AF887739

SOTHIS PRESS

Sothis Press aims to completely reimagine the public domain book with an emphasis on universal access to captivatingly designed print editions. We believe that technology will enable every single book ever published to be perpetually accessible in beautiful print form. Sothis is beauty in ubiquity.

www.sothis.press
1 Waterhouse Square, London, ECIN 2ST

The Heart of the Antarctic

Being the Story of the British Antarctic Expedition 1907-1909

E. H. Shackleton

SOTHIS PRESS

ISBN 978-1-0365-1097-8

The Heart of the Antarctic
is in the public domain.

This edition is published by
Sothis Press © 2023. The company
is registered in England and Wales.
The Company Number is 14900389.

THE HEART OF THE ANTARCTIC

BEING THE STORY OF THE BRITISH ANTARCTIC EXPEDITION 1907-1909

BY E. H. SHACKLETON, C.V.O.

WITH AN INTRODUCTION BY HUGH ROBERT MILL, D.Sc. AN ACCOUNT OF THE FIRST JOURNEY TO THE SOUTH MAGNETIC POLE BY PROFESSOR T. W. EDGEWORTH DAVID, F.R.S.

VOL. II

PHILADELPHIA
J. B. LIPPINCOTT COMPANY
1909

Contents of Volume Two

CONTENTS

CONTENTS

List of Illustrations of Volume Two

COLOURED PLATES

PLATES

LIST OF ILLUSTRATIONS

LIST OF ILLUSTRATIONS

LIST OF ILLUSTRATIONS

LIST OF ILLUSTRATIONS

LIST OF ILLUSTRATIONS

LIST OF ILLUSTRATIONS

ILLUSTRATIONS IN THE TEXT

LIST OF ILLUSTRATIONS

MAPS

In pocket of binding-case

NOTE

In the photographs the microscopic animals are magnified about 200 diameters

THE HEART OF THE ANTARCTIC

Chapter One

SOME NOTES ON THE SOUTHERN JOURNEY

WE brought back with us from the journey towards the Pole vivid memories of how it feels to be intensely, fiercely hungry. During the period from November 15, 1908, to February 23, 1909, we had but one full meal, and that was on Christmas Day. Even then we did not keep the sense of repletion for very long, for within an hour or two it seemed to us that we were as hungry as ever. Our daily allowance of food would have been a small one for a city worker in a temperate climate, and in our case hunger was increased by the fact that we were performing vigorous physical labour in a very low temperature. We looked forward to each meal with keen anticipation, but when the food was in our hands it seemed to disappear without making us any the less ravenous. The evening meal at the end of ten hours' sledging used to take us a long time to prepare. The sledges had to be unpacked and the camp pitched. Then the cooker was filled with snow and the Primus lamp lit, often no easy matter with our cold, frost-bitten fingers. The materials for the thin hoosh would be placed in the boiling-pot, with the addition, perhaps, of some pony maize, and the allowance of tea was placed in the outer boiler. The tea was always put in a strainer, consisting of a small tin in which we had punched a lot of holes, and it was removed directly the water had come to the boil. We used to sit round the cooker waiting for our food, and at last the hoosh would be ready and would be ladled into the pannikins by the cook of the week. The scanty allowance of biscuit would be distributed and we would commence the meal. In a couple of minutes the hot food would be gone, and we would gnaw carefully round the sides of our biscuits, making them last as long as possible. Marshall used sometimes to stand his pannikin of hoosh in the snow for a little while, because it got thicker as it cooled, but it was a debatable point whether this paid. One seemed to be getting more solid food, but there was a loss of warmth, and in the minus temperatures on the plateau we found it advisable to take our hoosh very hot. We would make the biscuits last as long as possible, and sometimes we tried to save a bit to eat in the sleeping-bag later on, but it was hard to do this. If one of us dropped a crumb, the others would point it out, and the owner would wet his finger in his mouth and pick up the morsel. Not the smallest fragment was allowed to escape.

We used to "turn backs" in order to ensure equitable division of the food. The cook would pour the hoosh into the pannikins and arrange the biscuits in four heaps. Perhaps some one would suggest that one pannikin had rather less in it than another, and if this view was endorsed by the others there would be a readjustment. Then when we were all satisfied that the food had been divided as fairly as possible, one man would turn his back, and another, pointing at one pannikin or group of biscuits, would say, "Whose?" The man who had his back turned, and therefore could not see the food, would give a name, and so the distribution would proceed, each of us always feeling sure that the smallest share had fallen to our lot. At lunch-time there would be chocolate or cheese to distribute on alternate days, and we much preferred the chocolate days to the cheese days. The chocolate seemed more satisfying, and it was more easily divided. The cheese broke up into very small fragments on the march, and the allowance, which amounted to two spoonfuls per man, had to be divided up as nearly as possible into four equal heaps. The chocolate could be easily separated into sticks of equal size. It can be imagined that the cook for the week had no easy task. His work became more difficult still when we were using pony-meat, for the meat and blood, when boiled up, made a delightful broth, while the fragments of meat sunk to the bottom of the pot. The liquor was much the better part of the dish, and no one had much relish for the little dice of tough and stringy meat, so the cook had to be very careful indeed. Poor old Chinaman was a particularly tough and stringy horse.

We found that the meat from the neck and rump was the best, the most stringy portions coming from the ribs and legs. We took all the meat we could, tough or tender, and as we went south in the days when horse-meat was fairly plentiful, we used to suck frozen, raw fragments as we marched along. Later we could not afford to use the meat except on a definite allowance. The meat to be used during the day was generally cut up when we took a spell in the morning, and the bag containing the fragments was hung on the back of the sledge in order that the meat might be softened by the sun. It cut more easily when frozen than when partially thawed, but our knives gradually got blunt, and on the glacier we secured a rock on which to sharpen them. During the journey back, when every ounce of weight was of great importance, we used one of our geological specimens, a piece of sandstone, as a knife-sharpener. The meat used to bulk large in the pot, but as fresh meat contains about 60 per cent. of moisture, it used to shrink considerably in the process of cooking, and we did not have to use very much snow in the pot.

We used the meat immediately we had started to kill the ponies in order to save the other food, for we knew that the meat contained a very large percentage of water, so that we would be carrying useless weight with it. The pemmican and biscuits, on the other hand contained very little moisture, and it was more profitable to keep them for the march further south, when we were likely to want to reduce the loads as far as possible. We left meat at each depôt, to provide for the march back to the coast, but always took on as much as possible of the prepared foods. The reader will understand that the loss of Socks, which represented so many pounds of meat, was a very severe blow to us, for we had after that to use sledging stores at the depôts to make up for the lost meat. If we had been able to use Socks for food, I have no doubt that we would have been able to get further south, perhaps even to the Pole itself, though in that case we could hardly have got back in time to catch the ship before she was forced to leave by the approach of winter.

When we were living on meat our desire for cereals and farinacious foods became stronger; indeed any particular sort of food of which we were deprived seemed to us to be the food for which nature craved. When we were short of sugar we would dream of sweet-stuffs, and when biscuits were in short supply our thoughts were concerned with crisp loaves and all the other good things displayed in the windows of the bakers' shops. During the last weeks of the journey outwards, and the long march back, when our allowance of food had been reduced to twenty ounces per man a day, we really thought of little but food. The glory of the great mountains that towered high on either side, the majesty of the enormous glacier up which we travelled so painfully, did not appeal to our emotions to any great extent. Man becomes very primitive when he is hungry and short of food, and we learned to know what it is to be desperately hungry. I used to wonder sometimes whether the people who suffer from hunger in the big cities of civilisation felt as we were feeling, and I arrived at the conclusion that they did not, for no barrier of law and order would have been allowed to stand between us and any food that had been available. The man who starves in a city is weakened, hopeless, spiritless, and we were vigorous and keen. Until January 9 the desire for food was made the more intense by our knowledge of the fact that we were steadily marching away from the stores of plenty.

We could not joke about food, in the way that is possible for the man who is hungry in the ordinary sense. We thought about it most of the time, and on the way back we used to talk about it, but always in the most serious manner possible. We used to plan out the enormous meals that we proposed to have when we got back to the ship and, later, to civilisation. On the outward march we did not experience really severe hunger until we got on the great glacier, and then we were too much occupied with the heavy and dangerous climbing over the rough ice and crevasses to be able to talk much. We had to keep some distance apart in case one man fell into a crevasse. Then on the plateau our faces were generally coated with ice, and the blizzard wind blowing from the south made unnecessary conversation out of the question. Those were silent days, and our remarks to one another were brief and infrequent. It was on the march back that we talked freely of food, after we had got down the glacier and were marching over the barrier surface. The wind was behind us, so that the pulling was not very heavy, and as there were no crevasses to fear we were able to keep close together. We would get up at 5 A.M. in order to make a start at 7 A.M., and after we had eaten our scanty breakfast, that seemed only to accentuate hunger, and had begun the day's march, we could take turns in describing the things we would eat in the good days to come. We were each going to give a dinner to the others in turn, and there was to be an anniversary dinner every year, at which we would be able to eat and eat and eat. No French chef ever devoted more thought to the invention of new dishes than we did.

It is with strange feelings that I look back over our notes, and see the wonderful meals that we were going to have. We used to tell each other, with perfect seriousness, about the new dishes that we had thought of, and if the dish met with general approval there would be a chorus of, "Ah! That's good." Sometimes there would be an argument as to whether a suggested dish was really an original invention, or whether it did not too nearly resemble something that we had already tasted in happier days. The "Wild roll" was admitted to be the high-water mark of gastronomic luxury. Wild proposed that the cook should take a supply of well-seasoned minced meat, wrap it in rashers of fat bacon, and place around the whole an outer covering of rich pastry, so that it would take the form of a big sausage-roll. Then this roll would be fried with plenty of fat. My best dish, which I must admit I put forward with a good deal of pride as we marched over the snow, was a sardine pasty, made by placing well-fried sardines

* The Hut in Summer Time; Coal Bags at the Left

Another View of the Hut in Summer. The Meteorological Station can be seen on the Extreme Right inside pastry. At least ten tins of sardines were to be emptied on to a bed of pastry, and the whole then rolled up and cooked, preparatory to its division into four equal portions. I remember one day Marshall came forward with a proposal for a thick roll of suet pudding with plenty of jam all over it, and there arose quite a heated argument as to whether he could fairly claim this dish to be an invention, or whether it was not the jam roll already known to the housewives of civilisation. There was one point on which we were all agreed, and that was that we did not want any jellies or things of that sort at our future meals. The idea of eating such elusive stuff as jelly had no appeal to us at all.

On a typical day during this backward march we would leave camp at about 6.40 A.M., and half an hour later would have recovered our frost-bitten fingers, while the moisture on our clothes, melted in the sleeping-bags, would have begun to ablate, after having first frozen hard. We would be beginning to march with some degree of comfort, and one of us would remark, "Well, boys, what are we going to have for breakfast to-day?" We had just finished our breakfast as a matter of fact, consisting of half a pannikin of semi-raw horse-meat, one biscuit and a half and a pannikin of tea, but the meal had not taken the keenness from our appetites. We used to try to persuade ourselves that our half-biscuit was not quite a half, and sometimes we managed to get a little bit more that way. The question would receive our most serious and careful consideration at once, and we would proceed to weave from our hungry imaginations a tale of a day spent in eating. "Now we are on board ship," one man would say. "We wake up in a bunk, and the first thing we do is to stretch out our hands to the side of the bunk and get some chocolate, some Garibaldi biscuits and some apples. We eat those in the bunk, and then we get up for breakfast. Breakfast will be at eight o'clock, and we will have porridge, fish, bacon and eggs, cold ham, plum pudding, sweets, fresh roll and butter, marmalade and coffee. At eleven o'clock we will have hot cocoa, open jam tarts, fried cods' roe and slices of heavy plum cake. That will be all until lunch at one o'clock. For lunch we will have Wild roll, shepherd's pie, fresh soda-bread, hot milk, treacle pudding, nuts, raisins and cake. After that we will turn in for a sleep, and we will be called at 3.45, when we will reach out again from the bunks and have dough-nuts and sweets. We will get up then and have big cups of hot tea and fresh cake and chocolate creams. Dinner will be at six, and we will have thick soup, roast beef and Yorkshire pudding, cauliflower, peas, asparagus, plum pudding, fruit, apple-pie with thick cream, scones and butter, port wine, nuts, and almonds and raisins. Then at midnight we will have a really big meal, just before we go to bed. There will be melon, grilled trout and butter-sauce, roast chicken with plenty of livers, a proper salad with eggs and very thick dressing, green peas and new potatoes, a saddle of mutton, fried suet pudding, peaches *à la Melba,* egg curry, plum pudding and sauce, Welsh rarebit, Queen's pudding, angels on horseback, cream cheese and celery, fruit, nuts, port wine, milk and cocoa. Then we will go to bed and sleep till breakfast time. We will have chocolate and biscuits under our pillows, and if we want anything to eat in the night we will just have to get it." Three of us would listen to this programme and perhaps suggest amendments and improvements, generally in the direction of additional dishes, and then another one of us would take up the running and sketch another glorious day of feeding and sleeping.

I daresay that all this sounds very greedy and uncivilised to the reader who has never been on the verge of starvation, but as I have said before, hunger makes a man primitive. We did not smile at ourselves or at each other as we planned wonderful feats of over-eating. We were perfectly serious about the matter, and we noted down in the back pages of our diaries details of the meals that we had decided to have as soon as we got back to the places where food was plentiful. All the morning we would allow our imaginations to run riot in this fashion. Then would come one o'clock, and I would look at my watch and say, "Camp!" We would drop the harness from our tired bodies and pitch the tent on the smoothest place available, and three of us would get inside to wait for the thin and scanty meal, while the other man filled the cooker with snow and fragments of frozen meat. An hour later we would be on the march again, once more thinking and talking of food, and this would go on until the camp in the evening. We would have another scanty meal, and turn into the sleeping-bags, to dream wildly of food that somehow we could never manage to eat.

The dysentery from which we suffered during the latter part of the journey back to the coast was certainly due to the meat from the pony Grisi. This animal was shot one night when in a greatly exhausted condition, and I believe that his flesh was made poisonous by the presence of the toxin of exhaustion, as is the case with animals that have been hunted. Wild was the first to suffer, at the time when we started to use Grisi meat with the other meat, and he must have been unfortunate enough to get the greater part of the bad meat on that occasion. The other meat we were using then came from Chinaman, and seemed to be quite wholesome.

A few days later we were all eating Grisi meat, and we all got dysentery. The meat could not have become affected in any way after the death of the pony, because it froze hard within a very short time. The manner in which we managed to keep on marching when suffering, and the speed with which we recovered when we got proper food, were rather remarkable, and the reason, no doubt, was that the dysentery was simply the result of the poison, and was not produced by organic trouble of any sort. We had a strong wind behind us day after day during this period, and this contributed in a very large measure to our safety, for in the weakened condition we had then reached we could not have made long marches against a head-wind, and without long marches we would have starved between the depôts. We had a sail on the sledge, formed of the floorcloth of a tent, and often the sledge would overrun us, though at other times it would catch in a drift and throw us heavily.

When we were travelling along during the early part of the journey over the level Barrier surface, we felt the heat of the sun severely, though as a matter of fact the temperature was generally very low, sometimes as low as zero Fahr. though the season was the height of summer. It was quite usual to feel one side of the face getting frozen while the other side was being sunburned. The ponies would have frozen perspiration on their coats on the sheltered side, while the sun would keep the other side hot and dry, and as the day wore on and the sun moved round the sky the frosted area on the animals would change its position in sympathy. I remember that on December 4 we were marching stripped to our shirts, and we got very much sunburned, though at noon that day the air temperature showed ten degrees of frost. When we started to climb the glacier and marched close to the rocks, we felt the heat much more, for the rocks acted as radiators, and this experience weighed with me in deciding to leave all the spare clothing and equipment at the Upper Glacier Depôt, about seven thousand feet up. We did not expect to have to climb much higher, but, as the reader knows, we did not reach the plateau until we had climbed over ten thousand feet above sea-level, and so we felt the cold extremely. Our wind-proof Burberry clothing had become thin by this time, and had been patched in many places in consequence of having been torn on the sharp ice. The wind got in through a tear in my Burberry trousers one day and I was frost-bitten on the under part of the knee. This frost-bite developed into an open wound, into which the wool from my underclothing worked, and I had finally to perform a rather painful operation with a knife before the wound would heal. We were continually being frost-bitten up on the plateau, and when our boots had begun to give out and we were practically marching on the senne-grass inside the finnesko, our heels got frost-bitten. My heels burst when we got on to hard stuff, and for some time my socks were caked with blood at the end of every day's march. Finally Marshall put some "New-skin" on a pad, and that stuck on well until the cracks had healed. The scars are likely to remain with me. In the very cold days, when our strength had begun to decrease, we found great difficulty in hoisting the sail on our sledge, for when we lifted our arms above our heads in order to adjust the sail, the blood ran from our fingers and they promptly froze. Ten minutes or a quarter of an hour sometimes elapsed before we could get the sledge properly rigged. Our troubles with frost-bite were no doubt due in a measure to the lightness of our clothing, but there was compensation in the speed with which we were able to travel. I have no doubt at all that men engaged in polar exploration should be clothed as lightly as is possible, even if there is a danger of frost-bite when they halt on the march.

The surface over which we travelled during the southern journey changed continually. During the first few days we found a layer of soft snow on top of a hard crust, with more soft snow underneath that again. Our weight was sufficient to break through the soft snow on top, and if we were pulling the increased pressure would cause the crust to break also, letting us through into the second layer of soft snow. This surface made the travelling very heavy. Until we had got beyond Minna Bluff we often passed over high, sharp sastrugi, and beyond that we met with ridges four to six feet high. The snow generally was dry and powdery, but some of the crystals were large, and showed in reflected light all the million colours of diamonds. After we had passed latitude 80° South the snow got softer day by day, and the ponies would often break through the upper crust and sink in right up to their bellies. When the sun was hot the travelling would be much better, for the surface snow got near the melting-point and formed a slippery layer not easily broken. Then again a fall in the temperature would produce a thin crust, through which one broke very easily. Between latitude 80° South and 83° South there were hard sastrugi under the soft snow, and the hoofs of the horses suffered in consequence. The surface near the land was broken up by the pressure from the glaciers, but right alongside the mountains there was a smooth plain of glassy ice, caused by the freezing of water that had run off the rocky slopes when they were warm under the rays of the sun. This process had been proceeding on the snow slopes that we had to climb in order to reach the glacier. Here at the foot of the glacier there were pools of clear water round the rocks, and we were able to drink as much as we wanted, though the contact of the cold water with our cracked lips was painful.

The glacier itself presented every variety of surface, from soft snow to cracked and riven blue ice, by-and-by the only constant feature were the crevasses, from which we were never free. Some were entirely covered with a crust of soft snow, and we discovered them only when one of us broke through, and hung by his harness from the sledge. Others occurred in mazes of rotten ice, and were even more difficult to negotiate than the other sort. The least unpleasant of the crevasses were those that were wide open and easily seen, with firm ice on either side. If these crevasses were not too wide, we would pull the sledges up to the side, then jump over, and pull them after us. This was more difficult than it sounds from the fact that the ice gave only a very uncertain footing, but we always had the harness as a safeguard in case of a fall. If the crevasses were wide we had to make a *détour*. The sledges, owing to their length, were not liable to slip down a crevasse, and we felt fairly safe when we were securely attached to them by the harness. When the surface was so bad that relay work became necessary we used to miss the support of a sledge on the back journeys. We would advance one sledge half a mile or a mile, put up a bamboo pole to mark the spot, and then go back for the other. We were roped together for the walk back to the second sledge, but even then we felt a great deal less secure than when harnessed to one of the long, heavy sledges. On some days we had to travel up steep slopes of smooth ice, and often it became necessary to cut steps with our ice-axes, and haul the sledges after us with the Alpine rope. When we had gone up about sixty feet, the length of the rope, we would haul up the sledge to which we had attached the lower end, and jamb it so that it could not slide back. Then one of us would slide down in order to fix the rope to the other sledge.

One of the curious features of the glacier was a yellow line, evidently an old moraine, extending for thirty or forty miles. The rocks of the moraine had gradually sunk in out of sight, the radiation of the sun's heat from them causing the ice to melt and let them through, and there had remained enough silt and dust to give the ice a dirty yellow appearance. The travelling along this old moraine was not so bad, but on either side of it there was a mass of pressure ice, caused by the constriction of the glacier between the mountains to the east and west. Unfortunately we brought back no photographs of this portion of the glacier. The number of plates at our disposal was limited, and on the outward march we decided not to take many photographs in case we found interesting land or mountains in the far south nearer the Pole. We thought that we would be able to secure as many photographs of the glacier as we wanted on the way back if we had the plates to spare, but as a matter of fact when we did get on to the glacier a second time we were so short of food that we could not afford the time to unpack the camera, which had to be stowed away carefully on the sledge in order to avoid damage to it.

Many nights on the glacier there was no snow on which to pitch the tents, and we had to spend perhaps an hour smoothing out a space on a rippled, sharp-pointed sea of ice. The provision bags and sledges had to be packed on the snow cloths round the tents,

Bay of Whales, or Balloon Bight
(*Sketch by G. Marston*)

* THE MOTOR-CAR IN SOFT SNOW AFTER THE RETURN OF THE SHIP

and it was indeed fortunate for us that we did not meet with any bad weather while we were marching up the glacier. Had a blizzard come on while we were asleep, it would have scattered our goods far and wide, and we would have been faced with a very serious position. All the time that we were climbing the glacier we had a northerly wind behind us, although the direction of the sastrugi showed clearly that the prevailing wind was from the south; when we were coming back later in the season the wind was behind us all the time. We encountered a strong wind on the outward journey when near the top of the glacier, and as the ice slopes were covered with snow it was difficult to pull the sledges up them. When we reached the same slopes on the way back, the summer sun had cleared the snow from them, leaving clear ice, and we simply glissaded down all but the steepest slopes, although one of the sledge runners was very badly torn. We had to travel carefully on the steep slopes, for if we had let the sledge get out of hand it would have run away altogether, and would probably have been smashed up hundreds of feet below.

The Upper Glacier Depôt was overhung by great cliffs of rock, shattered by the frosts and storms of countless centuries, and many fragments were poised in such a fashion that scarcely more than a touch seemed needed to bring them hurtling down. All around us on the ice lay rocks that had recently fallen from the heights, and we wondered whether some boulder would come down upon us while we were in camp. We had no choice of a camping-ground, as all around was rough ice. The cliffs were composed largely of weathered sandstone, and it was on the same mountains, higher up the glacier, that the coal was found, at a point where the slope was comparatively gentle.

Looking down from this height, we could see the glacier stretching away to the point of junction with the Barrier, the mountains rising to east and west. Many of the mountains to the west of the glacier were more or less dome-shaped, but there were some sharp conical peaks to the westward of the particular mountain under which the Upper Glacier Depôt had been placed. There were three distinct peaks, as the photographs show, and the plateau ice sweeping down made a long moraine on the west side of the glacier. To the eastward there was a long ridge of high mountains, fairly uniform in shape and without any sharp peaks, but with ridges, apparently of granite, projecting towards the west and so constricting the glacier. The mountains were distant about twenty-five miles, but well-defined stratification lines could plainly be seen. Below us, as we looked from the depôt, could be seen the cumulus clouds that always hung above the " Cloud-Maker."

When we looked to the south from this depôt we saw no clouds; there was nothing but hard clear sky. The sky gave no indication of the blizzard winds that were to assail us when we reached the plateau, and after we had gone as far south as we could and retraced our footsteps to the depôt, we looked back and saw the same clear sky, with a few wisps of fleecy cloud in it. We had no doubt that below those clouds the pitiless gale was still raging across the great frozen plain, and that the wind which followed us during our march back to the coast was coming from the vicinity of the Pole. As we advanced from the Upper Glacier Depôt we came upon great ice falls. The surface looked smooth from a distance, and we thought that we were actually on the plateau, but as we advanced we saw that before us lay enormous ridges rising abruptly. We had to relay our gear over these ridges, and often at the tops there would be a great crevasse, from which would radiate smaller crevasses fringed with crystals and showing ghastly depths below. We would creep forward to see what lay on the other side, and perhaps would find a fall of fifty feet, with a grade of about 1 in 3. Many times we risked our sledge on very severe slopes, allowing it to glissade down, but other times the danger of a smash was too great, and we had to lower the sledge slowly and carefully with the rope. The ice was safe enough to walk upon at this time except at the ridges, where the crevasses were severe, for the smaller crevasses in the hollows and slopes could be passed without difficulty.

The ice falls delayed us a good deal, and then we got into soft snow, over which the sledge dragged heavily. We thought that we were finally on the plateau level, but within a few days we came to fresh ridges and waves of pressure ice. The ice between the waves was very rotten, and many times we fell through when we put our weight on it. We fastened the Alpine rope to the sledge harness, and the first man pulled at a distance of about eighteen feet from the sledge, while the whole party was so scattered that no two men could fall into a crevasse together. We got on to better ground by steering to the westward, but this step was rather dangerous, for by taking this course we travelled parallel with the crevasses and were not able to meet them at right angles. Many times we nearly lost the sledge and ourselves when the ice started to break away into an unseen crevasse running parallel with our course. We felt very grateful to Providence that the weather remained clear, for we could not have moved a yard over this rotten ice in thick weather without courting disaster. I do not know whether the good weather we experienced in that neighbourhood was normal. We generally had about seven miles of easy going after we had passed one ridge in this area, and then another ridge would rise up ahead of us, and we would start to climb again. There were always crevasses at the top of the ridges, suggesting that the ice was moving over land at no great depth.

We passed the last ridge at last, and reached the actual plateau, but instead of hard névé, such as the *Discovery* expedition had encountered in the journey to the plateau beyond the mountains west of McMurdo Sound, we found soft snow and hard sastrugi. All the sastrugi pointed to the south, and the wind blew strongly nearly all the time from the south or south-east, with an occasional change to the south-west. Sometimes we marched on hard sastrugi, and at other times we had soft snow under our feet, but could feel the sastrugi on which the snow was lying. I formed the opinion that during the winter on the plateau the wind must blow with terrible violence from the south, and that the hard sastrugi are produced then. Still further south we kept breaking through a hard crust that underlay the soft surface snow, and we then sank in about eight inches. This surface, which made the marching heavy, continued to the point at which we planted the flag. After the long blizzard, from the night of January 6 until the morning of January 9, we had a better surface over which to make our final march southwards, for the wind had swept the soft snow away and produced a fairly hard surface, over which, unencumbered with a sledge as we were, we could advance easily.

We found the surface generally to be improved on the march back. The blizzard winds had removed the soft surface snow, and incidentally uncovered many of the crevasses. We were following our outward tracks, and often I noticed the tracks lead us to the edge of a crevasse which had been covered previously and over which we had passed in ignorance of our danger on the march southwards. When we got to the head of the glacier we tried to take a short cut to the point where we had left the Upper Glacier Depôt, but we got enmeshed in a maze of crevasses and pressure ridges to the eastward, and so had to steer in a westerly direction again in order to get clear. The dangers that we did know were preferable to those that we did not know. On the way down the glacier we found all the snow stripped away by the wind and sun for nearly one hundred miles, and we travelled over slippery blue ice, with innumerable cracks and sharp edges. We had many painful falls during this part of the journey. Then when about forty miles from the foot of the glacier we got into deep soft snow again, over which rapid progress was impossible. There had evidently been a heavy snowfall in this area while we were further south, and for days, while our food was running short, we could see ahead of us the rocks under which the depôt had been placed. We toiled with painful slowness towards the rocks, and as the reader has already learned, we were without any food at all for the last thirty hours of that march. We found the Barrier surface to be very soft when we got off the Glacier, but after we had passed Grisi Depôt there was an improvement. The surface remained fairly good until we reached the winter quarters, and in view of our weakened condition it was fortunate for us that it did so.

In reviewing the experience gained on the southern journey, I do not think that I could suggest any improvement in equipment for any future expedition. The Barrier surface evidently varies in a remarkable fashion, and its condition cannot be anticipated with any degree of certainty. The traveller must be prepared for either a hard surface or a very soft one, and he may get both surfaces in the course of one day's march. The eleven-foot sledge is thoroughly suitable for the work, and our method of packing the stores and hauling the sledges did not develop any weak points. We would have been glad to have had crampons for use on the glacier; what would be better still would be heavy Alpine boots with nails all round, for very often the surface would give little grip to crampons, which would only touch the rough ice at one or two points. The temperature is too cold to permit of the explorer wearing ordinary leather boots, and some boot would have to be designed capable of keeping the feet warm and carrying the nails all round. A mast consisting of a bamboo lashed to the forward oil-box proved as efficient as could be required for use in connection with a sail on the sledges. It was easily rigged and had no elaborate stays. I would suggest no change in the clothing, for the light woollen underclothing, with thin windproof material outside, proved most satisfactory in every way. We could certainly not have travelled so fast had we been wearing the regulation pilot cloth garment generally used in polar exploration. Our experience made it obvious that a party which hopes to reach the Pole must take more food per man than we did, but how the additional weight is to be provided for is a matter for individual consideration. I would not take cheese again, for although it is a good food, we did not find it as palatable as chocolate, which is practically as sustaining. Our other foods were all entirely satisfactory.

Each member of the Southern Party had his own particular duties to perform. Adams had charge of the meteorology, and his work involved the taking of temperatures at regular intervals, and the boiling of the hypsometer, sometimes several times in a day. He took notes during the day, and wrote up the observations at night in the sleeping-bag. Marshall was the cosmographer and took the angles and bearings of all the new land; he also took the meridian altitudes and the compass variation as we went south. When a meridian altitude was taken, I generally had it checked by each member of the party, so that the mean could be taken.

Marshall's work was about the most uncomfortable possible, for at the end of a day's march, and often at lunch-time, he would have to stand in the biting wind handling the screws of the theodolite. The map of the journey was prepared by Marshall, who also took most of the photographs. Wild attended to the repair of the sledges and equipment, and also assisted me in the geological observations and the collection of specimens. It was he who found the coal close to the Upper Glacier Depôt. I kept the courses and distances, worked out observations and laid down our directions. We all kept diaries. I had two, one my observation book, and the other the narrative diary, reproduced in the first volume.

Chapter Two

SUMMER AT THE WINTER QUARTERS

WE were distant about thirty-two miles from Hut Point when I decided to send the supporting-party back. The men watched us move off across the white plain until we became mere dots on the wide expanse, and then loaded up their gear and started north. Joyce was left in charge of the party, and he decided to make one forced march to Hut Point. They had to cross a good deal of crevassed ice, but a special effort would enable them to make their next camp under shelter. They got under way at 7 A.M. and marched till noon, making good progress in spite of the surface. In the afternoon they marched from two till five o'clock, and then a final march, from 7 P.M. till 1.30 A.M., took them to the old *Discovery* hut. The only incident of the day had been the succumbing of Brocklehurst's feet to another attack of frost-bite, he having worn ski-boots when the other men had put on finnesko. The damage was not serious, although the sufferer himself had trouble with his feet for some time after. The party had covered thirty-two miles in fourteen hours and a half, very good marching in view of the soft and broken character of the surface.

The party left Hut Point on the morning of November 12, and had a hard pull to Glacier Tongue. They at first thought of camping on the southern side of the Tongue, but, fortunately, kept on, for on the other side they met Day, Murray and Roberts, who had brought out stores with the motor-car. I had left orders that about 1800 lb. of provisions and gear should be taken to the depôt there, as a provision for the sledging-parties, in case they should be cut off from Cape Royds by open water on their return. Day had succeeded in running the car right up to the Tongue, about twelve miles from winter quarters. After a good meal of biscuits, jam, lobscouse, tongue and cods' roe, the two parties joined in getting the stores up to the depôt. Then they all went back to the winter quarters on the car and the light sledges it had in tow, leaving the heavy sledge that had been used by the supporting-party to be brought in at some later opportunity. They reached the hut in the small hours of the morning, and after another meal turned in for a good sleep.

Routine work occupied the men at the hut for some time after the return of the supporting-party. The scientific members were more than a little grieved to find that during the days when the hut had been untenanted, for Murray, Day and Roberts had been away too on a small expedition, some of the dogs had managed to get loose, and had killed thirty or forty penguins. We had from the first tried very hard to avoid any accidents of this sort, for we did not want to cause any unnecessary destruction of animal life. The penguins were now laying, and the men found that the eggs were very good to eat. The egg of the penguin is about the same size as that of a duck, and it has a transparent, jelly-like white and a small yolk. It takes about eight minutes' boiling to cook the egg nicely, and ten minutes if it is required set hard to the centre. The shell is the most beautiful dark-green inside, while the outer shell is chalky and white, though generally stained prettily by guano. Murray set aside a certain portion of the rookery for the supply of eggs for "domestic purposes," partly in order to ensure freshness and partly in order to ascertain how many eggs the penguins would lay. The other portion of the rookery was left untouched in order that the development and education of the young penguins might be studied.

The scientific work in its various branches was carried on by the men at the winter quarters, and they made a series of small expeditions to points of interest in the surrounding country. "To-day we motored to Tent Island *viâ* Inaccessible Island," wrote Priestley on November 14. "The main object of the expedition was to enable Joyce to kill and skin some young seals, but we did geological work as well. Day, Joyce, Murray and myself were the party, and when the motor was pulled up opposite Inaccessible Island three of us strolled over to look at its western slopes. We did not have time to climb, but the island from that side consists entirely of a flow of massive basalt, with small porphyritic felspars, which show out best in the weathered specimens. The sheet of basalt appeared to be dipping to the south. Day endeavoured to join us, but he chose a bad place, and got so deep in the drift that his axle was aground, so he was obliged to reverse engines and back out. From there we proceeded to Tent Island, and after Joyce had picked out a young seal and started operations, Murray, Day and I climbed up a water-worn gully on the island and had a cursory look at the rocks, which are an agglomerate with very coarse fragments; capping the agglomerate there is a massive flow of kenyte. . . . Day photographed the lower slopes of the gully while Murray and I climbed the rock-slopes till they ended, and then cut steps up a snow slope, at the top of which I came across a snow cornice and nearly got into trouble getting through it. On reaching the top we walked along the ridge, and

*

Glacier South of Cape Barne, with Motor travelling on Sea Ice

A Seal destined for the Larder

Fetching Snow for Cooking Purposes

Transporting a Sledge over bare Rocks for the Summer Journey to the Slopes of Mount Erebus

Parasitic Cone on the Slopes of Mount Erebus

* Smoke Streaming from the Crater of Erebus

photographed a splendid weathered kenyte boulder, hollowed out like a summer-house, and studded with felspars as an old-fashioned church door is studded with nails. After taking these photographs we climbed down the other side of the island, and walked round to join the others. The rock-climbing here, on any slopes at all steep, is very difficult because of the weathered fragments, which, owing to lack of powerful natural agents of transportation and to the fact that the wind carries all the lightest soil away, are left lying just at their angle of repose; a false step may send mountaineer and mountain surface hurtling down fifty or a hundred feet—no agreeable sensation, as I know from frequent experience. The sun was very hot to-day, and the gully was occupied by a little stream which was carrying quite a quantity of light soil down with it.

Day had an exciting experience with the car during this journey. He encountered a big crack in the ice near Cape Barne, and steering at right angles to its course, put on speed in order to "fly" it in the usual way. When only a few yards from it and travelling at a speed of about fifteen miles an hour, he found that the crack made a sudden turn, so as to follow the line he was taking, and an instant later his right-hand front wheel dropped in. Any weak points in the car would have been discovered by the sudden strain, but happily nothing broke, and the crack making another turn, the wheel bounded out at the elbow, and the car was on sound ice again.

On November 16 Priestley made an interesting trip up the slopes of Erebus. Beyond the lower moraines and separated from them by a snow-field of considerable size, he found a series of kenyte ridges and cones, covered by very little *débris*. The ridges continued for some distance to the edge of the main glacier, where they terminated in several well-marked nunataks. "One which I visited, and which was the nearest to the large parasitic cone, was eighty feet high, of massive kenyte of brown colour and close texture, jointed into very large cubical joints by a very complete series of master-joints. From this nunatak I obtained nine kinds of lichen, including four or five new species, and one piece of moss. One of the lichens was so much larger than the others and branched so much that it might well be called a forest-lichen, and Murray considers it to be very closely allied to the reindeer-moss, or ice-moss."

Joyce was engaged at this time in making zoölogical collections, and with the aid of the motor-car he was able to cover a great deal of ground. The motor-car, driven by Day, would take him fifteen or sixteen miles over the sea-ice to some suitable locality, generally near the Cathedral Rocks on the north side of Glacier Tongue, and leave him there to kill seals and penguins. In order to kill young seals, some specimens of which were required, he had first to drive the mothers away, and this often took a long time, as the female seal becomes aggressive when interfered with in this manner. The work was not at all pleasant, but Joyce killed and prepared for preservation five young Weddell seals and four adult specimens. He had taken lessons in taxidermy before leaving England in order to be ready for this duty. Joyce and Day also killed and skinned twenty Emperor penguins, twelve Adelie penguins and twelve skua gulls, and all the men at the winter quarters assisted in collecting eggs.

Murray was looking after the scientific work, paying special attention to his own particular domain, that of biology, and Marston was devoting as much time as he could to sketching and painting. He had taken oils,

THE RETURNING SUN

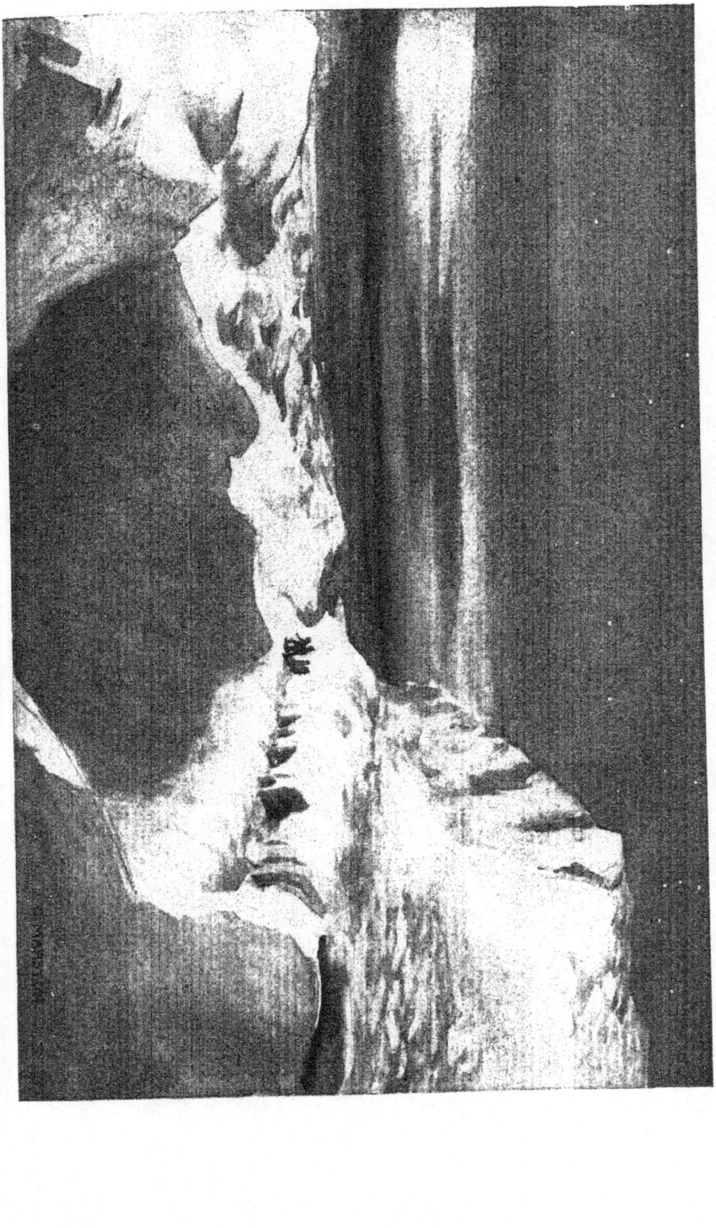

water-colours and pastels to the south with him. He found that the water-colours could not be used in the open at all; for they froze at once. Oils could be used fairly comfortably in the summer, though it was always chilly work to sit still for any length of time; during the spring the oils froze after they had been in the open air for about an hour, so that steady work was not possible. The pastels could always be used for making "colour notes," and they were also used for some of the colour-sketches that are reproduced in this book. Mits had to be worn for all outside work, and this made sketching difficult.

Marston found, as other artists have found, that Nature's color-schemes in the Antarctic are remarkably crude, though often wondrously beautiful. Bright blues and greens are seen in violent contrast with brilliant reds, and an accurate record of the colours displayed in a sunset, as seen over broken ice, would suggest to many people an impressionistic poster of the kind seen in the London streets. Words fail one in an attempt to describe the wildly bizarre effects observed on days when the sky was fiery red and pale green, merging into a deep blue overhead, and the snow-fields and rocks showed violet, green and white under the light of the moon. Marston used to delight in the "grey days," when there was no direct sunlight and the snow all around showed the most subtle tones of grey; there would be no shadows anywhere, perhaps light drifts of snow would be blowing about, and the whole scene became like a frozen fairyland. The snow-bergs and snow-fields were white under direct light, but any hollows showed a vivid blue, deepening almost to black in the depths. There was an unlimited amount of interesting work for an artist, and Marston suffered to some extent, as did the other specialists on the expedition, from the fact that the number of men available was so small that every one, in addition to his own work, had to take a share in the routine duties.

Joyce devoted what spare time he could find to the completion of the volumes of the "Aurora Australis." Practice had made him more skilful in the handling of type, and he was able to make a good deal of progress, Day assisting with the preparation of the Venesta boards in which the volumes were to be bound. Some of the contributions towards the literary part of the work had come in late, so that there was plenty of work left to do. Marston went on with the lithographing for the illustrations.

Instructions had been left for a geological reconnaisance to be made towards the northern slopes of Mount Erebus, to examine, if possible, some parasitic cones and the oldest main crater of the mountain. Threatening weather prevented the carrying out of this plan for some time, yet for nearly a fortnight after the return of the southern supporting-party the expected blizzard did not come, while the weather was not propitious for the journey. At length no further delay was possible if the trip was to be made, as Priestly, the geologist, had to leave for the western mountains, so on November 23 the trip was begun, though with misgivings as to the long overdue blizzard.

The party consisted of Priestley, Marston, Joyce, Murray and Brocklehurst, and they took seventy pounds of food—a week's supply on the ordinary basis of thirty-two ounces per day for each man—but carried only one tent, intended to hold three men, their idea being that one or two men could sleep in the bags outside the tent. The weather was fine when they left the hut, but in the afternoon a strong southerly wind sprang up, and they had to march through low drift. They camped for the night close to a steep nunatak about five miles from the hut and nearly two thousand feet above sea-level. There was difficulty in getting a good snowy camping-ground, and they had to put up the tent on smooth blue glacier ice, having a thin coating of snow, and sloping gently down till it terminated in an ice-cliff overlooking the sea not many hundreds of yards below. After dinner Priestley, Murray and Joyce climbed over the nunataks, and found several new lichens, but the specimens collected were lost in the blizzard later on. Priestley also found a number of very perfect felspar crystals weathered out of the kenyte, and collected a couple of handfuls of the best. The members of the party retired to their sleeping-bags at eight o'clock on Monday night, and before midnight a blizzard swept down upon them, and proved to be an exceptionally severe one, with dense drift. Priestley had volunteered to sleep outside that night, and had taken his sleeping-bag to a nook in the rocks some distance away. When the other men heard the roaring of the blizzard they looked out, and were reassured to find that he had come down while there was time and had lain down close by the tent. The first night the light snow round the tent was blown away leaving one side open to the wind, but the occupants were able to find a few bits of rock close by, and secured it with those.

"Inside the tent for the next three days we were warm enough in our sleeping-bags," wrote Murray in his report. "Though we could not cook anything we ate the dry biscuit and pemmican. The little snow under the floorcloth was squeezed in the hand till it became ice, and we sucked this for drink. We were anxious about Priestley, and occasionally opened the door-flap and hailed him, when he always replied that he was all right. Joyce had managed to pass him some food early in the storm, so there was no fear of starving, but as we learned afterwards he could get nothing to drink and so could not eat. No one could offer to change places with him, as in doing so the sleeping-bag would have filled with snow, and might have blown away. On Wednesday Marston dressed in his Burberries and crawled down to Priestley, who reported " All well," but he had had no food for twenty-four hours. Marston gave him some biscuits and chocolate. On Thursday morning he replied to the hail, but he was getting further and further from the tent, as every time he moved he slipped a little bit down the smooth glacier. At mid-day there was no reply to our hail, and we thought of the precipitious ice-foot and imagined things. Joyce and I dressed and went out to seek him. The drift was so thick that nothing whatever could be seen, and when the head was lifted to try and look the whole face and eyes were instantly covered by a sheet of ice. We crept about on hands and knees looking for the lost man. The only chance of getting back to the tent again was to steer by the wind, down the wind looking for Priestley, up the wind home again. At one side the sledge lay, forming a landmark, and Priestley had been not very far from the far-away end. Creeping along the sledge to where he had lain, I found that he was not there. Joyce went a little further to the right and came upon him, all alive."

Priestley's experiences during this period are related in his diary. " I had volunteered to sleep in the bag outside the tent," he wrote, " and by the time I was ready to turn in the drift had started again pretty badly, and the only sheltered spot I could find was at the top of the hill, so I told Joyce where he would find me in the morning and camped down, first luckily taking the precaution to put a few cubic feet of kenyte on my Burberry trousers and jacket outside the bag. A few hours later I woke up to find that the wind had increased to the dimensions of a blizzard, and that the drift was sweeping in a steady cloud over my head. I realised that those in the tent would have trouble in reaching me in the morning, so I got out of the bag and dressed, getting both the bag and my clothes full of snow in the process. Then, after some trouble, I got the bag down the steep slope of the nunatak to the sledge, where I wrapped myself up in the tent-cloth and lay athwart the wind. In about two hours I got drifted up so close that I was forced to get my shoulders out of the bag and lever myself out of the drift, and I then tried the experiment of lying head to wind on the opening of the sleeping-bag. This answered very well, and it was in this position that I spent the next seventy-two hours, getting shifted down a yard or two at a time at every change in the direction of the wind, and being gradually pushed along the wind-swept surface of the glacier until I was some twenty or thirty yards from the tent, and in some danger of getting swept, as the wind increased in violence, either on to some rocks a quarter of a mile below or else straight down the glacier and over a hundred-foot drop into Horseshoe Bay.

"Three times the people in the tent managed to pass me over some biscuits and raw pemmican, and Marston got my chocolate from the rücksack and brought it to me. My chief difficulty, however, was want of water. I had had a little tea before I turned in, but from that time for nearly eighty hours I had nothing to drink but some fragments of ice that I could prise up with the point of a small safety-pin. The second time Joyce came down, I believe about the beginning of the third day, he reported that the lashings at the top of the tent-poles had given way and that a rent had been torn in the material by the corner of a biscuit-tin. He added that it was impossible to keep any snow on the skirt of the tent, and that, as the snow-cloth was kept down only by a few rocks, the occupants of the tent were in constant expectation of seeing the tent leave them altogether. When Joyce left me on this occasion the drift was so thick that he could see nothing, and had to find his way back by shouting and listening for the return shouts of his tent-mates. He had gone only a quarter of the distance when both his eyes were filled with drift and immediately choked with ice, and when he reached the tent his face was a mask of ice and both feet were frost-bitten. He was helped inside and his feet brought round with rubbing, but no further attempt could be made to reach me. He had brought me some biscuit and raw pemmican. Cooking was not possible in the tent owing to the impossibility of reaching the sledge to get the oil-filler. It may sound like an exaggeration to say that we could not reach the sledge, which was four yards or less from the tent, but it must be remembered that we were lying on the slopes of a clean-swept glacier, on which finnesko could get no hold. The snow that had covered the ice when we pitched camp had all disappeared before the fury of the blizzard. Our spiked ski-boots were on the ice-axes round the sledge, where they had been hung to dry, but in any case it would have been impossible to wear them in a blizzard when feet were getting badly frost-bitten even in finnesko. A slip on the ice meant very serious danger of destruction.

"A slight decrease in the wind at the close of the third day gave me hope of getting up to the tent, and I prepared to move by putting on my outdoor clothing, no easy task in a sleeping-bag; then, rolling over on my side, I tried to get out. I found that there was less wind and less drift, and that I was able for the first time to see where I was with regard to surrounding objects. I was unable, however, to get out of the bag without being blown further down the slippery glacier, and I could see that it would be impossible to crawl up the slopes with the cumbrous bag. If I lost the bag I might as well have let myself slide."

About two hours after this Marston ventured forth from the tent in one of the remarkable intervals of calm occasionally experienced in the course of an Antarctic blizzard. On either side of the spot on which the camp had been pitched he could see the drift flying along with the full force of the wind, but he was able to make his way down to Priestley before the blizzard swept down on them again. They dragged the bag up the glacier by kneeling on it and jerking it along, and both got into the tent. "Four men in a three-men tent is a big squeeze," continued Priestley, "but five was fearful, and it was some time before I managed to get even sitting room. The first thing to do was to examine and attend frost-bitten feet, and the examination showed as big a crop as could be expected, for Marston and I each had both feet frost-bitten. A course of massage brought them round, and I got into Marston's bag while he made tea. . . . After tea I got into my own bag and lay down on top of Murray and Marston, and by dint of much wriggling we managed to get fairly settled, though our positions were so cramped that sleep was impossible.

"At about half-past four in the morning we cooked some pemmican in the tent and had a proper breakfast, as for the first time the wind had really begun to die away. Owing to the cold, the long period of semi-starvation in our cramped quarters, and the fact that oil had got mixed up with the food, we were unable to do justice either to the hoosh or to the cocoa which followed it, and were still fairly empty when the drift ceased and we turned out to face the blizzard, pack the sledge and start for home. The ascent of the mountain had, of course, to be abandoned. I put on my damp finnesko and went out to help, but in less than five minutes, though the temperature was plus 22° Fahr., I was back in the tent with the front portions of both feet frozen, and we took half an hour to bring them round by beating, massaging and rubbing with snow. This latter remedy, Marston's favourite, is a very drastic one, and as painful as any I know, for the Antarctic snow is invariably in small sharp crystals, very brittle and hard. We all chafed very much at the unavoidable delay, as there was every sign of a renewal of the blizzard and the drift, but fortunately we got under way before any drift rose, and the wind was rather in our favour. We left all the provisions there, and unanimously named the nunatak 'Misery Nunatak,' and we were about as glad to leave the place as a soul would be to leave purgatory. We also left a tin of biscuits and some oil with a view to a future attempt at an ascent, to be made by Murray, Day, Marston and Joyce.

"There was a remarkable contrast between the wind-swept surface of the glacier and the surface over which we had toilsomely dragged the sledges during our day's journey outward. Instead of a uniform carpet, six inches deep, of soft snow, varied with drifts up to one's knees, we found patches of glacier ice, larger stretches of névé, and hard drifts of snow, on which neither our weight nor the weight of the sledge made the slightest impression; these drifts were deeply

START OF WESTERN PARTY FROM CAPE ROYDS

THE MOTOR NEAR THE WINTER QUARTERS

A HAUL OF FISH

undercut on the south-east side, and were frequently a foot to eighteen inches in height. It was no easy matter to direct the sledge across the strong wind then blowing, although we had two men pulling and two others guiding the sledge, and we ascended about half a mile to the north of Horseshoe Bay in what was, for a long time, totally unfamiliar country, and through a series of moraines which had not yet been explored. I was, unfortunately, of no use in the pulling, being only just able to get along myself, and we were all extremely glad to get the sledge to the Back Door Bay end of Blue Lake, where it was left till the next day. We reached the hut and started on a course of feeding and recuperating, having been five days out."

Mount Erebus was noticed to be in eruption when the party was maching back to the hut on November 27. Huge diverging columns of steam were rising from the crater, and behind could be seen curious clouds of feathery cirrus. The temperature during the blizzard had not fallen below 12° Fahr., and been above 20° Fahr. during most of the time, so that the frost-bites sustained by the men must have been due mainly to lowered vitality, caused by the cramped situation and the lack of hot food.

The experience had been rather a severe one, but the men were none the worse for it after a day or two at the winter quarters, and they commenced at once to make preparations for the western journey. I had left instructions that on December 1 Armytage, Priestley and Brocklehurst should start for Butter Point with 600 lb. of stores in order to lay a depôt for the Northern Party which might be expected to reach that point on its journey back from the Magnetic Pole. Then the three men were to secure what stores they required for their own purposes, and proceed up the Ferrar Glacier as far as the Depôt Nunatak in order that Priestley might search for fossils in the sandstones of the western mountains. They were to get back to Butter Point early in January in order to meet Professor David, Mawson and Mackay, and if a junction was effected, Mawson, Priestley and Brocklehurst were to carry on geological work in Dry Valley and the surrounding country, while Professor David, Armytage and Mackay were to return to the winter quarters. The fact that the Northern Party did not arrive upset this arrangement to some extent, but the other three men did some very useful work. The mountains to the west of McMurdo Sound had been explored by Lieutenant Armytage and Captain Scott during the *Discovery* expedition, Armitage having climbed the mountains and penetrated west to an altitude of 9000 ft. on the ice-cap, while Scott had reached longitude 146° 33′ East, on the western plateaux. Further information was required, however, in regard to the geology of the mountains.

Armytage, Priestley and Brocklehurst accordingly left the winter quarters on December 1, taking with them about 1200 lb. of gear and stores. The motor-car carried them for the first sixteen miles, although the sea ice was by this time in a very bad condition. The season was well advanced, the sun was above the horizon all the time, and there were cracks and pools in all directions. Day and Marston took the car out, and when they were coming back after leaving the Western Party the car got stuck firmly in a crack that ran across the course. They spent two hours cutting away the ice sufficiently to get the car out, and then had to make a *détour* of five miles in order to get round the crack. This was the last journey of the car in the Antarctic, for it was laid up when it got back to the hut.

The Western Party, after some heavy sledging, camped on December 4 at the foot of the Ferrar Glacier. Armytage was, at this time, suffering from an attack of snow-blindness. Priestley found moss and a species of fungus at the stranded moraines and also some kenyte. The men had been looking forward with pleasurable anticipation to securing skuas' eggs, which would have been a welcome change from pemmican and biscuit, but the birds had apparently not begun to lay and no eggs were secured. "A good deal of water denudation and transportation is taking place along the sea-cliffs of these moraines," wrote Priestley in his diary. "Quite a thick alluvial deposit, bearing a strong resemblance to a series of miniature deltas, is to be seen along the ice-foot awaiting the breaking up of the ice and its removal to the sea. The dust from the moraines had made a remarkable surface for two miles this side of them. Some winds had evidently been strong enough to remove a considerable quantity of the gravel with the snow, and the drifts which had contained this gravel had melted away, undercutting the edges of the cleaner snow-drifts, and thus giving a surface of bare ice with patchy snowdrifts undercut on all sides."

The party reached Butter Point, about thirty-five miles as the crow flies from the winter quarters, on December 5, and found a small depôt left there by the Northern Party on its way to the Magnetic Pole. Professor David and his companions had placed some final letters in a milk-tin. The stores brought for the purpose were placed at the depôt, and then Armytage, Priestley and Brocklehurst proceeded back to the winter quarters, arriving there on December 7 at 11.30 P.M. On December 9 they started for Butter Point again, taking five weeks' provisions for three men, in order to proceed up the Ferrar Glacier, and later to try to effect a junction with the Northern Party.

Only five men—Murray, Joyce, Day, Marston and Roberts—were now at the winter quarters. The heat of the Antarctic summer being at its height, the snow-drifts were melting rapidly, and the trickling of running water was everywhere to be heard. A large drift remained on the hill behind the hut, leading up to Mawson's anemometer. On December 1 it was melting in several little trickles, and next day it was found that one of these had got under the hut and made a pool about a foot in depth at the lower end. Many valuable things were stored under the hut, and the only opening was occupied by the pool of water. A hole had to be made at one side of the house, where the ground was higher, and into this Joyce crawled and spent some hours wriggling about in a space hardly more than one foot in height, rescuing valuable boxes of printing material and printed matter.

In the succeeding days the men at the hut had an illustration of the contrasts which the Antarctic climate presents. The heat of the sun melted the snow, and indeed made the weather oppressively warm, yet the water which ran below the house where the sunshine could not penetrate and the air temperature never rose above 32° Fahr., froze at night and never thawed again, so that the water each day added a layer to the accumulation beneath the hut, till it reached nearly up to the floor.

After the final departure of the Western Party on December 9, life at the winter quarters was uneventful until the arrival of the *Nimrod*. The members of the expedition remaining at Cape Royds were busy collecting skua eggs, preparing skins, carrying on the routine scientific observations, and watching the doings of the Adelie penguins. Many photographs were taken, especially by Day, of penguins in every variety of attitude, and of other subjects of interest. Experiments were made in photographing microscopic animals, and many pictures of them from life were obtained.

Chapter Three

RETURN OF THE *NIMROD*

AFTER leaving us on February 22, the *Nimrod* had an uneventful voyage back to New Zealand. Fair winds were encountered all the way, and the ice gave no difficulty, the coast of New Zealand being sighted twelve days after the departure from Cape Royds. During the winter the *Nimrod* had been laid up in Port Lyttelton waiting till the time arrived to bring us back to civilisation. The little ship had been docked and thoroughly overhauled, so that all effects of the severe treatment she received during the first voyage down to the ice had been removed, and she was once more ready to battle with the floes. Towards the end of the year stores were taken on board, for there was a possibility that a party might have to spend a second winter at Cape Royds, if the men comprising one of the sledging expeditions had not returned, and, of course, there was always the possibility of the *Nimrod* herself being caught in the ice and frozen in for the winter. Sufficient stores were taken on board to provide for any such eventualities, and as much coal as could be stowed away was also carried. Captain P. F. Evans, who had commanded the *Koonya* at the time she towed the *Nimrod* down to the Antarctic Circle, was appointed master of the *Nimrod* under my power of attorney, Captain England having resigned on account of ill-health after reaching New Zealand earlier in the year.

The *Nimrod* left Lyttelton on December 1, 1908,

and encountered fine weather for the voyage southwards. On the evening of the 3rd, the wind being favourable, the propeller was disconnected, and the vessel proceeded under sail alone until the 20th, when she was in latitude 66° 30′ South, longitude 178° 28′ West. The "blink" of ice was seen ahead and the ship was hove to until steam had been raised and the propeller connected. Then Captain Evans set sail again, and proceeded towards the pack. The vessel was soon in brash ice, and after pushing through this for a couple of hours reached the pack, and made her way slowly through the lanes. Numerous seals were basking on the floes, regarding the ship with their usual air of mild astonishment. On the following day the pack was more congested, and the progress southward was slow, so much so that the crew found time to kill and skin several crabeater seals. Open water was reached again that evening, and at noon on the 22nd the *Nimrod* was in latitude 68° 20′ South, longitude 175° 23′ East, and proceeding under sail through the open water of Ross Sea. The belt of pack-ice had been about sixty miles wide.

On December 26 the *Nimrod* reached latitude 70° 42′ South, longitude 173° 4′ West, the position, in which, in 1843, Sir James Ross sighted "compact, hummocky ice," but found only drift ice, with plenty of open water. A sounding gave no bottom with 1575 fathoms of wire, so that the theory that the ice seen by Ross was resting on land was completely disproved. At noon on the 27th the *Nimrod,* which was proceeding in a south-east direction, was brought up by thick floes in latitude 72° 8′ South, longitude 173° 1′ West. Progress became possible again later in the day, and at four o'clock on the following morning the *Nimrod* was in open water, with the blink of pack to the eastward.

Captain Evans had kept east with the hope of sighting King Edward VII Land, but the pack seemed to be continuous in that direction, and on the 30th he therefore shaped a course for Cape Bird, and on January 1, 1909, Mount Erebus was sighted. The experience of Captain Evans on this voyage confirms my own impression that, under normal conditions, the pack that stretches out from the Barrier to the eastward of the Ross Sea is not penetrable, and that the *Discovery* was able to push to within sight of King Edward VII Land in 1902 for the reason that the ice was unusually open that season.

The progress of the *Nimrod* towards the winter quarters was blocked by ice off Beaufort Island, and after manœuvring about for three hours Captain Evans made the vessel fast to a floe with ice anchors. The next morning he cast off from the floe, and with the help of the current, which seems to set constantly to the west between Cape Bird and Beaufort Island, and by taking advantage of lanes of open water, gradually proceeded in two days to a point only twenty-eight miles from Cape Royds. Some heavy bumps against the floes tested the strength of the vessel, and finally what appeared to be fast ice was encounterd, so that no further progress towards the south was possible for the time.

There seemed to be no immediate possibility of the *Nimrod* reaching Cape Royds, and Captain Evans therefore decided to send Mackintosh with three men to convey a mail-bag and the news of the ship's arrival to the winter quarters. The party was to travel over the sea ice with a sledge, and it did not seem that there would be any great difficulties to be encountered. A start was made at 10.15 A.M. on January 3, the party consisting of Mackintosh, McGillan, Riches and Paton,

SERRATED EDGE OF GLACIER SOUTH OF CAPE BARNE, ROSS ISLAND

*

* View from High Hill after second arrival of the "Nimrod." The Ship in loose pack

with one sledge, a tent, sleeping-bags, cooking equipment and a supply of provisions. The distance to be covered was about twenty-five miles. In the afternoon Mackintosh sent Riches and Paton back to the ship, and he reduced the load on the sledge by leaving fifty pounds of provisions in a depôt. The travelling became very rough, the two men encountering both bad ice and soft snow. They camped at 7.50 P.M., and started for Cape Royds again at 1.55 A.M. on the following day. They soon got on to a better surface, and made good progress until 5.30 A.M., when they met with open water, with pressure ice floating past. This blocked the way. They walked for two hours in a westerly direction to see how far the open water extended, but did not reach the end of it. The whole of the ice to the southward seemed to be moving, and the stream at the spot at which they were then standing was travelling at the rate of about three miles an hour. They breakfasted at 7.30 A.M., and then started back for the ship, as there seemed to be no chance of reaching Cape Royds in consequence of the open water.

Presently Mackintosh found that there was open water ahead, blocking the way to the ship, and a survey of the position from a hummock revealed the unpleasant fact that the floe-ice was breaking up altogether, and that they were in most serious danger of drifting out into the sound. Safety lay in a hurried dash for the shore to the east, and they proceeded to drag their sledge across rough ice and deep snow with all possible speed. At places they had to lift the sledge bodily over the ice-faces, and when, after an hour's very heavy work, they arrived off the first point of land, they found an open lane of water barring their way. "We dragged on to the next point, which appeared to be safe," wrote Mackintosh in his diary. "The floes were small and square in shape. Every two hundred yards we had to drag our sledge to the edge of a floe, jump over a lane of water, and then with a big effort pull the sledge after us. After an hour of this kind of work our hands were cut and bleeding, and our clothes, which, of course, froze as stiff as boards, had been wet through to the waist, for we had frequently slipped and fallen when crossing from floe to floe. At 2.30 P.M. we were near to the land, and came to a piece of glacier ice that formed a bridge. The floe that we were on was moving rapidly, so we had to make a great effort and drag our sledge over a six-foot breach. Our luck was in, and we pulled our sledge a little way up the face of the fast ice, and unpacked it. We were in a safe position again, and none too soon, for fifteen minutes later there was open water where we had gained the land."

Mackintosh decided to go into camp near the spot where they had landed, as a journey across the rocks and the glaciers of the coast was not a thing to be undertaken lightly, and would probably be impossible unless the mail-bag was left behind. McGillan, moreover, had developed snow-blindness, and both men were very tired. I will quote from Mackintosh's report on the subsequent experience of this little party.

"Early the next morning I found McGillan in great pain," wrote Mackintosh. "His eyes were closed up completely, and his face was terribly swollen. The only remedy I could apply was to bathe them, and this seemed to give him some relief. From an elevated position I had a good look round for the ship, and could not see a trace of her. As the day wore on my own eyes became painful. I fervently hoped I was not going to be as bad as my companion, for we would then be in a very difficult position. The morning of January 6 found us both blind. McGillan's face was frightfully swollen, and his eyes completely and tightly shut, so that he did not know that I was attacked too. At first I refrained from telling him, but the pain was very severe, and I had to tell him. By the painful process of forcing my eyelids apart with my fingers I could see a little, but I was not able to do this for long. I continued to bathe McGillan's eyes, and then suffered six hours' agony, ending in a good long sleep, from which I awoke refreshed and much better. I was able to see without effort. McGillan was also much better, and our relief, after the anxiety we had felt, was very great. By midnight we had improved so much that we walked to the penguin rookery, where we had great fun with the birds and found several eggs."

The men stayed in camp for several days, seeing no sign of the ship, and after their eyes were better spent a good deal of time studying the neighbourhood and especially the bird-life. They cut down their food to two meals a day, as their supply of food was not large. Finally, Mackintosh decided that he would leave the mail-bag in the tent, it being too heavy to carry for any distance, and march into Cape Royds. They made a start on the morning of January 11, carrying forty pounds each, including food for three meals, and expected to be able to reach the winter quarters within twenty-four hours. The first portion of the journey lay over hills of basaltic rock, at the base of Mount Bird, and they thought it best to get as high as possible in order to avoid the valleys and glaciers. They went up about five thousand feet, and had fairly easy travelling over slopes until they got well on to the glaciers. Then their troubles commenced. They were wearing ski-boots without spikes, and had many heavy falls on the slippery ice. "We were walking along, each picking his own tracks, and were about fifty yards apart, foolishly not roped, when I happened to look round to speak to my companion, and found that he had disappeared," wrote Mackintosh. "Suddenly I heard my name called faintly from the bowels of the glacier, and immediately rushed towards the place from which the sound proceeded. I found McGillan in a yawning chasm, many feet beneath me, and held up on a projection of ice. I took off my straps from my pack and to them tied my waist lashing, and lowered this extemporised rope down to him. It just reached his hand, and with much pulling on my part and knee-climbing on his, he got safely to the surface of the glacier again. The Primus stove and our supply of food had gone further down the crevasse. We tried to hook them up, and in doing so I lost my straps and line which I had attached to a ski-stick, so we were left almost without equipment. As soon as McGillan had recovered from the shock he had received we started off again, with the spare strap tying the two of us together. We crossed over many snow-bridges that covered the dangers underneath, but soon we were in a perfect hot-bed of crevasses. They were impassable and lay right across our path, so that we could look down into awful depths. We turned and climbed higher in order to get a clear passage round the top. We were roped together and I was in the lead, with McGillan behind, so that when I fell, as I often did, up to my waist in a crevasse, he could pull me out again. We found a better surface higher up, but when we began to descend we again got into crevassed regions. At first the crevasses were ice-covered gaps, but later we came to huge open ones, whose yawning depths made us shudder. It was not possible to cross them. We started to ascend again, and soon came to a bridge of ice across a huge crevasse about twenty feet wide. We lashed up tighter, and I went off in the lead, straddle-legged across the narrow bridge. We both reached the other side in safety, but one slip, or the breaking of the bridge would have precipitated us into those black depths below."

The two men found their way blocked by crevasses in whichever direction they turned, and at last reached a point from which ascent was out of the question, while below lay a steep slope running down for about three thousand feet. They could not tell what lay at the bottom of the slope, but their case was desperate, and they decided to glissade down. Their knives, which they attempted to use as brakes, were torn from their grasp, but they managed to keep their heels in the snow, and although they passed crevasses, none lay directly in their path. They reached the bottom in safety at 4 P.M. on the 11th. They were very hungry and had practically no food, but they could get forward now, and at 6 P.M. they could see Cape Royds and were travelling over a smooth surface. They ate a few biscuit crumbs and half a tin of condensed milk, the only other food they had being a little chocolate. Soon snow commenced to fall, and the weather became thick, obscuring their view of the Cape. They could not see two yards ahead, and for two hours they stumbled along in blinding snow. They rested for a few minutes, but their clothes were covered with ice, icicles hung from their faces, and the temperature was very low. In a temporary clearing of the blizzard Mackintosh thought that he could make out the Cape and they dashed off, but at lunch-time on the 12th they were still wandering over the rocks and snow, heavy snow cutting off all view of the surrounding country. Soon after this the snow ceased to fall, though the drift-snow, borne along by the blizzard wind, still made the weather thick. Several times they thought that they saw Cape Royds, but found that they had been mistaken. As a matter of fact they were quite close to the winter quarters when, at about 7 P.M., they were found by Day. They were in a state of complete exhaustion, and were just managing to stagger along because they knew that to stop meant death. Within a few minutes they were in the hut, where warm food, dry clothes and a good rest soon restored them. They had a narrow escape from death, and would probably have never reached the hut had not Day happened to be outside watching for the return of the ship.

Mackintosh and McGillan reached the hut on January 12, but in the meantime the *Nimrod* had arrived at Cape Royds and had gone north again in search of the missing men. Murray had sailed in the *Nimrod,* and as events turned out, he was not able to get back to the hut for about ten days. "We were having tea on the afternoon of January 5, and Marston happening to open the door, there was the *Nimrod* already moored to the edge of the fast ice, not more than a mile away," wrote Murray in a report on the summer work. "We ran towards the ship, over the rotten sea ice, in boots or slippers as chanced, with the one idea that is uppermost in these circumstances—to get 'letters from home.' We were doomed to disappointment. Before we had finished greeting our old friends, the officers asked us, 'Has Mackintosh arrived?' and we learned to our horror that he and a companion had left the ship two days before and thirty miles north of Cape Royds, to try to bring the letters sooner to us over the sea ice, over the bay where only a few days ago we saw a broad

Group in the Hut in the Summer: Joyce at the Sewing Machine

Bluff Depot Party on the Barrier

sheet of open water to the horizon, and which was even now only filled with loose pack! So we got no home letters, and had good reason to believe that our friends had lost their lives in the endeavour to bring them. We knew that they must have embarked on a large floe, and little expected to see them again. On January 7 the *Nimrod* left Cape Royds to seek for the lost men, on the chance that they might have got ashore near Cape Bird. Within a few hours she was caught by the pack which was drifting rapidly southward along the shore of Ross Island. Driven almost on shore near Horseshoe Bay, the ship, by dint of hard steaming, got a little way off the land, and was there beset by the ice and so remained from the 7th to the 15th, with only a few hours' ineffectual steaming during the first day or two. At length she was rigidly jammed and was carried helplessly by a great eddy of the pack away towards the western side of the sound, and gradually northward.

"On January 12 she was as tight as though frozen in for the winter. In the afternoon sudden pressure affected all the ice from the *Nimrod* as far as we could see. Great blocks of ice, six or eight feet in thickness, were tossed and piled on the surface of the floes. These pressure heaps were formed on each side of the ship's bow, but she took no harm, and in about an hour the pressure ceased. On the morning of January 15 there was not the slightest sign of slacking of the pack, but in the early afternoon Harbord, from the crow's-nest, saw lanes of water at no great distance to the east. Steam was got up and in a few hours we had left our prison and got into a broad lane, with only thin ice which the ship could charge, and the open water was in sight. Shortly after midnight we got clear of the ice. When released we were not very far from the Nordenskjold Ice-Barrier.

"The deceptive appearance of loose pack was impressed upon us. For many hours there was blue water apparently only a mile or two ahead, but it never appeared to get any nearer for hours, and we could not be sure it was really near till we were within a few hundred yards of the edge. All this time in the pack we were in doubt as to the fate of Mackintosh, or rather, we had not much doubt about it, for we had given him up for lost, but we were helpless to do anything. On the afternoon of the 16th, on which day we cleared the ice, we had passed Beauford Island and were approaching through very loose pack the only piece of shore on which there was any chance of finding the lost men. Near the end of this stretch of beach, where it is succeeded by hopeless cliffs, a small patch of greenish colour was seen, and the telescope showed the details of a deserted camp, a tent torn to ribbons and all the camp gear lying around. A boat was sent ashore in charge of Davis, who found the bag of letters, and a note from Mackintosh pinned to the tent, telling of his risky attempt to cross the mountains nearly a week before. Knowing the frightfully crevassed character of the valley between Mount Bird and Mount Erebus, there seemed to us little hope that they would get through. The crevassed slope extends right to the top of Mount Bird, and is very steep towards the Erebus side. When we reached Cape Royds about midnight, only two men came out to meet the ship. One of the men was Mackintosh's comrade in all his adventures, and we soon learned that all had ended well."

In the meantime the Bluff Depôt party had started off to place a supply of provisions off Minna Bluff in readiness for the return of the Southern Party. The crew of the *Nimrod* proceeded to take on board the geological and zoölogical specimens collected by the expedition and stored at the hut, so that all might be in readiness for the final departure when the parties had been picked up. Then followed weeks of uncertainty as to the fate of the men who were away.

Chapter Four

THE BLUFF DEPÔT JOURNEY

I HAD left instructions at the winter quarters that a party should proceed to Minna Bluff at the beginning of the new year, and place at a point opposite the Bluff a depôt of stores for the use of the Southern Party on its return journey. Joyce was to take charge of this work, and it was of very considerable importance, since we four of the Southern Party would be depending on the depôt to supply us with the provisions necessary for the last hundred miles or so of the journey back to the winter quarters. Joyce was accompanied by Mackintosh, Day and Marston, and he found that as the snow surface was very soft it would be necessary to make two journeys to the Bluff, one with ordinary sledging provisions, and the other with special luxuries from the ship. The party left the winter quarters at Cape Royds on January 15, with one sledge and 500 lb. of provisions, drawn by eight dogs. Early in the afternoon they encountered soft ice, sticky with salt, and the travelling became very heavy. They kept well away from the land, but Joyce, Mackintosh and Marston all fell through at different times, the soft surface giving away under them, and they got wet up to their waists. Their clothing froze stiff at once. They camped for the night at Glacier Tongue, and the next morning found the weather so bad that they were unable to march. There was a strong southerly wind, with drift, and this soon turned into a howling blizzard. A calm succeeded at mid-

A Dog Team with Loaded Sledge going South to lay a Depot for the Return of the Southern Party

*

Depot Party pitching a Tent

The Bluff Depot

night, and on the morning of the 18th they got under way again. The dogs had been buried under the drift by the blizzard, only their noses showing at the surface, and it was necessary to dig them out before they could be harnessed up. A seven-foot sledge was loaded with 300 lb. of store from the depôt at the Tongue, and the four men took on the two sledges, with a total weight of 800 lb. They had a heavy day's work, over soft ice and snow-drift, but reached the old *Discovery* winter quarters at Hut Point at midnight. The dogs pulled very well, and seemed to be enjoying their work after the long spell of semi-idleness at Cape Royds.

On the morning of the 19th the party proceeded on to the Barrier. The surface was fairly good, and the dogs ran practically all the time, Joyce finding it necessary to put two men on the sledge in order to reduce the speed, for the men would not travel at the pace set by the dogs. The weight per dog was well over 100 lb., though only one sledge had been taken on from Hut Point. The temperature was low during the days that followed, and the men's beards were constantly coated with ice, but their progress was rapid. On January 23, when they were travelling over a deep snow surface covering sastrugi, they sighted a depôt about three miles to the west of their course. This was the depôt at which some pony fodder had been left in the spring. Soon after this the party came upon crevasses running at right angles to their course, and the travelling became difficult. Joyce had the members of the party roped together, as the crevasses were hidden by treacherous snow lids and were therefore dangerous. The crevasses became worse in the following two days. Some of the pressure ridges were over thirty feet in height, running in an east-south-east and west-north-west direction, with enormous crevasses between them, and they all had the experience of falling through, to be hauled out again by means of the rope, after they had dropped to the length of their harness with a heavy jerk. On one occasion the four centre dogs fell through a snow-lid into a crevasse, and were got out with great difficulty. Day and Joyce, with two leaders, were on one side of the crevasse, and Mackintosh and Marston, with the two rear dogs, were on the other side. Day and Joyce had to unharness and ease the dogs, while the other two men pulled them back to the sledge. The dogs meanwhile were hanging over the abyss, and evidently did not like their position. Joyce had to keep altering his course in order to avoid these crevasses, but after steering in a south-west direction for about six hours he reached a better surface. The crevasses were getting smaller, although the surface of névé caused many falls. An attempt to steer south, straight for the spot at which the depôt was to be laid, resulted in the party getting into the badly crevassed area again, and once more Joyce had to steer east-south-east. Finally they got clear of the crevasses, and at midnight on January 25, reached the spot at which it had been decided to place the depôt, about fourteen miles off Minna Bluff.

An early start was made on the 26th, and for seven hours the party laboured erecting a mound of snow ten feet high. On top of the mound they put two eleven-foot bamboos, lashed together and carrying three black flags. The total height of the depôt was twenty-two feet, and it could be seen at a distance of eight miles. The bearing of this depôt I had arranged with Joyce during the spring depôt journey, before my departure for the southern journey. It was on a line drawn through a sharp peak on the Bluff, well-known to Joyce, and the top of Mount Discovery, with a cross bearing secured by getting the centre peak of White Island in line with a peak of Mount Erebus.

The party started north again on the morning of the 27th, and after they had travelled a short distance Day sighted a pole projecting from the snow, some distance to the west of their course. Joyce was able to identify this as the depôt laid out for the *Discovery's* Southern Party in the spring of 1902. There was a bamboo pole about eight feet high projecting from the snow, with a tattered flag attached to it, and a food tin on top. The guys to which the pole was attached were completely buried under the snow. The men dug down for about five feet with the idea of ascertaining how deeply the depôt had been covered by snow, but as the bottom had not been reached and time was limited, they put fresh flags on the pole and proceeded on their way, intending to visit the depôt on the second journey. A fresh southerly wind was blowing, and rapid progress was made to the north towards Cape Crozier. A sail was hoisted on the sledge, and this assisted the dogs so much that three men were able to sit on the sledge, while a pace of about four miles an hour was maintained. Soon the area of crevasses, caused by the impinging of the Barrier ice on the land to the west, was reached again, and for thirty-seven miles the party twisted and turned in making a course past the obstacles; Joyce counted the crevasses that were passed, and he reported that he had seen one hundred and twenty-seven, ranging from two feet to thirty feet in width. The larger ones were open, and therefore easily detected, but the smaller ones had the usual snow-lids. On the 30th the men were held up by another blizzard, which completely buried the dogs and sledge, but they reached Hut Point at 11 P.M. on January 31.

A second load of stores was secured from the depôt, including some luxuries, such as apples and fresh mutton, brought by a party from the ship, and on February 2 the party started south again. Joyce decided to take a new course in order to avoid the crevasses. He kept a course towards Cape Crozier for two days, and then marched south on the 5th and reached the depôt without having seen any crevasses at all. I think that the crevasses run right across to Cape Crozier from the district around White Island, but they are evidently more snow covered along the outer course. When the party was close to the depôt a blizzard came up from the south, and there was just time to get the tents up before the drift became thick. The tents were completely snowed up before the weather cleared, and the men had some difficulty in getting out again. The dogs were covered, but they seemed to be quite happy in their "nests" deep in the drift. When dogs and sledge had been dug out the party started again, and at 2 A.M. on the 8th they reached the Bluff depôt for the second time.

"We expected to find the Southern Party camped there, and to surprise them with the luxuries we had brought out for them," wrote Joyce in his report, "but they were not there. As our orders were to return on the 10th if the Southern Party did not turn up, we began to feel rather uneasy. It came on to blow again from the south, and presently the wind turned into a howling blizzard, and did not ease down until the 11th. During every lull we climbed the depôt and looked round the horizon with the glasses, expecting every minute to see the Southern Party loom up out of the whiteness, but they did not appear. On the 11th, after a consultation, we decided to lay depôt flags in towards the Bluff, so that there would be no chance of

THE DEPÔT PARTY AMONGST CREVASSES

the other party missing the food depôt. We knew that they would be run out of provisions, as they were then eleven days overdue, and the position caused us great anxiety. After we had laid the flags, three miles and a half apart, with directions where to find the depôt, we decided to march due south to look for the Southern Party. At every rest we would get on the sledge with the glasses, and look around, thinking that each snow hummock was a man or a tent. On the 13th Day sighted some marks in the snow that looked unusual, and on examination we found them to be the hoof-prints of the ponies, evidently made on the outward march of the Southern Party three months before. The tracks of the four sledges showed distinctly. We followed these tracks for seven hours, and then we lost them. We camped that night at 10 P.M., and early the next morning proceeded south again, thinking all the time that we would see something appear out of the loneliness. It is curious what things one can see in circumstances like these, especially with a bad light. We started back to the depôt with all sorts of fears for the Southern Party."

They reached the depôt again at noon on the 16th, and Joyce states that as they approached the mound they were all sure that they could see a tent up and men walking about. When they got close, however, they found that everything was just as they had left it. They put all the provisions on top of the mound, lashed everything securely, and examined the flags to the eastward, and started on the march back to the coast, full of gloomy thoughts as to the fate of the Southern Party, which was now eighteen days overdue.

They proceeded first to the old *Discovery* depôt found on the first journey, Joyce wishing to take some measurements in order to ascertain the movement of the Barrier ice, and the amount of the snowfall. The depôt had been laid six years previously on bearings off the Bluff, and after its original position had been ascertained as exactly as possible, the distance to the bamboo pole was measured off by Day and Marston with a forty-foot length of rope, which had been measured off with a tape-measure. The distance was found to be 9600 ft., and the direction of the movement was about east-north-east. The Barrier ice at this point must therefore be moving forward at the rate of about 1500 ft. a year. The party then worked till 1 A.M. digging down in order to find what depth of snow had been deposited on top of the depôt during the six years. It was found that the level at which the stays of the depôt pole had been made fast was eight feet three inches down in hard compressed snow. A measured quantity of this snow was melted in order to ascertain the actual amount of the snowfall. The interesting points involved in these investigations will be dealt with in the reports on the scientific work of the expedition.

The party started north again on the following day, and covered a distance of thirty-three miles. The dogs pulled splendidly, and three men were able to ride on the sledge. On the second day crevasses were encountered again, and several times men fell through to the length of their harness. The general direction of the crevasses was east-south-east and west-south-west. The party had a narrow escape from complete disaster at this stage. "We were going at a good trot over a very hard surface," wrote Joyce, "when I felt my foot go through. I called out 'Crack!' and rushed the dogs over, and as the sledge touched the other side of the hard ridge, the whole snow-bridge over which we had passed fell in. Marston, who was running

* Digging to ascertain the depth of snow covering a depot left by the "Discovery" Expedition

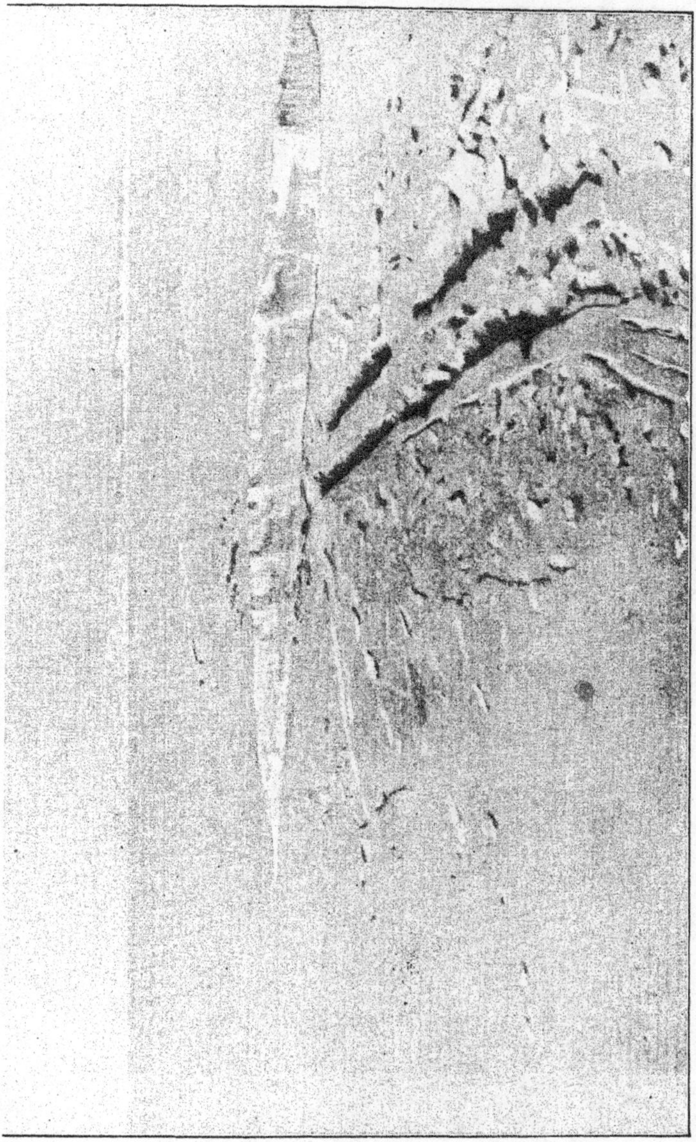

The Winter Quarters, with Mount Erebus in the background. The Dog-kennels are seen on the left

astern of the sledge, felt himself falling through space, but the pace of the dogs brought him over the crevasse, at the length of his harness. We found ourselves standing on the edge of a yawning gap that would easily have swallowed up sledge, dogs, and the whole party, and on the far side we could see our sledge tracks leading right up to the edge. It seemed almost a miracle how we had managed to escape. Day took a photograph, and we altered the course for Cape Crozier, getting out of the crevasses about 5 P.M. Then we camped for the night, having all had a good shaking up."

A long march the next day over a good surface brought the party to Cape Armitage at midnight. Joyce found that the ice in the sound had gone out, and it was therefore necessary to climb through the gap at Observation Hill. A blizzard came up, and with great difficulty the party reached the old *Discovery* hut at Hut Point at 2 A.M. The distance covered during the day had been forty-five miles, an unusually good performance. The surface had been good, and the wind favourable, and the dogs had pulled splendidly. Joyce speaks very highly of the work of the dogs on this journey. They were pulling over 100 lb. per dog, and yet ran most of the time. They suffered a good deal from snow-blindness, and then they used to dig a hole in the snow and bury their faces right in; this method of treatment seemed to ease their eyes and they recovered from the attacks very quickly. "One day I released Tripp, because he had a chafed leg," wrote Joyce, "and for the whole day he ran in his place in the team, as if he had been harnessed up. He slept about half a mile from camp that night, and when I tried to coax him over in the morning he would not come, but as soon as we got under way he came running up to his old place. I fed the dogs on one pound of biscuit a day each, and this seemed to satisfy them; as they went on their condition improved. The dog harness was generally satisfactory, but could have been improved with a few more swivels, in order to avoid tangling when the dogs jumped over their traces. I think that all dog teams should be taught to be driven, as a man cannot keep pace with dogs, and holding them back in order that a man may go ahead causes them to get fagged out. If they were let go at their best pace, one could, with a light load, say 80 lb. per dog, get forty miles a day out of them over a good surface."

Chapter Five

THE WESTERN JOURNEY

MEANWHILE the Western Party, which had left the winter quarters for the second time on December 9, had been working in the western mountains. The three men (Armytage, Priestley and Brocklehurst) reached the stranded moraines on December 13, and on this occasion succeeded in securing a large supply of skuas' eggs. The anticipated feast was not enjoyed, however, for only about a dozen of the eggs were "good enough for eating," to quote the words of a member of the party. The other eggs were thrown on to the snow near the tent, and the result was an invasion of skuas, which not only ate the eggs, but also made themselves a general nuisance by pulling about the sledge harness and stores. At this time the men were troubled by patches of thin ice, about an eighth to a quarter of an inch thick, forming a lenticle, the top of the middle being sometimes as much as five or six inches from the actual surface. When these patches of ice were trodden on they broke down, and not infrequently disclosed a puddle of salt water an inch or two deep. Priestley thought that they were the final product of the thawing of snowdrifts, and owed their character to the fact that the salt water worked faster from below than did the sun from above.

On December 15 the party started to ascend the Ferrar Glacier, Priestley examining the rocks carefully on the way with a view to securing fossils if any were to be found. The surface was bad for the most part, soft snow being encountered where ice had been expected. On December 19 they were held up by a blizzard, and then they got on to very slippery crevassed ice. On December 20 they camped near the Solitary Rocks, at the spot at which Captain Scott had camped after leaving Dry Valley. The idea of getting to Depôt Nunatak had to be abandoned, for a heavy snowfall made the travelling difficult, and the time at the disposal of the expedition was short. Priestley worked under the Bluff between Dry Valley and the east fork of the glacier without success, and then they moved over to Obelisk Mount. "I have examined block after block of unfossiliferous-looking sandstone without any success," wrote Priestley in his diary on the evening of the 21st. "The only thing I can find different from the ordinary quartz grains are a few seams of conglomerate with quartz pebbles, and a few lenticles of a soft clayey substance. The other rock I have collected here is a junction between granite and porphyry, which is common. The sandstone is very weathered, dropping to pieces in many cases at a single blow. I am faced with the necessity of examining for fossils rocks which I should carefully avoid if I were at home or anywhere else. I have never seen a sedimentary rock that looked more unfossiliferous. Many of the boulders are coated with a hard crust of white, opaque salt (probably calcium carbonate), and if there was ever any lime in the sandstone it has probably been dissolved out long ago. There are numerous interesting rocks about here, but I am debarred from collecting much by the difficulty of transport. . . . There are evidently serious defects in the map near this point. The whole of the bluff opposite is marked as Beacon sandstone, and from the face of the cliff here it is easily seen, for at least 3000 feet, to be granite; the very grain in the stones can

* Camp on December 17 on the Ferrar Glacier below Sentinel Rock

Rough Ice Surface near Windy Gully

* The Western Party Camped on the Ferrar Glacier on December 18. Heavy hanging Glaciers on the Hills

be seen. This granite is capped by the beacon sandstone on the tops of the hills, but the dolerite seems to have died out, with the exception of the upper flow. This formation is as it should be, taking into consideration the horizontal structure of the rocks, and it was the fact that I doubted the existence of Beacon sandstone so low down the series that brought me here, as much as the expectation of finding fossils if the mapping should be correct."

At this time thawing was proceeding rapidly on the glacier. The party made for the north wall of the glacier, but was stopped by a precipice of ice, 200 to 300 feet in height, with a stream of water flowing at its foot. The deeper ripple-cracks and potholes were filled with water, and water was streaming down the convex face of the glacier to the stream which was roaring beneath the ice-face. The scene was a magnificent one, but the conditions were unpleasant from the point of view of the party. The ice was separated from the rock at the sides of the glacier by a thaw-gully about fifty feet deep, with a stream of water flowing at the bottom. Then came the lateral moraine, which was still within the region in which the rock-heat was felt, and formed a depression some three to six feet below the main surface of the glacier, commencing as an abrupt rise, almost perpendicular, but somewhat convex. After this came the ordinary billowy surface, and the next stones met with formed a sub-medial moraine not sufficiently thick to effect any lowering in the general level of the surface, although each stone was surrounded by its own hollow, filled with thaw water. Along the middle meandered a small stream, a few feet wide, with the bottom of its channel filled with morainic gravel matter. In the evening freezing commenced in the potholes that were sheltered from the rays of the sun,

Long needles of ice formed from the small grains of gravel, and crossed and recrossed in the most beautiful pattern. In some holes hexagonal plates of ice were being formed on the surface.

An examination of the Solitary Rocks proved that the map was incorrect at this point. The previous expedition had thought that the rocks formed an island, with the glacier flowing down on either side, but a close examination showed that the rocks were in reality a peninsula, joined to the main north wall by an isthmus of granite at least one thousand feet high. The glacier surged round the peninsula on its way down Dry Valley, and just below the isthmus was a lake of some size, fed by streams from a glacier opposite. These streams were yellow with silt, and another stream, also much discoloured, was running from the lake down to Dry Valley. The Solitary Rocks are at an altitude of about two thousand feet above sea-level. Priestley proceeded with geological and survey work in the neighbourhood of the east fork, and made an extensive examination of the spots known as Kurki Kills, Knob Head Mount and Windy Gully. On December 24, close to the camp, they found the bleached skeleton of a crabeater seal. It is rather curious that one of these animals should proceed so far up the glacier. A new camp was pitched at the foot of Knob Head Mountain, just below the second gully east of Windy Gully, and here Armytage and Priestley climbed up the slope behind to an altitude of 4200 ft., finding a yellow lichen at 3100 ft., a black lichen at 3800 ft. and a green lichen or moss at 4200 ft. The altitude of the camp was 2470 ft.

Christmas Day was spent at this camp, and, as was the case with the other sledging expeditions that were out at the time, a special feast was provided. For breakfast they had hoosh, sardines in tomato sauce and raisins; for lunch Garibaldi biscuits and jelly; and for dinner potted boneless chicken and a small plum pudding. Armytage picked up a piece of sandstone with fernlike markings, but Priestley was not hopeful of finding fossils in the greatly altered sandstone. The day was spent in geological work. "We lose the sun here about 9.30 P.M.," noted Priestley in his diary, "and it is curious to observe the sudden change from bright light to darkness in the tent, while outside the thin surface of ice covering the thaw-water round the rocks immediately contracts with reports like a succession of pistol-shots, and sometimes breaks up and flies about in all directions, making a noise like broken glass. This is the effect of the quick cooling of the ice by the cold plateau wind immediately the sun's influence is withdrawn. The plum pudding was 'top hole.' Must remember to give one of the pot-holed sandstones to Wild for the New Zealand girl who gave him the plum pudding."

On December 27 the men proceeded down the glacier again in order to see whether the Northern Party had arrived at Butter Point. Priestley studied the moraines on the way down, and made an extensive collection of specimens, and on January 1 they arrived at the depôt. They had constant trouble with crevasses and "potholes" on the way down the glacier, but met with no serious accident. The snow-bridges many times let them through up to their knees or waists, but never broke away entirely. The weather was unpleasantly warm for the sort of work they were undertaking, since the snow was thawing, and they were constantly wet.

They found no sign of the Northern Party at Butter Point, and after waiting there until the 6th they proceeded to the "stranded moraines," a day's trek to the south, in order that geological specimens might be secured. The moraines, which were found by the *Discovery* expedition, and are relics of the days of more extensive glaciation, present a most varied collection of rocks, representative of the geological conditions to be found in the mountains to the west, and are of very great interest on that account. After spending two days at this spot, the party went back to Butter Point with about 250 lb. of specimens, and camped again till the 11th. Still there was no sign of the Northern Party, and on the 12th they went north to Dry Valley. There Priestley found a raised beach, about sixty feet above sea-level, and Brocklehurst climbed the mountain known as the Harbour Heights.

Numerous fragments of Pecten Colbecki, the shell at present common at Cape Royds, were found imbedded in the sand as far up as sixty feet, and Priestley thought that they would probably be found higher still. Writing of the moraines at this point, he said: "In their chief characteristics they are very similar to the stranded moraines. Large patches of gravel are mixed with boulders of every description and size, a chaos of sedimentary, volcanic, plutonic, hypabyssal and metamorphic rocks, segmented by watercourses, which are bordered by flats of gravel and spread out before reaching the sea over large, alluvial, fan-shaped mudflats. They differ from the stranded moraines in the presence of numerous specimens of now existing shells, imbedded in the gravel and sand of the moraines, but found in most cases under any steep declivity in a stream's bed where it has cut back through a gravel terrace. The remarkable part of the preservation of these shells is their extreme fragility. Of Pecten Colbecki I have seen thousands of specimens, and have secured many whole single valves. Of the Anatæna so common at

Western Party in Camp on December 20

WESTERN PARTY'S CAMP ON DECEMBER 28 BELOW A HANGING GLACIER AT THE CATHEDRAL ROCKS

Looking down the Eastern Arm of the Ferrar Glacier towards Dry Valley from Solitary Rocks

* Dry Valley

Cape Royds I came across several patches. At the head of one of the aluvial mudflats, about two feet above the present level of the sea-water, I secured many dried bodies of a small amphipod and a dried fish an inch long. The whole of the moraines so far as I have been are covered with seal bones, and I have seen two whole dried bodies, with the skin still on. One was a crabeater. Among several rock specimens secured was one of Beacon sandstone with the same curious markings as were found on two specimens secured by Armytage at Knob Head. The impression in the original stone was exactly as if the body of a wasp-like wingless insect several inches long had been pressed into clay."

They went back to the depôt on the 14th, and in accordance with the orders I had left, pitched camp in order to wait for the Northern Party until the 25th, when they were to make their way back to winter quarters, or signal for the ship by means of the heliograph. On January 24–25 this party had a very narrow escape from disaster. They were camped on the sea-ice at the foot of Butter Point, intending to move off on the return journey early on the morning of the 25th. Their position was apparently one of safety. Armytage had examined the tide-crack along the shore, and had found no sign of more than ordinary movement, and the ice in the neighborhood seemed to be quite fast. At 7 A.M. on the 24th Priestley was first out of the tent, and a few minutes later he came running back to his companions to tell them that the ice they were on had broken away and was drifting away north to the open sea. The other two men turned out promptly, and found that his statement was only too true. There were two miles of open water between the floe and the shore, and they were apparently moving steadily out to sea. "When we found that the ice had gone out," wrote Armytage in his report to me, "we struck camp, loaded up the sledge, and started away with the object of seeing whether we could get off the floe to the north. The position seemed to be rather serious, for we could not hope to cross any stretch of open water, there was no reasonable expectation of assistance from the ship, and most of our food was at Butter Point. We had not gone very far to the north before we came to an impassible lane of open water, and we decided to return to our original position. We went into camp, and had breakfast at 11 A.M. Then we held a consultation and agreed that it would be best to stop where we then were for a time, at any rate, on the off-chance of the ship coming along one of the lanes to pick us up on the following day, or of the current changing and the ice once more touching the shore. We waited till three o'clock in the afternoon, but there did not seem to be any improvement in the position. The Killer whales were spouting in the channels, and occasionally bumping the ice under us. Then we marched north again, but met with open water in every direction, and after we had marched right round the floe we got into camp at the old position at 10 P.M. We had a small meal of hoosh and biscuit. We had only four days' provisions on the floe with us, and I decided that we would have to go on short rations. We were encouraged by the fact that we had apparently ceased to move north, and were perhaps getting nearer the fast ice again. We got into our sleeping-bags in order to keep warm. At 11.30 P.M. Brocklehurst turned out to see whether the position had changed, and reported that we seemed to be within a few hundred yards of the fast ice, and still moving towards the land. I got out of my bag and put on my finnesko, and at midnight saw that we were very close to the fast ice, probably not more than two hundred yards away. I ran back as fast as I could, deciding that there was a prospect of an attempt to get ashore proving successful, and gave the other two men a shout. They struck the camp and loaded up within a very few minutes, while I went back to the edge of the floe at the spot towards which chance had first directed my steps. Just as the sledge got up to me, I felt the floe bump the fast ice. Not more than six feet of the edge touched, but we were just at that spot, and we rushed over the bridge thus formed. We had only just got over when the floe moved away again, and this time it went north to the open sea. The only place at which it touched the fast ice was that to which I had gone when I left the tent, and had I happened to go to any other spot we would not have escaped. We made our way to Butter Point, and at about three o'clock in the morning camped and had a good meal. Then we turned in and slept. When we got up for breakfast, there was open water where we had been drifting on the floe, and I sighted the *Nimrod* under sail, ten or twelve miles out. We laid the heliograph on to the vessel, and after flashing for about an hour got a reply. The *Nimrod* came alongside the fast ice at three o'clock in the afternoon of January 26, and we went on board with our equipment and specimens. We left a depôt of provisions and oil at Butter Point in case the Northern Party should reach that point after our departure."

On January 22 and 23 a fresh wind blew from the south and commenced to break up the ice-sheet in the neighbourhood of Cape Royds, compelling the ship to re-fasten further to the southward. From this point Davis took a sledge-party to Hut Point with despatches that the supporting-party was to convey to me at the Bluff Depôt. On the 25th the ice had broken up to such an extent that Captain Evans thought there would be a chance of getting far enough across McMurdo Sound to search the western coast-line for the party that had been exploring the western mountains, and also for the Northern Party, which might by that time have returned from the journey to the Magnetic Pole and reached Butter Point. The *Nimrod* stood out into the sound, and from a distance of ten or twelve miles a heliograph was seen twinkling near Butter Point. The ship was able to get right alongside the fast ice, and picked up Armytage, Priestley and Brocklehurst.

After this date fine weather was experienced only at short intervals, the season being advanced, and as a consequence the fast ice that remained in the sound commenced to break up rapidly, and took the form of pack trending northwards. When blizzards blew, as they did frequently, the *Nimrod* moored on the lee-side of a stranded iceberg in the neighbourhood of Cape Barne, with the object of preserving her position without the consumption of more coal than was absolutely necessary. After the ice had broken up sufficiently, shelter was found under Glacier Tongue.

The waiting was rather unpleasant for the remaining members of the shore-party and for those on board the ship, for the time was approaching when it would be necessary to leave for the north unless the *Nimrod* was to be frozen in for the winter, and two of the parties were still out. I had left instructions that if the Northern Party had not returned by February 1 a search was to be made along the western coast in a northerly direction. The party was three weeks overdue, and on February 1, therefore, the *Nimrod* went north, and

Captain Evans proceeded to make a close examination of the coast. The ship did not get back to the hut until February 11. During this time Murray and Priestley found work of scientific interest. Priestley tramped the country, and now that the snow had in great measure disappeared, was enabled to see various interesting geological deposits previously covered up. Beds of sponge spicules, enclosing various other fossils, were evidence of recent elevation of the sea bottom. A thick deposit of salts was found on a mound between two lakes, and some curious volcanic formations were discovered. The smaller ponds were entirely melted, and gave a chance to find some forms of life not evident in winter. The penguins continued to afford Murray material for study.

The *Nimrod's* search for the Northern Party was both difficult and dangerous. Captain Evans had to keep close to the coast, in order to guard against the possibility of overlooking a signal, which might consist only of a small flag, and the sea was obstructed by pack-ice. He was to go north as far as a sandy beach on the northern side of the Drygalski Barrier, and he performed his duty most thoroughly in the face of what he afterwards modestly described as "small navigational difficulties." The beach, which had been marked on the chart, was found to have no existence in fact, but the *Nimrod* reached the neighbourhood indicated, and then proceeded south again, still searching every yard of the coast. On the 4th a tent was sighted on the edge of the Barrier, and when a double detonator was fired the three men who had been to the Magnetic Pole came tumbling out and ran down towards the edge of the ice. Mawson was in such a hurry that he fell down a crevasse, and did not get out again until a party from the ship went to his assistance. "They were the happiest men I have ever seen," said Davis in describing the finding of the party. Their sledge, equipment and specimens were taken on the *Nimrod,* which was able to moor right alongside the fast ice, and then Captain Evans proceeded back to the winter quarters. In the chapters that follow Professor David tells the story of the Northern Party's journey.

Chapter Six

PROFESSOR DAVID'S NARRATIVE

ON September 19, 1908, Lieutenant Shackleton gave me his final instructions for the journey of the Northern Party to the Magnetic Pole. These he read over to me in the presence of Mawson and Dr. Mackay. The instructions were as follows:

"BRITISH ANTARCTIC EXPEDITION, 1907.
"CAPE ROYDS, *September* 19, 1908.

INSTRUCTIONS FOR NORTHERN SLEDGE-PARTY UNDER COMMAND OF PROFESSOR E. DAVID.

"DEAR SIR,—The sledge-party which you have charge of consists of yourself, Douglas Mawson and Alistair Mackay.

"You will leave winter quarters on or about October 1, 1908. The main objects of your journey are to be as follows:

"(1) To take magnetic observations at every suitable point with a view of determining the dip and the position of the Magnetic Pole. If time permits, and your equipment and supplies are sufficient, you will try and reach the Magnetic Pole.

"(2) To make a general geological survey of the coast of Victoria Land. In connection with this work you will not sacrifice the time that might be used to carry out the work noted in paragraph (1). It is unnecessary for me to describe or instruct you as to details *re* this work, as you know so much better than I do what is requisite.

"(3) I particularly wish you to be able to work at the geology of the Western Mountains, and for Mawson to spend at least one fortnight at Dry Valley to prospect for minerals of economic value on your return from the north, and for this work to be carried out satisfactorily you should return to Dry Valley not later than the first week of January. I do not wish to limit you to an exact date for return to Dry Valley if you think that by lengthening your stay up north you can reach the Magnetic Pole, but you must not delay, if time is short, on your way south again to do geological work. I consider that the *thorough* investigation of Dry Valley is of supreme importance.

"(4) The *Nimrod* is expected in the Sound about January 15, 1909. It is quite possible you may see her from the west. If so, you should try to attract attention by heliograph to winter quarters. You should choose the hours noon to 1 P.M. to flash your signal, and if seen at winter quarters the return signal will be flashed to you, and the *Nimrod* will steam across as far as possible to meet you and wait at the ice-edge. If the ship is not in, and if she is and your signals are not seen, you will take into account your supply of provisions and proceed either to Glacier Tongue or Hut Point to replenish if there is not a sufficient amount of provision at Butter Point for you.

"(5) *Re* Butter Point. I will have a depôt of at least fourteen days' food and oil cached there for you. If there is not enough in that supply you ought to return as mentioned in paragraph (4).

"(6) I shall leave instructions for the master of the *Nimrod* to proceed to the most accessible point at the

Picking up the Western Party

THE MOTOR-CAR IN THE GARAGE, AND MAIZE-CRUSHER ON THE RIGHT

west coast and there ship all your specimens. But before doing this, he must ship all the stores that are lying at winter quarters, and also keep in touch with the fast ice to the south on the look-out for the Southern Sledge-party. The Southern Party will not be expected before February 1, so if the ship arrives in good time you may have all your work done before our arrival from the south.

"(7) If by February 1 after the arrival of the *Nimrod*, there is no evidence that your party has returned, the *Nimrod* will proceed north along the coast, keeping as close to the land as possible, on the look-out for a signal from you flashed by heliograph. The vessel will proceed very slowly. The ship will not go north of Cape Washington. This is a safeguard in event of any accident occurring to your party.

"(8) I have acquainted both Mawson and Mackay with the main facts of the proposed journey. In the event of any accident happening to you, Mawson is to be in charge of the party.

"(9) Trusting that you will have a successful journey and a safe return.

"I am yours faithfully,

"(Sgd.) Ernest H. Shackleton,

"Commander."

"Professor David,

"Cape Royds,

"Antarctic."

"Cape Royds,

"British Antarctic Expedition, 1907.

"Professor David.

"Dear Sir,—If you reach the Magnetic Pole, you will hoist the Union Jack on the spot, and take possession of it on behalf of the above expedition for the British nation.

"When you are in the Western Mountains, please do the same at one place, taking possession of Victoria Land as part of the British Empire.

"If economic minerals are found, take possession of the area in the same way on my behalf as Commander of this expedition.

"Yours faithfully,

"(Sgd.) ERNEST H. SHACKLETON,

"Commander."

We had a farewell dinner that night, given in honour of the Southern Depôt Party who were about to start to lay out a depôt one hundred miles southerly from our winter quarters.

The following day, September 20, a strong south-easterly blizzard was blowing. In the afternoon the wind somewhat moderated, and there was less drift. Mackay had been making a sail for our journey to the Magnetic Pole, and we now tried the sail on two sledges lashed together on the ice at Backdoor Bay. We used the tent poles of one of the sledging-tents as a mast. The wind was blowing very strongly and carried off the two sledges with a weight on them of 300 lb., in addition to the weights of Mackay and myself, who were sitting on the sledges. We considered this a successful experiment.

The weather continued bad till the night of the 24th.

On September 25 we were up at 5.30 A.M., and found that the blizzard had subsided. Priestley, Day and I started in the motor-car, dragging behind us two sledges over the sea ice. One sledge, weighing altogether 606 lb. with its load, contained five of our fortnightly food-bags, six large tins of biscuits, and 60 lb. of oil. The other sledge, which with its load weighed about 250 lb., carried personal gear which we might have to use on the depôt laying trip in the event of being surprised by a blizzard. At first Day travelled on his first gear; he then found that the engine became heated, and we had to stop for it to cool down. He discovered while we were waiting that one of the cylinders was not firing. This he soon fixed up all right. He then remounted the car and he put her on to the second gear. With the increased power given by the repaired cylinder we now sped over the floe-ice at fourteen miles an hour, much to the admiration of the seals and penguins. When, however, we had travelled about ten miles from winter quarters, and were some five miles westerly from Tent Island, we encountered numerous sastrugi of softish snow, the car continually sticking fast in the ridges. A little low drift was flying over the ice surface, brought up by a gentle blizzard. We left the heavy sledge ten miles out, and then with only the light sledge to draw behind us, Day found that he was able to travel on his third gear at eighteen miles an hour. At this speed the sledge, whenever it took one of the snow sastrugi at right angles, leapt into the air like a flying fish and came down with a bump on the surface of the ice. As we had occasionally to make sharp turns in order to avoid sastrugi and lumps of ice, our sledge had one or two capsizes. Meanwhile, the blizzard was freshening, and we tore along in hopes of reaching our winter quarters before it became very violent. We had just reached Flagstaff Point, and were taking a turn in towards the shore opposite the penguin rookery when the blizzard wind caught the side of the sledge nearly broadside on, and capsized it heavily. So violent was the shock that the aluminium cooking apparatus was knocked out of its straps, and the blizzard wind immediately started trundling this metal cylinder over the smooth ice. Day stopped his car as soon as possible, Priestley and I jumped off, and immediately gave chase to the runaway cooker. Meanwhile, the cooker had fallen to pieces so to speak; the tray part came away from the big circular cover; the melter and the supports for the cooking-pot and for the main outer covering also came adrift as well as the cooking-pot itself. The lid of the last-mentioned fell off, and immediately dumped on to the ice the three pannikins and our three spoons. These articles raced one another over the smooth ice-surface in the direction of the open water of Ross Sea. The spoons were easily captured, as also were the pannikins, which, being conical in shape, could not be bowled by the wind in a straight line, but described arcs of circles. Priestley and I recovered also the cooking-pot, and with Brocklehurst's help (for he had run down to meet us) the aluminium supports, but the large snow melter, the main outer casing, and the tray kept revolving in front of us at a speed which was just sufficient to outclass our own most desperate efforts. Finally, when we were nearly upon them, they took a joyous leap over the low cliff of floe-ice and disappeared one after another most exasperatingly in the black waters of the Ross Sea.

This was a shrewd loss, as aluminium cookers were, of course, very scarce. Priestley and I returned disconsolate, and very much winded after our mile's run in vain.

The following day we had intended laying out our second depôt, but as some of the piston rings of the motor-car needed repair, we decided to postpone the departure until the day after. That afternoon, after the repairs had been completed, Day and Armytage went out for a little tobogganing before dinner. Late in the evening Armytage returned dragging slowly and painfully a sledge bearing the recumbent, though not inanimate form of Day. We crowded round to inquire what was the matter, and found that just when Armytage and Day were urging their wild career down a steep snow slope Day's foot had struck an unyielding block of kenyte lava, and the consequence had been very awkward for the foot. It was severely staved, so that he was quite unable to walk without assistance. As no one but Day could be trusted to drive the motor-car, this accident necessitated a further postponement of the laying of our second depôt.

On September 28 it was blowing. On the 29th the day was fairly fine, but Day's foot was not well enough for him to start in the motor-car.

On September 30 a mild blizzard commenced blowing, and on October 1, the day on which Lieutenant Shackleton had intended that we should start, it was still raging with increased force. That day was spent chiefly in nailing strips of tin, painted blue, on all the geological specimen boxes, and double-labelling them.

On October 2 the weather was still bad, so that we were unable to start. On October 3, the weather having cleared, Day, Priestley, Mackay and I started with two sledges to lay our second depôt. All went well for about eight miles out, then the carburetter played up. Possibly there was some dirt in the nozzle. Day took it all to pieces in the cold wind, and spent three-quarters of an hour in fixing it up. We then started off again gaily in good style. We crossed a large crack in the sea ice where there were numbers of seals and Emperor penguins. On the other side of this crack our wheels stuck fast in snow sastrugi. All hands got on to the spokes and started swinging the car backwards and forwards; when we got a good swing on, Day would suddenly snatch on the power and over we would go—that is, over one of the sastrugi—only to find, often, that we had just floundered into another one ahead. In performing one of these evolutions Priestley, who as usual, was working like a Trojan, got his hand rather badly damaged through its being jammed between the spokes of the car wheel and the framework. Almost immediately afterwards one of my fingers was nearly broken, through the same cause, the flesh being torn off one of my knuckles; and then Mackay seriously damaged his wrist in manipulating what Joyce called the "thumb-breaking" starter. Still we went floundering along over the sastrugi and ice cracks, Day every now and then getting out to lighten the car and limping alongside. At last we succeeded in reaching a spot amongst the snow sastrugi on the sea ice, fifteen miles distant from our winter quarters. Here we dumped the load intended for the Northern Party, and then Day had a hard struggle to extricate the car from the tangle of sastrugi and ice-cracks. At last, after two capsizes of the sledges, we got back into camp at 10 P.M., all thoroughly exhausted, all wounded and bandaged. Brocklehurst carried Day on his back for about a quarter of a mile from where we left the car up to our winter quarters. So thoroughly exhausted were we, that we had to take a day's rest on October 4, before making our final start.

The following are the details respectively of our permanent load and equipment and of our consumable load (food and oil) when we did eventually start:

* The Start of a Blizzard from the South; Drift coming round Mount Erebus

NORTHERN PARTY'S EQUIPMENT

Northern Party's Permanent Load

	Weight. Lb.	Oz.
2 11-ft. sledges	120	0
Tent, poles and floorcloth	30	0
Shovel	6	0
Primus and cooker	20	0
Three-man sleeping bag	26	0
3 dozen plates	3	0
¼-plate camera and case	4	13
Legs of camera	1	14
Lloyd Creak dip circle	23	0
Legs for dip circle	7	0
Spirit for Primus stove	9	0
1 ready bottle for spirit	0	8
Sail and yard	11	0
Venesta board for table Centimetre rule Horn protractor Pencils "Hints to Travellers" and Nautical Almanac	1	10
3-inch. theodolite and case	9	0
Legs of theodolite	5	4
Field-glasses	1	13
3 ice axes, 3 lb. each	9	0
Rücksack and 60 ft. Alpine rope	6	0
Haversack, hammer and chisel	3	0
Aneroid 2 prismatic compasses	3	0
2 pairs of sledge thermometers in cases 2 low-temperature thermometers	0	12
1 hypsometer in case	1	0
Labels and small bags for specimens	1	0
Repair kit	2	0
Copper wire	0	4
Cod-line	1	0
Leather for repairs	2	0
1 pair shooting-boots for depôt at Butter Point	3	8
1 pair ski-boots (Mawson)	2	8
1 pair ski-boots (David)	2	8

	Weight. Lb.	Oz.
3 pairs ski-boots	12	0
9 pairs finnesko, 2¼ lb. each	20	4
Charts and tin case	1	0
Dram case of paper	1	0
30 lb. of personal gear 6 lb. of bags. }	36	0
Prickers, nipples, and washers for Primus . . .	0	8
3 hanks sennegrass	4	8
3 bags for drying sennegrass	0	8
Medical bag	5	0
Depôt flags, jack, and poles	4	0

Sledge harness.
Sledge ropes and toggles.
Small set of tools.
Books:
Field note-books.
"Magnetic Memoir of *Discovery* Expedition."
Sketch-book.

NORTHERN PARTY'S CONSUMABLE LOAD

	Lb.	Oz.
Plasmon biscuit: 1 lb. per man per day = 3 lb. per day. 93 days × 3 = 279 lb. Substitute for oatmeal, 1 lb. 3 oz. for 3 men per week × 13 = 14 lb. 10 oz. . . . . . =	294	0
Pemmican: 7.5 oz. per man per day × 3 × 93 = 2092 oz. . . . . =	131	0
Emergency rations (checked by Marshall): 1½ oz. per man per day × 3 × 93 = 418½ oz. . . . . =	26	0
Sugar (lumps): 3.8 oz. per man per day × 3 × 93 . =	70	0
Tea (twice a day): a little less than half a tin per week =	9	0
Rowntree's Sweet Chocolate: 8 oz. per man per week = normal allowance. 4½ " do. = substitute for honey. ——— 12½ oz. do. 12½ × 3 × 13 = 487 oz. . . . . . =	32	0
Cocoa: 14½ oz. for three men per week (once a day for dinner). 14½ × 13 = 188 oz. . . . . . =	12	0

∴ Out of this plasmon cocoa available for 6 weeks.

A BLIZZARD ON THE BARRIER

G. MARSTON.

	Weight. Lb.	Oz.
Cheese: 2 oz. per man per day, 3 days per week = 18 oz. per week. 18 oz. × 13 = 239 oz. . . . =	15	0
Plasmon and dried milk . . . . =	17	7
Salt: 2 oz. per week for 2 men = 4 oz. per week × 13 = 52 oz. . . . . =	3	4
Paraffin oil in 10-lb. tins . . . . =	100	0
	709	lb.

October 4 was a Sunday, and after the morning service we took the ponies out for exercise. In the evening the gramophone discoursed appropriate music, such as "We parted on the Shore," "I and my true love will never meet again by the bonnie, bonnie banks of Loch Lomond," concluding with the universal favourite, "Lead, Kindly Light."

Meanwhile, Mackay had his damaged wrist attended to, and I put the question to him as to whether or not he was prepared to undertake the long journey to the Magnetic Pole under the circumstances. He said that he was quite ready, provided Mawson and I did not object to his going with his wrist damaged and in a sling. We raised no objection, and so the matter was settled. All that night Mawson and I were occupied in writing final letters, and packing little odds and ends.

The following morning, October 5, after an early breakfast, we prepared for the final start. It was quite wonderful what a lot of things had been forgotten until this last moment. The sledges were dragged down from our hut to the edge of the sea ice at the Penguin Rookery, a distance of a little over a quarter of a mile. Day was there with the motor-car, ready for the start. Every now and then some one of the pilgrims would remember that he had left something very important behind at the hut, and would go running back for it. These odd belongings had to be tied with bits of string on to the second sledge which we were going to take with us on our northern journey, consequently, by degrees, this sledge became hung over with boots, crampons, and all sorts of what Day called "gagdets." Murray, Brocklehurst and Armytage came down on the ice to bid us a final farewell. Brocklehurst took a photograph of us just before we started, then Day, Priestley, Roberts, Mackay, Mawson and I got aboard, some on the motor-car, some on the sledges. Those remaining behind gave us three cheers, Day turned on the power and away we went. A light wind was blowing from the south-east at the time of our start, bringing a little snow with it and another blizzard seemed impending.

After travelling a little over two miles, just beyond Cape Barne, the snow had become so thick that the coast-line was almost entirely hidden from our view. Under these circumstances I did not think it prudent to take the motor-car further, so Mackay, Mawson and I bid adieu to our good friends. Strapping on our harness we toggled on to the sledge rope, and with a one, two, three, and away, pulled off into the thick falling snow, which in a few minutes blotted out all view of the motor-car in our rear. As we slowly trudged along the signs of an approaching blizzard became more pronounced and we bore somewhat to our left so as to have Inaccessible Island as a lee under which to run for shelter, but after a time, as the threatened blizzard did not come up, we slewed our sledge more to the right, away from Inaccessible Island, heading up for our ten-mile depôt. At last, towards evening Mackay sighted the black flag over the depôt about a mile distant.

We reached the depôt about 7 P.M. and got up our

tent. A fairly strong wind was still blowing from the south-east, raising low drift. We slept that night on the floe-ice, with about three hundred fathoms of water under our pillow.

The following morning, October 6, we started our relay work. We dragged the Christmas Tree sledge on first, as we were specially liable to lose parcels off it, for a distance of from one-third to half a mile. Then we returned and fetched up what we called the Plum Duff sledge, chiefly laden with our provisions. The light was dull, and a certain amount of soft, newly-fallen snow made the sledging heavy. The weather may be described as thick, with snow falling at intervals. During the afternoon it cleared somewhat and the Western Mountains came into view at about 2 P.M. This was fortunate for us, as it enabled us later on to sight the flag over our fifteen-mile depôt. We camped that night amongst screw pack-ice within less than a mile of this depôt.

The following day, October 7, was beautifully fine and calm. We started about 9 A.M. and sledged over pressure ice ridges and snow sastrugi, reaching our fifteen-mile depôt in three-quarters of an hour. Here we camped and repacked our sledges. We took the whole-meal plasmon biscuits out of two of the biscuit tins and packed them into canvas bags. This saved us a weight of about 8 lb.

We started again in the afternoon, relaying with the two sledges. The sledging again was heavy on account of the fresh, soft snow, and small sastrugi. We had a glorious view of the Western Mountains, crimsoned in the light of the setting sun. We camped that night close to a seal hole which belonged to a fine specimen of Weddell seal. We were somewhat disturbed that night by the snorting and whistling of the seals as they came up for their blows. Evidently this seal hole was a syndicate affair. The sounds at times seemed right under our tent.

October 8 was a fine, clear day, with a beautiful sunset, and a magnificent mirage, in the direction of Beaufort Island. To the north of us, the curious hills. called by Captain Scott the " Stranded Moraines," were now beginning to show out very plainly in the direction in which we were travelling.

On the morning of October 9 we got under way soon after eight o'clock. It was a lovely, calm day but cold, the thermometer registering 30° Fahr. at 8 P.M. The surface was fairly good for sledging, but in places we came on patches of soft snow, and a small, lumpy structure of the ice-surface, resembling a newly raked garden bed, evidently due to the thawing down and refreezing of " ice flowers." This made travelling very heavy. The " Stranded Moraines " now showed up very clearly, and Butter Point itself became visible.

The following day, Saturday, October 10, we were awakened by the chatter of some Emperor penguins who had marched down on our tent during the night to investigate us. The sounds may be described as something between the cackle of a goose and the chortle of a kookaburra. On peeping out of the Burberry spout of our tent I saw four standing by the sledges. They were much interested at the sight of me, and the conversation between them became lively. They evidently took us for penguins of an inferior type, and the tent for our nest. They watched, and took careful note of all our doings, and gave us a good send-off when we started about 8.30 A.M.

On our journey that morning we passed close by a large bull seal of the Weddell species. A little further on we noticed a curious dark object on the ice in the

Marston and Murray at the Door of the Hut

* Daisy's Third Litter at the Winter Quarters

* The Motor hauling Stores for a Depot

distance, and on coming up to it found that it was a dead Weddell seal with its head, neck, and shoulders firmly frozen into the ice. Evidently it had stuck fast in a seal hole in the ice in trying to get down to the sea-water.

The sky was overcast, and light snow began to fall in the afternoon. A little later a mild blizzard sprang up from the south-east; we thought this a favourable opportunity for testing the sailing qualities of our sledges, and so made sail on the Plum Duff sledge. As Mackay put it, we "brought her to try with main course." As the strength of the blizzard increased, we found that we could draw both sledges simultaneously, which was, of course, a great saving in labour. We were tempted to carry on in the increasing strength of the blizzard rather longer than was wise, and consequently, when at last we decided that we must camp, had great difficulty in getting the tent up. We slipped the tent over the poles placed close to the ground in the lee of a sledge. While two of us raised the poles, the third shovelled snow on to the skirt of the tent, which we pulled out little by little, until it was finally spread to its full dimensions. We were glad to turn in and escape from the biting blast and drifting snow.

The following day, Sunday, October 11, a violent blizzard was still blowing, and we lay in our sleeping-bag until past noon, by which time the snow had drifted high upon the door side of our tent. As this drift was pressing heavily on our feet and cramping us, I got up and dug it away. The cooker and Primus were then brought in and we all got up and had some hoosh and tea. The temperature, as usually happens in a blizzard, had now risen considerably, being 8.5° Fahr. at 1.30 P.M. The copper wire on our sledges was polished and burnished by the prolonged blast against it of tiny ice crystals, and the surface of the sea ice was also brightly polished in places. As it was still blowing we remained in our sleeping-bag for the rest of that day as well as the succeeding night.

When we rose at about 2 A.M. on Monday, October 12, the blizzard was over. We found very heavy snow-drifts on the lee side of our sledges, and it took us a considerable time to dig these away and get the hard snow raked out of all the chinks and crannies among the packages on the sledges. We made a start about 4 A.M., and all that day meandered amongst broken pack-ice. It was evident that the south-east blizzards drive large belts of broken floe-ice in this direction across McMurdo Sound to the western shore. The fractured masses of sea ice, inclined at all angles to the horizontal, are frozen in later, as the cold of winter becomes more intense, and of course, constitute a very difficult surface for sledging.

In order to make up for the time we had lost in our sleeping-bags during the blizzard, we travelled altogether fourteen hours, and succeeded in doing about six statute miles, that is, eighteen miles of relay work, and all felt much exhausted when we turned in that evening. As a result of this we did not wake until after 8 A.M next morning.

We were now only about two miles from Butter Point. We got under way at 10 A.M., and a few hours later camped at the foot of a low ice cliff, about 600 yards south-south-east of Butter Point. Butter Point is merely an angle in this low ice cliff near the junction of the Ferrar Glacier valley with the main shore of Victoria Land. This cliff was from fifteen to twenty feet in height, and formed of crevassed glacier ice. It was covered by a hard snow crust, which every now and then gave away and let one down for a foot or so. This glacier ice was not part of the main Ferrar Glacier, but appeared to be simply a local piedmont glacier stretching along for some considerable distance between the base of the coast range and the sea ice, past the "Stranded Moraines," until still further south it became confluent with that Mr. H. G. Ferrar has described as the "pinnacled ice." It was evident that this piedmont ice was firmly attached to the land, as it was separated from the sea ice, by a well-marked tide-crack. With the help of our ice axes we crossed over this crack and got up the little ice cliff on to the glacier ice, and selected there a suitable spot for our depôt.

According to arrangements with Lieutenant Shackleton we were to leave a depôt flag at Butter Point with a letter giving an account of our doings, and stating approximately by what date we hoped to return there. But the progress of our journey had been so much slower than we had originally anticipated that we decided before reaching Butter Point that it would be imperatively necessary, in order to make the Magnetic Pole in the time available, to lighten the load on our sledges by leaving a portion of our equipment and food.

During the latter part of this day Mawson and Mackay were busy making a mast and boom for the second sledge, it being our intention to use the tent floorcloth as a sail. Meanwhile I sorted out the material to be left at the depôt.

The following day, Wednesday, October 14, we spent the morning in resorting the loads on our sledges. We depôted two tins of wholemeal plasmon biscuits, each weighing about 27 lb., also Mackay's mountaineering nail boots, and my spare head-gear material and mits. Altogether we lightened the load by about 70 lb. We sunk the two full tins of biscuits and a tin containing boots, &c., a short distance in the glacier ice to prevent the blizzards blowing them away. We then lashed to the tins a short bamboo flag-pole, carrying one of our black depôt flags, and securely fastened to its base one of our empty air-tight milk tins, in which we placed our letters. In these letters for Lieutenant Shackleton and R. E. Priestley, respectively, I stated that in consequence of our late start from Cape Royds, and also on account of the comparative slowness of our progress thence to Butter Point, it was obvious that we could not return to Butter Point until January 12, at the earliest, instead of the first week of January, as was originally anticipated. We ascertained months later that this little depôt survived the blizzards, and that Armytage, Priestley and Brocklehurt had no difficulty in finding it, and that they read our letters.

Leaving the depôt about 9 A.M. on October 14, we started sledging across New Harbour in the direction of Cape Bernacchi. In the afternoon a light southerly wind sprang up bringing a little snow with it, the fall lasting from about 12.30 to 2.30 P.M. We steered in the direction of what appeared to us to be an uncharted island. On arriving at it, however, we discovered that it was a true iceberg, formed of hard blue glacier ice with a conspicuous black band near its summit formed of fine dark gravel. The iceberg was about a quarter of a mile in length, and thirty to forty feet high. In addition to the coarser bands of gravel there was a great quantity of dust, and fine dust bands, near the surface of the berg. This dust absorbing the heat of the sun had thawed its way deep down into the berg, thus forming numerous dust wells and dust grooves. There were several large cracks in the sea ice in the neighbourhood of this iceberg, and having taken the bearing of the trend of these by a compass they helped us to keep direction when the air was thick with falling snow.

The following day, October 15, was beautifully fine and calm; the sky was slightly cloudy with long belts of cirrus-stratus and alto-stratus. Erebus, now over fifty miles distant, was cloud capped. We had a glorious view up the magnificent valley of the Ferrar Glacier; the spurless hills on either side of the valley, strongly faceted in a direction parallel to each side of the valley, spoke eloquently of intense abrasive glacial action in the immediate geological past. The hills in the foreground, formed of gneissic granite, were of a rich chocolate brown to warm sepia hue, fading in the distance to exquisite tints of reddish purple and violet. Towards evening we had a wonderful vision of several large icebergs close ahead of us; it seemed as though they were only a mile or so distant, as one could see clearly the re-entering angles and bright reflected sides of the bergs lit up in the rays of the setting sun. Suddenly, as if by magic, they all vanished. They had been momentarily conjured up to our view by a wonderful mirage. In the departing rays of the setting sun Mount Erebus and Mount Bird glowed with a glorious golden light. This was one of the most beautiful days we experienced during the whole of our journey. The cold was now less severe than it had been, the temperature being 9.5° Fahr. at 8 P.M.

Chapter Seven

PROFESSOR DAVID'S NARRATIVE (*Continued*)

ON October 16 we were up at 3.30 A.M., and got under way at 5.30. A cold wind was blowing from the south, and after some trouble we set sail on both sledges, using the green floorcloth on the Christmas Tree sledge, and Mackay's sail on the Plum Duff sledge. A short time after we set sail it fell nearly calm; thick clouds gathered; a light wind sprang up from the south-east, veering to east-north-east, then back again to south-east in the afternoon. Fine snow fell for about three hours, forming a layer nearly a quarter of an inch in thickness. Towards evening we reached one of the bergs that had been miraged up the night before. It was four hundred yards long, and eighty yards wide, and was a true iceberg formed of glacier ice; Mackay, Mawson and I explored this. Like the previous iceberg its surface was pitted with numerous deep dust wells. It was wonderful to see what a very small amount of dust sufficed to dig these wells to a depth of several feet. The cliff of the berg which faced towards the north-west was deeply grooved, the appearance in the distance reminding one of a number of large parallel stalactites. Climbing up one of these deep grooves I found numbers of small angular rock pebbles, up to one and a half inches in diameter, adhering to the bottom of the groove, and it seemed as though these grooves, like the dust wells, were formed by the warmth of these small fragments of rock which, as the process of thawing of the ice cliff progressed, gradually settled down into long rows or strings as they crept gradually downwards under the influence of gravitation to the level of the sea ice below. As the shore was high and rocky, and seemed not more than half a mile distant, I went over towards it after our evening meal. It proved to be somewhat further than it appeared.

On the way, for the first time, I met with a structure in the sea ice known as pancake ice. The surface of the ice showed a rounded polygonal structure something like the tops of a number of large weathered basaltic columns. The edges of these polygons were slightly raised, but sufficiently rounded off by thawing or ablation to afford an easy surface for the runners of our sledge. Later on, in the autumn of the following year, we noticed this pancake ice in process of formation. If, as was often the case, there was any wind when the sea began to freeze over, the water at first commenced to look soupy; little by little the small ice particles which caused this appearance aggregated and formed myriads of small structures which may be likened to small open jam tarts. These would then coalesce in groups at their edges and form pancake ice. These pancakes were from one foot up to about three feet in diameter. Later, the pancakes would cohere and so a continuous hard ice crust would be formed over the sea surface; later freezing simply had the effect of strengthening and thickening this ice-sheet from below. Close in shore the pancake ice was traversed by deep tidal cracks. After climbing over these I reached the shore, which was composed of a well-marked terrace of coarse gravel and large and small erratic boulders. The smaller rock fragments were from three to six inches in diameter, the boulders being as much as five feet in diameter. The lower terrace was about twenty yards wide and as many feet in height above the sea;

then followed a belt of coarsely crystalline white marble traversed by bands of grey gneiss and blackish rock. Capping this belt of ancient crystalline rocks was a terrace of angular gravel, from eighty to one hundred feet above sea-level, formed of small rock fragments from one to two inches in diameter. The belts of coarse marble, gneiss, &c., were stained green and reddish to ochreous brown in places, and appeared to have been much mineralised. The whole area seemed to promise well for economic minerals. One curious boulder specially attracted my attention; its large pinkish crystals were peppered over with small black crystals, the whole being enclosed in a greenish black base. A specimen of this boulder which we brought back with us will be described in detail in the geological notes.

On Saturday, October 17, Mawson, Mackay and I landed at Cape Bernacchi, a little over a mile north of our previous camp. Here we hoisted the Union Jack just before 10 A.M. and took possession of Victoria Land for the British Empire. Cape Bernacchi is a low rocky promontory, the geology of which is extremely interesting. The dominant type of rock is a pure white coarsely crystalline marble; this has been broken through by granite rocks, the latter in places containing small red garnets. The marble or talc schist contains graphite disseminated through it in small scales. A great deal of tourmaline and epidote are developed in the granite at its point of contact with the calcareous schists. It appeared that the granite had intruded the black tourmaline rocks.

After taking possession we resumed our sledging, finding the surface of pancake ice very good. The day being calm and clear and free from either falling or drifting snow, we were able, for the first time, to turn

Moraine Cone with raised Beach material. Mt. Larsen on the right. "Backstairs Passage" is behind the Cone

One of the Sledges taken by the Northern Party

*

* Taking Possession of Cape Bernacchi, Victoria Land

our sleeping-bag inside out and air it in the sun. Previous to this the reindeer fur inside the bag had become much encrusted with ice, chiefly the result of the freezing of our breath. Although the heat of the sun was insufficient to actually thaw the ice it evaporated it to a considerable extent, and we found the bag that night much more comfortable to sleep in than it had been for many nights previous.

The following day, one and a quarter miles north of our preceding camp, we reached an interesting rocky headland. Here we found a mother seal with a newly born baby seal, the latter about three and a half feet in length. The mother seal at short intervals made a sound like "Wa-a-a." After a close inspection the mother and baby were left undisturbed, and we turned our attention to the rocks. These were most interesting, bearing a general resemblance to those at Cape Bernacchi. Some of the quartz veins traversing this point Mawson thought very favourable for gold.

When we left this point the wind had considerably freshened. We had previously hoisted sail on both sledges, and the wind was now sufficiently strong to admit of our pulling both sledges together. The total distance travelled was seven statute miles. This was the most favourable wind we experienced during the whole of our journey to and from the Magnetic Pole.

Shortly after leaving Baby Seal Point we encountered heavy belts of screw pack-ice with high sastrugi between. The Christmas Tree sledge capsized badly when being dragged over one of these high snow ridges. We were much exhausted when we camped that night and had suffered somewhat from the cold wind, the temperature being about 10° Fahr.

That night I experienced a rather bad attack of

snow-blindness, through neglecting to wear my snow goggles regularly. Finding that my eyes were no better next morning, and my sight being dim, I asked Mawson to take my place at the end of the long rope, the foremost position in the team. Mawson proved himself on this occasion and afterwards so remarkably efficient at picking out the best track for our sledges, and steering a good course that by my request he occupied this position throughout the rest of the journey.

The next two days were uneventful, except for the fact that we occasionally had extremely heavy sledging over screw pack-ice and high and long sastrugi. The latter were from two to three feet high, bounded frequently by almost perpendicular sides, and as they trended from west to east and our course was from south to north they proved formidable obstacles to our progress, and capsizes of the sledges were frequent.

On the night of October 20, we camped on the sea ice about three-quarters of a mile off shore. To the north-east of us was an outward curve of the shoreline, shown as a promontory on the existing chart. Early the next morning I walked over to the shore to geologise, and found the rocky headland composed of curious gneissic granite veined with quartz. On ascending this headland I noticed to my surprise that what had been previously supposed to be a promontory was really an island separated by a narrow strait from the mainland. It was clear that by going through this strait we would save several miles. Accordingly, after breakfast we sledged into the strait. The western side of the strait was formed of glacier ice terminating eastward in an almost precipitous slope. Here and there masses of gneissic granite showed from beneath the ice. The eastern side of the strait was formed of terraced moraine gravels with large erratics embedded in the gravel of the top terrace, eighty feet above sea-level.

While Mawson determined the position of this island by taking a round of angles with the theodolite, Mackay and I crossed the strait and explored the island, pacing and taking levels. The rocks of which the erratics and boulder-bearing gravels were formed were almost without exception of igneous origin. One very interesting exception was a block of weathered clayey limestone. This was soft and yellowish grey externally but hard and blue on the freshly fractured surfaces inside. It contained traces of small fossils which appeared to be seeds of plants. Specimens of these were taken by us and were depôted later at another small island, which we called Depôt Island. It is much to be regretted that we were unable later to reach this depôt on account of dense belts of pack-ice, and so these very interesting specimens were lost. Two chips, however, of this rock were fortunately preserved, sufficient for chemical analysis and microscopic examination. There could be little doubt that this clayey limestone has been derived from the great sedimentary formation, named by H. T. Ferrar, the Beacon Sandstone.

The island which we had been exploring we named provisionally Terrace Island. It was approximately triangular in shape, and the side facing the strait, down which we travelled, measured one mile 1200 yards in length.

The following day, October 22, we sighted the first skua gull we had seen that season. Snow fell in the afternoon between 2.30 and 5 P.M., forming a layer three-quarters of an inch deep. The temperature rose to plus 6.5° Fahr. at 7 P.M., and a blizzard seemed impending.

The following day, October 23, we held a serious council as to the future of our journey towards the Magnetic Pole. It was quite obvious that at our present rate of travelling, about four statute miles daily by the relay method, we could not get to the Pole and return to Butter Point early in January. I suggested that the most likely means of getting to the Pole and back in the time specified by Lieutenant Shackleton would be to travel on half rations, depôting the remainder of our provisions at an early opportunity. They both agreed, after some discussion, to try this expedient, and we decided to think the matter over for a few days and then make our depôt.

We found, as the result of the fall of snow on the preceding day, that the runners of our sledge clogged, making it difficult to start the sledge after each halt. The temperature at 5 P.M. was now as high as plus 5° Fahr. There were numerous seals, mostly mother seals with young ones, on the ice near the course of our sledge, as many as seventeen seals being sometimes visible simultaneously.

The previous day we had observed a mother seal with twin baby seals. Mackay took up one of these in his arms and stroked it while it was nuzzling round. It somewhat resembled a large lizard. The mother snorted at him indignantly, meanwhile, but made no attempt to attack him.

We encountered some large cracks in the sea ice. The sea water between the opposite walls had been only recently frozen over so that the ice was not more than a few inches in thickness. One of these was eighteen feet wide, and we found that the ice bent under our weight when we tested it primarily. Mackay called it "The Bridge of the Beresina." We rushed the sledges over at a good speed, and although the ice bent under their weight it fortunately held. At about 3 P.M. the weather grew very thick and it began to snow; a mild blizzard developing later, we hoisted sail on both sledges.

The next day, October 24, we found it very warm in the sleeping-bag, the sky being thickly overcast with dense stratus cloud. A strong water sky showed up to the east of us, while over the mountains to the west it was moderately clear. The presence of this water sky, indicating open sea, warned us that it was unsafe to stand out far from the land. We reached that evening a long rocky point of gneissic granite, which we called Gneiss Point. After our evening hoosh we walked across to the point and collected a number of interesting geological specimens, including blocks of kenyte lava.

The following day, October 25, proved a very heavy day for sledging, as we had to drag the sledges over new snow from three to four inches deep. In places it had a tough top crust which we would break through up to our ankles. We met also several obstacles in the way of wide cracks in the sea ice, from six to ten feet in width, and several miles in length. The sea water between the walls of the cracks had only recently been frozen over, so that the ice was only just thick enough to bear the sledges. The vicinity of these great ice-cracks were perfect baby farms for young seals. It was a pretty sight to see one of these baby seals playing with its mother, whom it kept gently flicking over the nose with its small flippers, the mother every now and then gently boxing the baby's ear with one of her large flippers. One of these mothers charged down on Mackay, who was making an inspection of her baby at too close quarters to suit her fancy. Another mother was moaning in great distress over her baby, which had just died. Evidently the mother seal's affection for her young is very strong.

In pursuing our north-westerly course we were now crossing a magnificent bay which trended westwards some five or six miles away from the course we were steering. On either side of this bay were majestic ranges of rocky mountains parted from one another at the head of the bay by an immense glacier with steep ice falls. On examining these mountains with a field-glass it was evident that in their lower portions they were formed of granites and gneiss, producing reddish brown soils. At the higher levels, further inland, there were distinct traces of rocks showing horizontal stratification. The highest rock of all was black in colour, and evidently very hard, apparently some three hundred feet in thickness. Below this was some softer stratified formation, approximately one thousand feet in thickness. We concluded that the hard top layer was composed of igneous rock, possibly a lava, while the horizontal stratified formation belonged in all probability to the Beacon Sandstone formation. Some fine nunataks of dark rock rose from the south-east side of the great glacier. On either side of this glacier were high terraces of rock reaching back for several miles from a modern valley edge to the foot of still higher ranges. It was obvious that these terraces marked the position of the floor of the old valley at a time when the glacier ice was several thousand feet higher than it is now, and some ten miles wider than at present. The glacier trended inland in a general south-westerly direction.

We longed to turn our sledges shorewards and explore these inland rocks, but this would have involved a delay of several days—probably a week at least—and we could not afford the time. Mawson took a series of horizontal and vertical angles with the theodolite to all the upper peaks in these ranges. We were much puzzled to determine on what part of the charted coast this wide bay and great glacier valley was situated. We speculated as to whether it was Granite Harbour, but decided that it could not be in view of the distance recorded by our sledge meter, for, according to this, we must still be some twenty miles south of Granite Harbour proper. We were to find out much later that the point opposite which we had now arrived was in reality Granite Harbour, and that its position was not shown correctly on the chart. Of course in pioneering work occasional mistakes such as these are inevitable.

The following day the sledge still proved very heavy on account of the soft snow—two to four inches deep—which was continually clogging the runners of our sledges. It was also difficult to steer a good course amongst the hummocky pack-ice on account of the day being dull and overcast. There was much low stratus cloud, and a light south-easterly wind.

The weather of October 27 was beautifully clear and sunshiny, and we had a glorious view of the great mountain ranges on either side of Granite Harbour. The rich colouring of warm sepia brown and terra cotta in these rocky hills was quite a relief to the eye. Wind springing up in the south-east we made sail on both sledges, and this helped us a good deal over the soft snow and occasional patches of sharp-edged brash ice. Occasionally the runners of our sledge would catch on one of these sharp fragments, and there would be a harsh rasping sound as a shaving was peeled off the runner. We feared that the wind would develop into a true blizzard, but it proved to be only what Joyce used to call a "carpet sweeper," driving along the newly fallen snow in white gossamer-like films over the sea ice.

Towards evening we fetched up against some high ice-pressure cracks with the ice ridged up six to eight feet high in huge tumbled blocks. We seemed to have got into a labyrinth of these pressure ridges from which there was no outlet. At last, after several capsizes of the sledges and some chopping through the ice ridges by Mackay, we got the sledges through, and camped on a level piece of ice. We were much helped in crossing the ridges by the long steep sastrugi of hardened snow. In places these ran like ramps up to the top of the pressure ridges, and were just wide enough and strong enough to bear our sledges. Mawson and I at this time were still wearing finnesko, while Mackay had taken to ski boots.

The following day, October 28, the sledging was again very heavy over sticky, soft snow alternating with hard sastrugi and patches of consolidated brash ice. Shavings of wood were being constantly rasped off the runners of our sledges. Mackay lost one of his finnesko off the sledge, but walked back a couple of miles in the evening and recovered it. Our course had taken us past a number of snowbergs; these were mostly about forty feet in height and from a quarter to half a mile in length. They were rigidly embedded in the sea ice. Occasionally we met with a true iceberg of blue ice amongst the snowbergs.

After our evening hoosh, Mawson and I went over to the shore, rather more than half a mile distant, in order to study the rocks. These we found were composed of coarse red granite; the top of the granite was much smoothed by glacier ice, and strewn with large erratic blocks. In places the granite was intersected by black dykes of basic rocks. One could see that the glacier ice, about a quarter of a mile inland from this rocky shore, had only recently retreated and laid bare the glaciated rocky surface. We found a little moss here amongst the crevices in the granite rock.

October 29 was beautifully fine, though a keen and fresh wind, rather unpleasantly cold, was blowing from off the high mountain plateau to our west. It blew from a direction west by south and caused a little low drift in the loose snow on the surface of the sea ice. There was still a great deal of deep, soft snow alternating with hard sastrugi and small patches of consolidated brash ice, so that the sledging was very heavy.

We were all thoroughly done up at night after completing our four miles of relay work. That evening we discussed the important question of whether it would be possible to eke out our food-supplies with seal meat so as to avoid putting ourselves on half rations, and we all agreed that this should be done. We made up our minds that at the first convenient spot we would make a depôt of any articles of equipment, geological specimens, &c., in order to lighten our sledges, and would at the same time, if the spot was suitable, make some experiments with seal meat. The chief problem in connection with the latter was how to cook it without the aid of paraffin oil. We could not afford more paraffin for this purpose, as we estimated that even with the utmost care the supply for our Primus, which we used for brewing tea, cocoa and hoosh, would become exhausted before we could hope to reach the Magnetic Pole, unless some kind of substitute for paraffin could be found.

The following day, October 30, was full of interest for us, as well as hard work. In the early morning, between 2.30 A.M. and 6.30 A.M., a mild blizzard was blowing. We got under way a little later and camped at about 10.30 A.M. for lunch alongside a very interesting rocky point. Mawson got a good set of theodolite angles from the top of this point. The point was formed of coarse porphyritic grey gneiss, traversed by black dykes of rock, apparently tinguaite, and another variety containing an abundance of sparkling black crystals of hornblende, which may be termed provisionally a hornblende lamprophyre.

After lunch we passed close by a mother seal and her baby. The mother charged us and we had to skid along quick and lively past her with the sledges. That day was the first occasion that we tried the experiment of strengthening the brew of the tea by using the old tea-leaves of a previous meal mixed with the new ones. This was Mackay's idea, and Mawson and I at the time did not appreciate the experiment. Later on, however, we were very glad to adopt it.

The sledging that afternoon was about the heaviest we had experienced up to date. The weather was now daily becoming warmer and the saline snow on the sea ice became sticky in consequence. It gripped the runners of the sledges like glue, and we were only able with our greatest efforts to drag the sledges over this at a snail's pace. We were all thoroughly exhausted that evening when we camped at the base of a rocky promontory about 180 ft. high. This cliff was formed of coarse gneiss, with numerous dark streaks, and enclosures of huge masses of greenish-grey quartzite.

After our evening hoosh we walked over to a very interesting small island about three-quarters of a mile distant. It was truly a most wonderful place geologically, and was a perfect elysium for the mineralogist. The island, which we afterwards called Depôt Island, was accessible on the shoreward side, but rose perpendicularly to a height of 200 ft. above sea-level on the other three sides. There was very little snow or ice upon it, the surface being almost entirely formed of

Dark Enclosures of Horneblende-Rock in Gneiss, Depot Island

Seals on Coast of Victoria Land

gneissic granite. This granite, as shown in the photograph, was full of dark enclosures of basic rocks, rich in black mica and huge crystals of hornblende. It was in these enclosures that Mawson discovered a translucent brown mineral, which he believed to be monazite, but which has since proved to be titanium mineral. Patches of a crystalline, milky-white mineral were to be seen amongst the large platy crystals of dark green hornblende. These white crystals we thought might be scapolite. We returned to camp and slept soundly after the severe work of the day.

We were up at 6 A.M. next morning, and after breakfast Mackay and Mawson went in pursuit of some seals which we had sighted further back on the previous day, while I climbed up an adjacent granite slope with the field-glasses, watching for a signal from them, if they were successful in their hunting, to bring up an empty sledge. They were, however, unsuccessful in their quest, and after some time returned to camp.

We packed up and made for the island at 9.30 A.M. The sledging was extremely heavy, and we fell into a tide-crack on the way, but the sledge was got over safely. Mackay sighted a seal about six hundred yards distant from the site of our new camp near the island, and just then we noticed that another seal had bobbed up in the tide-crack close to our old camp. Mackay and Mawson at once started off in the direction where the first seal had been sighted. It proved to be a bull seal in very good condition, and they killed it by knocking it on the head with an ice-axe. Meanwhile, I unpacked the Duff sledge and took it out to them. Returning to the site of our camp I put up the tent, and on going back to Mawson and Mackay found that they had finished fletching the seal. We loaded up the empty sledge with seal blubber, resembling bars of soap in its now frozen condition, steak and liver, and returned to camp for lunch.

After lunch we took some blubber and seal meat on to the island, intending to try the experiment of making a blubber fire in order to cook the meat. We worked our way a short distance up a steep, rocky gulley, and there built a fireplace out of magnificent specimens of hornblende rock. It seemed a base use for such magnificent mineralogical specimens, but necessity knows no laws. We had brought with us our Primus lamp in order to start the fire. We put blubber on our iron shovel, warmed this underneath by means of the heat of the Primus lamp so as to render down the oil from it, and then lit the oil. The experiment was not altogether successful. Mawson cooked for about three hours, closely and anxiously watched by Mackay and myself. Occasionally he allowed us to taste small snacks of the partly cooked seal meat, which were pronounced to be delicious.

While the experiment was at its most critical stage, at about 6 P.M., we observed sudden swirls of snowdrift high up on the western mountains, coming rapidly down to lower levels. For a few minutes we did not think seriously of the phenomenon, but as the drift came nearer we saw that something serious was in the air. Mackay and I rushed down to our tent, the skirt of which was only temporarily secured with light blocks of snow. We reached it just as it was struck by the sudden blizzard which had descended from the western mountains. There was no time to dig further blocks of snow, all we could do was to seize the heavy food-bags on our sledges, weighing sixty pounds each, and rush them on to the skirt of the tent. The blizzard struck our kitchen on the island simultaneously with our tent, and temporarily Mawson lost his mits and most of the tit-bits of seal meat, but these were quickly recovered, and he came rushing down to join us in securing the tent. While Mawson in frantic haste chopped out blocks of snow and dumped them on to the skirt of the tent, Mackay, no less frantically, struggled with our sleeping-bag, which had been turned inside-out to air, and which by this time was covered with drift snow. He quickly had it turned right side in again, and dashed it inside the tent. At last everything was secured, and we found ourselves safe and sound inside the tent. The Primus was quickly got going, and soon we had some hot cocoa and hot seal pottage, together with some small pieces of charred but delicious seal blubber. The blizzard continued until past our bedtime. We turned in with a determination of making further experiments on the cooking of seal meat on the following morning.

The following day, November 1, we breakfasted off a mixture of our ordinary hoosh and seal meat. After some discussion we decided that our only hope of reaching the Magnetic Pole lay in our travelling on half rations from our present camp to the point on the coast at the Drygalski Glacier, where we might for the first time hope to be able to turn inland with reasonable prospect of reaching the Magnetic Pole. Mawson was emphatic that we must conserve six weeks of full rations for our inland journey to and from the Pole. This necessitated our going on half rations from this island to the far side of the Drygalski Glacier, a distance of about one hundred statute miles. In order to supplement the regular half rations we intended to take seal meat.

While I was busy in calculating the times and distances for the remainder of our journey, and proportioning the food rations to suit our new programme,

Mawson and Mackay conducted further experiments on the cooking of seal meat with blubber. While at our winter quarters, Mackay had made some experiments on the use of blubber as a fuel. He had constructed a blubber lamp, the wick of which kept alight for several hours at a time, feeding itself on the seal oil. He had tried the experiment of heating up water over this blubber lamp, and was partly successful at the time when we left winter quarters for our present sledging journey. But his experiments at the time were not taken very seriously, and the blubber lamp was left behind, a fact which we now much regretted. An effective cooking-stove was, however, evolved, as the result of a series of experiments this day, out of one of our large empty biscuit tins. The lid of this was perforated with a number of circular holes for the reception of wicks. Its edges were bent down, so as to form supports to keep the wick-holder about half an inch above the bottom of the biscuit tin. The wick-holder was put in place; wicks were made of pieces of old calico food-bags rolled in seal blubber, or with thin slices of seal blubber enfolded in them, the calico being done up in little rolls for the purpose of making wicks, as one rolls a cigarette, the seal blubber taking the place of the tobacco in this case. Lumps of blubber were laid round the wick-holder. Then, after some difficulty, the wicks were lighted. They burned feebly at first, as seal blubber has a good deal of water in it. After some minutes of fitful spluttering, the wicks got fairly alight, and as soon as the lower part of the biscuit tin was raised to a high temperature, the big lumps of blubber at the side commenced to have the water boiled out of them and the oil rendered down. This oil ran under the wick-holder and supplied the wicks at their base. The wicks, now fed with warm, pure seal oil, started to burn brightly, and even fiercely, so that it became necessary occasionally to damp them down with chips of fresh blubber. We tried the experiment of using lumps of salt as wicks, and found this fairly successful. We also tried small pieces of our brown rope for the same purpose, using the separated strands of these cut in pieces of about one and a half inches long. These made excellent wicks, but we could not spare much rope. We also tried the lamp-wick that we had brought with us for binding on our finnesko, but in this case also rigid economy was an absolute necessity. We decided to rely for wicks chiefly on our empty food-bags, and thought possibly that if these ran out we might have recourse to moss. But the empty food-bags supplied sufficient wick for our need.

That day, by means of galvanised iron wires, we slung the inner pot from our aluminium cooker over the lighted wicks of our blubber cooker, thawed down snow in it, added chips of seal meat and made a delicious bouillon. This had a rich red colour and seemed very nutritious, but to me was indigestible. While Mawson was still engaged on further cooking experiments, Mackay and I ascended to the highest point of the island, selected a spot for a cairn to mark our depôt, and Mackay commenced building the cairn. Meanwhile, I returned to camp and wrote a number of letters, including one to the commander of the *Nimrod.* The latter was accompanied by a sketch plan taken from the Admiralty chart to show the proper position of our final depôt before we were to turn inland " on the low sloping shore " to the north-west of the Drygalski Glacier. The other letters were to Lieutenant Shackleton and to my family.

The letter to the commander of the *Nimrod* contained the following statement of our plans:

"CAMP, GRANITE HARBOUR.*

"DEAR SIR,—I beg to inform you that we intend leaving here to-morrow in continuation of our journey towards the Magnetic Pole. We have to work our two sledges by relays, which, of course, means slow progress—only about four miles per day. At this rate we hope to reach the north side of the Drygalski Ice Barrier at the point where 'low sloping shore' is marked on the Admiralty Chart of the Antarctic Sheet III. (please see sketch on opposite page), by about December 15. We propose to make a depôt there marked by a black flag similar to the one we are leaving here at the island at south side of entrance to Granite Harbour. We propose to travel inland from the 'low sloping shore,' and if possible reach the Magnetic Pole and return to depôt. We estimate that this may take six weeks, so that we may not return to the coast at the low sloping shore depôt until about January 25. We propose to wait there until the *Nimrod* calls for us at the beginning of February."

The letter concluded with detailed instructions regarding the course to be pursued in searching for the party.

---

* At this time we were under the impression that this island was on the south side of Granite Harbour. We did not know that we had already left Granite Harbour about twelve miles to the south of us.

Chapter Eight

PROFESSOR DAVID'S NARRATIVE (*Continued*)

The old dragon under ground
In straiter limits bound,
Not half so far casts his usurped sway,
And wroth to see his kingdom fail,
Swinges the scaly horror of his folded tail.
MILTON.

IT had, of course, become clear to us before this letter was written, in view of our experience of the already cracking sea ice near the true Granite Harbour, as well as in view of our comparatively slow progress by relay, that our retreat back to camp from the direction of the Magnetic Pole would in all probability be entirely cut off through the breaking up of the sea ice. Under these circumstances we determined to take the risk of the *Nimrod* arriving safely on her return voyage at Cape Royds, where she would receive the instructions to search for us along the western coast, and also the risk of her not being able to find our depôt and ourselves at the low sloping shore. We knew that there was a certain amount of danger in adopting this course, but we felt that we had got on so far with the work entrusted to us by our Commander, that we could not honourably now turn back. Under these circumstances we each wrote farewell letters to those who were nearest and dearest, and the following morning, November 2, we were up at 4.30 A.M. After putting all the letters into one of our empty dried-milk tins, and fitting on the air-tight lid, I walked with it to the island and climbed up to the cairn. Here, after carefully depôting several bags of geological specimens at the base of the flagstaff, I lashed the little post office by means of cord and copper-wire securely to the flagstaff, and then carried some large slabs of exfoliated granite to the cairn, and built them up on the leeward side of it in order to strengthen it against the southerly blizzards. A keen wind was blowing, as was usual in the early morning, off the high plateau, and one's hands got frequently frost-bitten in the work of securing the tin to the flagstaff. The cairn was at the seaward end of a sheer cliff two hundred feet high.

On returning to camp I put some chopped seal meat into the cooking-pot on our blubber stove, which Mawson had meanwhile lighted, and about three-quarters of an hour later we partook of some nourishing, but no less indigestible seal bouillon. It was later than usual when we started our sledges, and the pulling proved extremely heavy. The sun's heat was thawing the snow surface and making it extremely sticky. Our progress was so painfully slow that we decided, after, with great efforts, doing two miles, to camp, have our hoosh, and then turn in for six hours, having meanwhile started the blubber lamp. At the expiration of that time we intended to get out of our sleeping-bag, breakfast, and start sledging about midnight. We hoped that by adopting nocturnal habits of travelling, we would avoid the sticky ice-surface which by daytime formed such an obstacle to our progress.

We carried out this programme on the evening of November 2, and the morning of November 3. We found the experiment fairly successful, as at midnight and for a few hours afterwards, the temperature remained sufficiently low to keep the surface of the snow on the sea ice moderately crisp.

On November 3 and 4 the weather was fine, and we made fair progress. At noon Mawson cleaned out the refuse from our blubber lamp. Amongst this were a few dainty bits; Mackay was what he called "playing the skua," picking these over, when he accidentally transferred to his mouth and swallowed one of the salt wicks which we had been using in the blubber lamp. Mawson and I were unaware of this episode at the time. Later on, towards evening, he complained much of thirst, and proferred a gentle request, when the snow was being thawed down preparatory to making hoosh, that he might be allowed to drink some of the water before the hoosh was put into it; at the same time he gave us the plausible explanation above mentioned as the cause of his exceeding thirst. After debating the matter at some length, it was decided, in view of the special circumstances surrounding the case, and without creating a precedent—which otherwise might become a dangerous one—that he might be allowed on this occasion to take a drink. Mackay, however, considered that this water gift was given grudgingly, and of necessity, and accordingly he sternly refused to accept it. Just then, the whole discussion was abruptly terminated through the pot being accidentally capsized when being lifted off the blubber lamp, and the whole of the water was lost.

On the following day, November 5, we were opposite a very interesting coastal panorama, which we thought belonged to Granite Harbour, but which was really over twenty miles to the north of it. Magnificent ranges of mountains, steep slopes free from snow and ice, stretched far to the north and far to the south of us, and finished away inland, towards the heads of long glacier cut valleys, in a vast upland snow plateau. The rocks which were exposed to view in the lower part of these ranges were mostly of warm sepia brown to terracotta tint, and were evidently built up of a continuation of the gneissic rocks and red granites which we had previously seen. Above these crystalline rocks came a belt of greenish-grey rock, apparently belonging to some stratified formation and possibly many hundreds of feet in thickness; the latter was capped with a black rock that seemed to be either a basic plateau lava, or a huge sill. In the direction of the glacier valleys, the plateau was broken up into a vast number of conical hills of various shapes and heights, all showing evidence of intense glacial action in the past. The hills were here separated from the coast-line by a continuous belt of piedmont glacier ice. This last terminated where it joined the sea ice in a steep slope, or low cliff, and in places was very much crevassed. Mawson at our noon halts for lunch, continued taking the angles of all these ranges and valleys with our theodolite.

The temperature was now rising, being as high as 22° Fahr. at noon on November 5. We had a very heavy sledging surface that day, there being much consolidated brash ice, sastrugi, pie-crust snow, and numerous cracks in the sea ice. As an offset to these troubles we had that night, for the first time, the use of our new frying-pan, constructed by Mawson out of one of our empty paraffin tins. This tin had been cut in half down the middle parallel to its broad surfaces, and loops of iron wire being added, it was possible to suspend it inside the empty biscuit tin above the wicks of our blubber lamp. We found that in this frying-pan we could rapidly render down the seal blubber into oil, and as soon as the oil boiled we dropped into the pan small slices of seal liver or seal meat. The liver took about ten minutes to cook in the boiling oil, the seal meat about twenty minutes. These facts were ascertained by the empirical method. Mawson discovered by the same method that the nicely browned and crisp residue from the seal blubber, after the oil in it had become rendered down, was good eating, and had a fine nutty flavour. We also found, as the result of later experiments, that dropping a little seal's blood into the boiling oil produced eventually a gravy of very fine flavour. If the seal blood was poured in rapidly into the boiling oil, it made a kind of gravy pancake, which we also considered very good as a variety.

We had a magnificent view this day of fresh ranges of mountains to the north of Depôt Island. At the foot of these was an extensive terrace of glacier ice, a curious type of piedmont glacier. Its surface was strongly convex near where it terminated seawards in a steep slope or low cliff. In places this ice was heavily crevassed. At a distance of several miles inland it reached the spurs of an immense coastal range, while in the wide gaps in this range the ice trended inland as far as the eye could see until it blended in the far distance with the skyline high up on the great inland plateau.

A little before 9 P.M. on November 5 we left our sleeping-bag, and found snow falling, with a fresh and chilly breeze from the south. The blubber lamp, which we had lighted before we had turned in, had got blown out. We built a chubby house for it of snow blocks to keep off the wind, and relighted it, and then turned into the sleeping-bag again while we waited for the snow and chips of seal meat in our cooking-pot to become converted into a hot bouillon; the latter was ready after an interval of about one hour and a half. Just before midnight we brought the cooker alight into the tent in order to protect it from the blizzard which was now blowing and bringing much falling snow with it. Mawson's cooking experiments continued to be highly successful and entirely satisfactory to the party.

We waited for the falling snow to clear sufficiently to enable us to see a short distance ahead, and then started again, the blizzard still blowing with a little low drift. After doing a stage of pulling on both sledges to keep ourselves warm in the blizzard we set sail—always a chilly business—and the wind was a distinct assistance to us. We encountered a good deal of brash ice that day, and noticed that this type of ice surface was most common in the vicinity of icebergs, which just here were very numerous. The brash ice is probably formed by the icebergs surging to and fro in heavy weather like a lot of gigantic Yermaks, and crunching up the sea ice in their vicinity. The latter, of course, re-freezes, producing a surface covered with jagged edges and points.

We were now reduced to one plasmon biscuit each for breakfast and one for the evening meal, and we were unanimous in the opinion that we had never before fully realised how very nice these plasmon biscuits were. We became exceedingly careful even over the crumbs. As some biscuits were thicker than others, the cook for the week would select three biscuits, place them on the outer cover of our aluminium cooker, and get one of his mates to look in an opposite direction while the messman pointed to a biscuit and said, "Whose?" The mate with averted face, or shut eyes, would then state the owner, and the biscuit was ear-marked for him, and so with the other two biscuits. Grievous was the disappointment of the man to whose lot the thinnest of the three biscuits had fallen. Originally, on this sledge journey, when biscuits were more plentiful, we used to eat them regardless of the loss of crumbs, munching them boldly, with the result that occasional crumbs fell on the floorcloth. Not so now. Each man broke his biscuit over his own pannikin of hoosh, so that any crumbs produced in the process of fracture fell into the pannikin. Then, in order to make sure that there were no loose fragments adhering to the morsel we were about to transfer to our mouths, we tapped the broken chip, as well as the biscuit from which it had been broken, on the sides of the pannikin, so as to shake into it any loose crumbs. Then, and then only, was it safe to devour the precious morsel. Mackay, who adopted this practice in common with the rest of us, said it reminded him of the old days when the sailors tapped each piece of broken biscuit before eating it in order to shake out the weevils.

Mawson and I now wore our ski-boots instead of finnesko, the weather being warmer, and the ski-boot giving one a better grip on the snow surface of the sea ice. The rough leather took the skin off my right heel, but Mackay fixed it up later in the evening, that is, my heel, with some "Newskin." As we found the sharp iron spikes of the ski-boots made holes in our waterproof floorcloth we made a practice of always changing into our finnesko before entering the tent.

We sledged on uneventfully for the remainder of November 6, and during the 7th, and on November 8 it came on to blow again with fresh-falling snow. The blizzard was still blowing when the time came for us to pitch our tent. We had a severe struggle to get the tent up in the high wind and thick falling snow. At last the work was accomplished, and we were all able to turn into our sleeping-bag, pretty tired, at about 12.30 P.M.

The weather was still bad the following day, November 9. After breakfast off seal's liver, and digging out the sledges from the snow drift, we started in the blizzard, the snow still falling. After a little while we made sail on both sledges. The light was very bad on account of the thick falling snow, and we were constantly falling up to our knees in the cracks in the sea ice. It seemed miraculous that in spite of these very numerous accidents we never sprained an ankle.

That day we saw a snow petrel, and three skua gulls visited our camp. At last the snow stopped falling and the wind fell light, and we were much cheered by a fine, though distant view of the Nordenskjold Ice Barrier to the north of us. We were all extremely anxious to ascertain what sort of a surface for sledging we should meet with on this great glacier. According to the Admiralty chart, prepared from observations by the *Discovery* expedition, this glacier was between twenty-four and thirty miles wide, and projected over twenty miles from the rocky shore into the sea. We hoped that we might be able to cross it without following a circuitous route along its seaward margins.

We started off on November 10, amongst very heavy sastrugi and ridges of broken pack-ice. Cracks in the sea ice were extremely numerous. The morning was somewhat cloudy, but as the midnight sun got higher in the heavens, the clouds dispersed and the weather become comparatively warm, the temperature being up to plus 3° Fahr. at 8 A.M. That day when we pitched camp we were within half a mile of the southern edge of the Nordenskjold Ice Barrier.

The following day, November 11, as Mawson wished to get an accurate magnetic determination with the Lloyd-Creak dip circle, we decided to camp, Mackay and I exploring the glacier surface to select a suitable track for our sledges while Mawson took his observations. After breakfast we removed everything containing iron several hundred yards away from the tent, leaving Mawson alone inside it in company with the dip circle. We found that the ascent from the sea ice to the Nordenskjold Ice Barrier was a comparatively easy one. The surface was formed chiefly of hard snow glazed in places, partly through thawing and re-freezing, partly through the polishing of this windward surface by particles of fresh snow driven over it by the blizzards. Hummocky masses, apparently of the nature of large sastrugi, projected here and there to a height of six feet above the general level. The latter were something like elongated white ants' nests. In places the snow surface showed pie-crust structure, a bad surface for sledging. On the whole this Barrier was fairly free from crevasses, although Mackay and I crossed a few in our short pioneering excursion.

The surface ascended gradually to a little over one hundred feet above the level of the sea ice, passing into a wide undulating plain which stretched away to the north as far as the eye could see.

We returned to Mawson with the good news that the Nordenskjold Ice Barrier was quite practicable for sledging, and would probably afford us a much more easy surface than the sea ice over which we had previously been passing. Mawson informed us, as the result of his observations with the dip circle, that the Magnetic Pole was probably about forty miles further inland than the theoretical mean position calculated for it from the magnetic observations of the *Discovery* expedition seven years ago.

Early on the morning of November 12 we packed up, and started to cross the Nordenskjold Ice Barrier. We noticed here that there were two well-marked sets of sastrugi, one set, nearly due north and south, formed by the strong southerly blizzards, the other set, crossing nearly at right angles, coming from the west and formed by the cold land winds blowing off the high plateau at night on to the sea.

We were surprised to observe that this ice barrier was almost completely isolated from the shore by deep inlets, and for a time we speculated as to whether after all it might not be a gigantic tabular iceberg run aground In view, however, of what we observed later there can be little doubt that it is of the nature of a large piedmont glacier, afloat at its seaward end and central portions. It is now practically inert, having no forward movement from the land towards the sea. It is just the vanishing remnant of what at one time was no doubt a large active glacier, vigorously pushed out seaward, the overflow ice from the vast snow-fields of the inland plateau. The supply, however, of ice near the coast has dwindled so enormously that there is no longer sufficient pressure to move this ice barrier.

This day, November 12, was an important one in the history of Mawson's triangulation of the coast, for he was able in the morning to sight simultaneously Mount Erebus and Mount Melbourne, as well as Mount Lister. We were fortunate in having a very bright and clear day on this occasion, and the round of angles obtained by Mawson with the theodolite were in every way satisfactory.

The following day, November 13, we were still on the Nordenskjold Ice Barrier. The temperature in the early morning, about 3 A.M., was minus 13° Fahr. Mawson had provided an excellent dish for breakfast consisting of crumbed seal meat and seal's blood, which proved delicious. We got under way about 2 A.M. It was a beautiful sunshiny day with a gentle cold breeze off the western plateau. When we had sledged for about one thousand yards Mawson suddenly exclaimed that he could see the end of the barrier where it terminated in a white cliff only about six hundred yards ahead. We halted the sledge, and while Mawson took some more theodolite angles Mackay and I reconnoitred ahead, but could find no way down the cliff. We returned to the sledge and all pulled on for another quarter of a mile. Once more we reconnoitred, and this time both Mawson and I found some steep slopes formed by drifted snow which were just practicable for a light sledge lowered by an Alpine rope. We chose what seemed to be the best of these; Mackay tied the Alpine rope around his body, and taking his ice-axe descended the slope cautiously, Mawson and I holding on to the rope meanwhile. The snow slope proved fairly soft, giving good foothold, and he was soon at the bottom without having needed any support from the Alpine rope. He then returned to the top of the slope, and we all set to work unpacking the sledges. We made fast one of the sledges to the Alpine rope, and after loading it lightly lowered it little by little down the slope, one of us guiding the sledge while the other two slacked out the Alpine rope above. The man who went down the sledge to the bottom would unload it there on the sea ice and then climb up the slope, the other two meanwhile pulling up the empty sledge. This manœuvre was repeated a number of times until eventually the whole of our food and equipment, including two sledges, were safely down on the sea ice below.

We were all much elated at having got across the Nordenskjold Ice Barrier so easily and so quickly. We were also fortunate in securing a seal; Mackay went off and killed this, bringing back seal steak, liver and a considerable quantity of seal blood. From the last Mackay said he intended to manufacture a black pudding. Usually, I believe, a black pudding is manufactured from the part of the blood which does not contain the fibrin, but on this occasion the black pudding was wholly formed of fibrin, so that it may be described as a negative rather than a positive black pudding. This fibrin was boiled up in seal oil, and though rather tasteless was at all events nourishing, and was certainly filling.

While Mackay had been in pursuit of the seal meat Mawson had taken a meridian altitude while I kept the time for him. After our hoosh we packed the sledges, and Mawson took a photograph showing the cliff forming the northern boundary of the Nordenskjold Ice Barrier. This cliff was about forty feet in height. We had some discussion as to whether or not there was a true tidal crack separating the sea ice from this ice barrier. Certainly, on the south side there was no evidence of the presence of any such crack, but on the north side there were small local cracks; yet it could hardly be said that these were of sufficient importance to be termed true tide-cracks. In one of these cracks most beautiful filagree ice crystals, fully one inch across, lined the sides of the walls of the crack in the sea ice. There can be little doubt, I think, that the greater part of this Nordenskjold Ice Barrier is afloat.

The sun was so warm this day that I was tempted before turning in to the sleeping-bag to take off my ski-boots and socks and give my feet a snow bath, which was very refreshing.

The following day, November 14, we were naturally anxious to be sure of our exact position on the chart, in view of the fact that we had come to the end of the ice barrier some eighteen miles quicker than the chart led us to anticipate we should. Mawson accordingly worked up his meridian altitude, and I plotted out the

Cliff down which the Sledges were Lowered on the North Side of the Nordenskjold Ice Barrier Tongue

A Pause by the Way

angular distances he had found, respectively, for Mount Erebus, Mount Lister and Mount Melbourne. As the result of the application of our calculations to the chart it became evident that we had actually crossed the Nordenskjold Ice Barrier of Captain Scott's survey, and were now opposite what on his chart was termed Charcot Bay. This was good news and cheered us up very much, as it meant that we were nearly twenty miles further north than we previously thought we were.

The day was calm and fine, and the surface of the sea ice was covered with patches of soft snow with nearly bare ice between, and the sledging was not quite as heavy as usual. In the evening two skua gulls went for our seal meat during the interval that we were returning for the second sledge after pulling on the first sledge. It was wonderful how quickly these gulls made their appearance from distant parts of the horizon as soon as any fresh meat was available. The previous day one of them had actually attempted to eat the seal meat out of our frying-pan when the meat was being cooked in boiling oil. We could see as we came up from a distance that the heat of the savoury dish puzzled him a good deal, as each time he dipped his beak into our hot mince he jerked it out again very suddenly and seemed a very surprised bird.

We had a magnificent view of the rocky coast-line, which is here most impressive. The sea ice stretched away to the west of us for several miles up to a low cliff and slope of piedmont glacier ice, with occasional black masses of rock showing at its edge. Several miles further inland the piedmont glacier ice terminated abruptly against a magnificent range of mountains, tabular for the most part but deeply intersected. In the wide gaps between this coast range were vast glaciers fairly heavily crevassed, descending by steep slopes from an inland plateau to the sea.

On November 15, there was a fresh wind from the west-south-west. The weather was overcast, and a few flakes of snow were falling. We killed two young seals to replenish our food-supply. Mackay took over the blubber cooking apparatus so as to set Mawson free for his theodolite observations. The sky was dull and leaden for most of the day, with occasional glimpses of light over the western mountains. On the whole it looked as if a blizzard were approaching.

We were still doing our travelling by night and sleeping during the afternoon. When we arose from our sleeping-bag at 8 P.M. on the night of November 15, we found that the signs of the blizzard had more or less passed away. There was a beautifully perfect "Noah's Ark" in the sky; the belts of cirrus-stratus composing the ark stretched from south-south-west to north-north-east, converging towards the horizon in each of these directions. Fleecy sheets of frost smoke arose from over the open water on Ross Sea, and formed dense cumulus clouds. This, of course, was a certain indication to us that open water was not far distant, and impressed upon us the necessity of making every possible speed if we hoped to reach our projected point of departure on the coast for the Magnetic Pole before the sea ice entirely broke up.

This had been a truly glorious day, bright and sunny, and as this was the end of a food week and the messman for the week had kept a little food up his sleeve, so to speak, we fared sumptuously. The cocoa was extra strong, milky and sweet. Mackay's opinion was that such cocoa much reminded him of better days, and was absolutely uplifting.

The following day, November 17, after a very heavy sledging over loose powdery snow six inches deep we reached a low glacier and ice cliff. We were able to get some really fresh snow from this barrier or glacier, the cliffs of which were from thirty to forty feet high. It was a great treat to get fresh water at last, as since we had left the Nordenskjold Ice Barrier the only snow available for cooking purposes had been brackish.

The following day was also bright and sunny, but the sledging was terribly heavy. The sun had thawed the surface of the saline snow and our sledge runners had become saturated with soft water. We were so wearied with the great effort necessary to keep the sledges moving that at the end of each halt we fell sound asleep for five minutes or so at a time across the sledges. On such occasions one of the party would wake the others up, and we would continue our journey. We were even more utterly exhausted than usual at the end of this day.

By this time, however, we were in sight of a rocky headland which we took to be Cape Irizar, and we knew that this cape was not very far to the south of the Drygalski Glacier. Indeed, already, a long line was showing on the horizon which could be no other than the eastward extension of this famous and, as it afterwards proved, formidable glacier.

On November 19, we had another heavy day's sledging, ankle deep in the soft snow with occasional thin patches of sludgy saline ice from which ice flowers had recently disappeared through thawing. We only did two miles of relay work this day and yet were quite exhausted at the end of it.

The following day, November 20, being short of meat we killed a seal calf and cow, and so replenished our larder. At the end of the day's sledging I walked over about two miles to a cliff face, about six miles south of Cape Irizar. The rocks all along this part of the shore were formed of coarse gneissic granite, of which I was able to collect some specimens. The cliff was about one hundred feet high where it was formed of the gneiss, and above this rose a capping of from seventy to eighty feet in thickness of heavily crevassed blue glacier ice. There were here wide tide-cracks between the sea ice and the foot of the sea cliff. These were so wide that it was difficult to cross them. The whole shore line was literally alive with seals and seal calves; there were over fifty of them in a stretch of about three hundred yards. At a distance of two miles our tent was, of course, quite out of sight, and one had to be guided back, on this as on other similiar occasions, chiefly by one's footprints.

The following day, November 21, the sledging was painfully heavy over thawing saline snow surface, and sticky sea ice. We were only able to do two and two-thirds miles.

On November 22, on rounding the point of the low ice barrier, thirty to forty feet high, we obtained a good view of Cape Irizar, and also of the Drygalski Ice Barrier.

On November 23 we found that a mild blizzard was blowing, but we travelled on through it as we could not afford to lose any time. The blizzard died down altogether about 3 A.M., and was succeeded by a gentle westerly wind off the plateau. That evening, after our tent had been put up and we had finished the day's meal, I walked over a mile to the shore. The prevailing rock was still gneissic granite with large whitish veins of aplitic granite. A little bright green moss was growing on tiny patches of sand and gravel, and in some of the cracks in the granite. The top of the cliff was capped by blue glacier ice. With the help of steps cut by my ice-axe I climbed some distance up this in order to try and get some fresh ice for cooking purposes, but close to the top of the slope I accidentally slipped and glissaded most unwillingly some distance down before I was able to check myself by means of the chisel edge of the ice-axe. My hands were somewhat cut and bruised, but otherwise no damage was done. The whole of this ice was slightly bitter; no doubt sea spray in heavy weather when the sea was open during summer time, had dashed over the headland, and so flavoured the ice with sea salts. At last I obtained some fairly fresh ice in the form of large ice stalactites depending from an overhanging cliff of glacier ice. With these and my geological specimens I trudged back to the camp.

On November 24, a strong keen wind was blowing off the plateau from the west-south-west. This died down later on in the morning at about 2 A.M. and the temperature at 9 A.M. rose as high as plus 20° Fahr. We were all suffering somewhat from want of sleep, and although the snow surface was better than it had been for some little time we still found the work of sledging very fatiguing. A three-man sleeping-bag, where you are wedged in more or less tightly against your mates, where all snore and shin one another and each feels on waking that he is more sinned against than shinning, is not conducive to real rest; and we rued the day that we chose the three-man bag in preference to the one-man bags. That afternoon and evening we slept a little longer than usual, and felt much refreshed on the early morning of November 25.

It was interesting to watch at lunch time the anxious face of the messman for the week as he sat with his nose close to the outer cover of the aluminium cooker in order to catch the first whiff of the delicious aroma which told that the tea in the water of the inner cooking-pot had been just brought to the boil. With the first sniff of the aroma the messman would immediately unscrew the brass valve of the Primus, so as to let the air in and the Primus lamp down, with a view to saving paraffin oil.

Chapter Nine

PROFESSOR DAVID'S NARRATIVE (*Continued*)

ON the following day, November 26, we saw on looking back that the rocky headland, where I had collected the specimens of granite and moss, was not part of the mainland, but a small island. This day was rather a memorable one in our journey, as we reached a large rocky promontory, which we supposed at the time to be Cape Irizar. Subsequent observations, however, proved that we must already have passed Cape Irizar, which was in all probability the small island just referred to.

We had some good sledging here over pancake ice nearly free from snow and travelled fast. While Mackay secured some seal meat Mawson and I ascended the rocky promontory, climbing at first over rock, then over glacier ice, to a height of about six hundred feet above the sea. The rock was a pretty red granite traversed by large dykes of black rocks, apparently of an alkaline character, belonging to the phonolites or tinguaites.

From the top of the headland to the north we had a magnificent view across the level surface of sea ice far below us. We saw that at a few miles from the shore an enormous iceberg, frozen into the floe, lay right across the path which we had intended to travel in our northerly course on the morrow. To the north-west of us was Geikie Inlet and beyond that stretching as far as the eye could follow was the great Drygalski Glacier. Beyond the Drygalski Glacier were a series of rocky hills. One of these was identified as probably being Mount Neumaer. Several mountains could be seen further to the north of this, but the far distance was obscured from view by cloud and mist so that we were unable to make out the outline of Mount Nansen. It was evident that the Drygalski Glacier was bounded landwards on the north by a steep cliff of dark, highly jointed rock, and we were not a little concerned to observe with our field-glasses that the surface of the Drygalski Glacier was wholly different to that of the Nordenskjold Ice Barrier. It was clear that the surface of the Drygalski Glacier was formed of jagged surfaces of ice very heavily crevassed, and projecting in the form of immense séracs separated from one another by deep undulations or chasms. It at once suggested to my mind some scaly dragon-like monster and recalled the lines of Milton quoted at the beginning of this chapter. The "Scaly horror of his folded tail" did not seem enchanting even at this distance of ten to fifteen miles. We could see that much of this glacier was absolutely impossible for sledging, but it appeared that further eastward the inequalities of the ice-surface became less pronounced, and at the extreme eastward extension, at a distance of some twenty-five to thirty miles from where we stood, the surface appeared fairly smooth. After taking these observations from our point of vantage we retraced our steps. Mawson, in his spiked ski-boots, got down the sloping ice-surface with comparative ease, but as I had finnesko on I found it necessary to cut steps with my ice-axe all the way down the glacier ice.

It was obvious from what we had seen looking out to sea to the east of our camp that there were large bodies of open water trending shorewards in the form of long lanes at no great distance. The lanes of water were only partly frozen over, and some of these were interposed between us and the Drygalski Glacier. Clearly not a moment was to be lost if we were to reach the Drygalski Glacier before the sea ice broke up. A single strong blizzard would now have converted the whole of the sea ice between us and the Drygalski Glacier into a mass of drifting pack. We obtained from this rocky promontory a fine collection of geological specimens, and here, as elsewhere, got abundant evidence of former much greater extension of the inland ice sheet.

The following day, November 27, we decided to run our sledges to the east of the large berg, which we had observed on the previous day, and this course apparently would enable us to avoid a wide and ugly looking tide-crack extending northwards from the rocky point at our previous camp. The temperature was now as high as from plus 26° to plus 28° Fahr. at mid-day, consequently, the saline snow and ice were all day more or less sticky and slushy. We camped near the large berg.

On the morning of November 28 a mother seal with a well-grown baby came up to our tent and sniffed and snorted around its skirt. It seemed about to enter the tent when I hunted it off, and mother and baby, meanwhile, made tracks, in every sense of the word, for the open water. Then we packed up and started our sledges, and pulled them over a treacherous slushy tide-crack, and then headed them round an open lead of water in the sea ice. At 3 A.M. we had lunch near the east end of the big berg. Near here Mackay and Mawson succeeded in catching and killing an Emperor penguin, and took the breast and liver. This bird was caught close to a lane of open water in the sea ice. We found that in the direction of the berg this was thinly frozen over, and for some time it seemed as though our progress further north was completely blocked. Eventually we found a place where the ice might just bear our sledges. We strengthened this spot by laying down on it slabs of sea ice and shovelfuls of snow, and when the causeway was completed—not without Mackay breaking through the ice in one place and very nearly getting a ducking—we rushed our sledges over safely, although the ice was so thin that it bent under their weight. We were thankful to get them both safely to the other side.

We now found ourselves amongst some very high sastrugi and hard tough snow. We had to drag the sledges over a great number of these, which were nearly at right angles to our course. This work proved extremely fatiguing. The sastrugi were from five to six feet in height. As we were having dinner at the end of our day's sledging we heard a loud report which we considered to be due to the opening of a new crack in the sea ice. We thought it was possible that this crack was caused by some movement of the great active Drygalski Glacier, now only about four miles ahead of us to the north.

We got out of our sleeping-bag soon after 8 P.M. on the evening of the 28th, and started just before midnight. The ice-surface over which we were sledging this day had a curious appearance resembling rippling stalagmites, or what may be termed ice marble. This opacity appeared to be due to a surface enamel of partly thawed snow. This surface kept continually cracking as we passed over it with a noise like that of a whip being cracked. It was evidently in a state of tension, being contracted by the cold which attained its maximum soon after midnight, for, although of course we had for many weeks past been having the midnight sun it was still so low in the heavens towards midnight that there was an appreciable difference in the temperature between midnight and the afternoon. This difference in our case was further accentuated by the cold nocturnal wind from the high plateau to our west. This wind was of the nature of a land breeze on a large scale.

There were here two sets of sastrugi, the principal set parallel to the plateau wind and trending here from nearly north-west to south-east; the other set, caused by the blizzard winds, trended from south to north. We were now getting very short of biscuits, and as a consequence were seized with food obsessions, being unable to talk about anything but cereal foods, chiefly cakes of various kinds and fruits. Whenever we halted for a short rest we could discuss nothing but the different dishes with which we had been regaled in our former lifetime at various famous restaurants and hotels.

The plateau wind blew keenly and strongly all day on November 29. As we advanced further to the north the ice-surface became more and more undulatory, rising against us in great waves like waves of the sea. Evidently these waves were due to the forward movement, and consequent pressure of the Drykalski Glacier. We had a fine view from the top of one of these ridges over the surface of the Drygalski Glacier to the edge of the inland plateau. Far inland, perhaps forty or fifty miles away, we could see the great névé fields, which fed the Drygalski Glacier, descending in conspicuous ice falls, and beyond these loomed dim mountains. At the end of this day we hardly knew whether we were on the edge of the sea ice or on the thin edge of the Drygalski Glacier. Probably, I think, we were on very old sea ice, perhaps representing the accumulations of several successive seasons.

It fell calm at about 9 P.M., but just before midnight,

November 29–30, the plateau wind returned, blowing stronger than ever. As the sun during the afternoon had now considerable heating power we tried the experiment of putting snow into our aluminium cooking-pot, the exterior of which by this time was permanently coated with greasy lamp-black from the blubber lamp, and leaving the pot exposed in the evening to the direct rays of the sun. The lamp-black, of course, formed an excellent absorbent of the sun's heat-rays. On getting out of the sleeping-bag at 9 P.M. on November 29 I found that about half the snow I had put into the cooking-pot had been thawed down by the sun's heat. This, of course, saved both paraffin and blubber. It takes, of course, as much energy to thaw ice or snow at a temperature of 32° Fahr. to form a given volume of water as it does to raise that water from 32° Fahr. up to boiling-point. As our snow and ice used for domestic purposes frequently had a temperature of many degrees below zero, the heat energy necessary to thaw it was greater than that required to raise the water from freezing-point to boiling-point.

As we advanced with our sledge on the early morning of November 30, the ice ridges fronting us became higher and steeper, and we had much ado straining with all our might on the steep ice slopes to get the sledges to move, and they skidded a good deal as we dragged them obliquely up the slopes. The plateau wind, too, had freshened, and was now blowing on our port bow at from fifteen to twenty miles an hour, bringing with it a good deal of low drift. At last, about 10 A.M., the plateau wind dropped and with it the drift, and the weather became warm and sunny.

The glacier now spread before us as a great billowy sea of pale green ice, with here and there high embankments of marble-like névé resembling railway embankments. Unfortunately for our progress, the trend of the latter was nearly at right angles to our course. As we advanced still further north the undulations became more and more pronounced, the embankments higher and steeper. These embankments were now bounded by cliffs from forty to fifty feet in height, with overhanging cornices of tough snow. The cliffs faced northwards. The deep chasms which they produced formed a very serious obstacle to our advance, and we had to make some long *détours* in order to head them off. On studying one of these chasms it seemed to me that their mode of origin was somewhat as follows: In the first place the surface of the ice had become strongly ridged through forward movement of the glacier, with perhaps differential frictional resistance; the latter causing a series of undulations, the top of each ice undulation would then be further raised by an accumulation of snow partly carried by the west-north-west plateau wind, partly by the southerly blizzard wind. These two force components produced these overhanging cliffs facing the north. For some reason the snow would not lie at the bottoms of the troughs between the undulations. Probably they were swept bare by the plateau wind. It was hardly to be wondered at that we were unable to advance our sledges more than about one mile and a half that day.

The next day, December 1, the hauling of our sledges became much more laborious. For half a day we struggled over high sastrugi, hummocky ice ridges, steep undulations of bare blue ice with frequent chasms impassible for a sledge, unless it was unloaded and lowered by Alpine rope. After struggling on for a little over half a mile we decided to camp, and while Mawson took magnetic observations and theodolite angles, Mackay and I reconnoitred ahead for between two and three miles to see if there was any way at all practicable for the sledge out of these mazes of chasms, undulations and séracs. Mackay and I were roped together for this exploratory work, and fell into about a score of crevasses before we returned to camp, though in this case we never actually fell with our head and shoulders below the lids of the crevasses, as they were mostly filled at the surface with tough snow. We had left a black signal flag on top of a conspicuous ice mound as a guide to us as to the whereabouts of the camp, and we found this a welcome beacon when we started to return, as it was by no means an easy task finding one's way across this storm-tossed ice sea, even when one was only a mile or two from the camp. On our return we found that Mawson was just completing his observations. He found that the dip of the needle here was $2\frac{1}{2}°$ off the vertical. We brought the tent down from where he had been taking magnetic observations, and treading warily, because of crevasses, set it up again close to our sledge, and had lunch.

We noticed in the case of the snow lids over crevasses that they were covered by a very pretty moss-like growth of pointed ice crystals. This growth was apparently due to a slow upward steaming of moist air from the spaces between the walls of the crevasses below. Possibly during the day the air beneath the snow lids may become slightly warmed, and as the temperature falls at night, particularly under the influence of the plateau wind, a slow percolation of the warm air through the snow lid may take place, and the small amount of moisture in it is deposited on the surface of the lid on coming in contact with the colder air outside. This process, continued from day to day, gradually builds up these moss-like crystals.

That afternoon we discussed the situation at some

length. It appeared that the Drygalski Glacier must be at least twenty miles in width. If we were to cross it along the course which we were now following at the rate of half a mile every half day it would obviously take at least twenty days to get to the other side, and this estimate did not allow for those unforseen delays which experience by this time had taught us were sure to occur. The view which Mackay and I had obtained of the glacier ice ahead of us showed that our difficulties, for a considerable distance, would materially increase. Under these circumstances we were reluctantly forced to the conclusion that our only hope of ultimate success lay in retreat. We accordingly determined to drag the sledges back off the glacier on to the sea ice by the way along which we had come.

Early on the morning of December 2 the retreat began. Just before midnight it had been clear and sunny, but as midnight approached a thick fog suddenly came up and obscured everything. Consequently we had great difficulty in picking up our old sledge tracks as we retreated over the glacier ice. The weather was still very thick and foggy at 3 A.M., but a little before 6 A.M. the fog cleared off and the sun shone through. We had now reached the southern edge of the glacier, and were back on the old undulating sea ice. We turned our sledges eastwards following parallel with the glacier edge. Immediately on our left rose large rounded hummocky masses of ice belonging to the Drygalski Glacier, and from fifty to sixty feet in height. The sky, meanwhile, had become again overcast with dense cumulus which drifted across rapidly from a south-easterly direction.

The following day we still travelled eastwards parallel to the southern edge of the Drygalski Glacier. The sledging was chiefly over soft snow, ankle deep, with occasional high snow sastrugi, and here and there a patch of rippled sea ice. There could be no doubt about it being sea ice this time because in one place, at the foot of one of these steep snow ridges, we noticed a pool of water only thinly frozen over, and on breaking the ice I tasted the water and found it was very salt.

Towards the end of this day's sledging we passed a long inlet trending north-westerly. This inlet was floored with sea ice, and made a long, deep indentation in the glacier ice. After our hoosh, and before turning into the sleeping-bag, Mawson and I went on to the north over some high hummocky ridges of the Dryalski Glacier to look ahead. Mawson, after a while, returned to camp, while I turned north-westwards to explore the inlet. After falling into a few crevasses which traversed the great billowy hummocks of blue glacier ice in all directions, I got down into the inlet, and on following it north-westwards, found that it gradually passed into a definite glacial river channel, and became quite unpracticable for sledging. It was quite clear from the steep banks of this channel, cut out of the hard snow névé and glacier ice, that during the few weeks of thaw in this part of the Antarctic, great volumes of thaw water must rush down off the higher parts of the glacier towards the sea, and in their passage they tear out deep canyon-like channels in the glacier ice and névé. This channel trended at first exactly in the direction in which we wished to make, but it was obvious that it was an impossible route for the sledges.

We decided on December 4 that we had better go on an extended reconnoitring expedition before we again risked landing our sledge in a labyrinth of pressure ridges and crevasses. After hauling our sledges for a little less than a mile, and meeting with steep slopes of snow dunes, we halted. While Mackay sewed one of the tent-poles,

Skua Gulls at the Ice-edge

which had become loosened, back into its place in the canvas crown which held the tops of the tent-poles together, Mawson and I climbed on to some hummocks a little north of the camp to see which route would be best to follow on our reconnoitring journey. After lunch we all three started with our ice-axes and the Alpine rope. We travelled up a broad bottomed snow valley for about two miles trending in the direction of Mount Larsen. Then for a little over a mile beyond it trended more to the right in the direction of Mount Nansen. Here we got into difficult country, the snow surface being succeeded by steep-sided hummocks, rolls and ridges of blue glacier ice, with occasional deep chasms and very numerous crevasses. We fell into numbers of the latter up to our thighs, but the snow lids as yet were just strong enough to stop us going deeper. Mawson opened up one of these snow lids with his ice-axe, and we noticed that the lid was from one to one and a half feet thick, while the crevasse was thirty feet wide and of vast depth. Much of the ground over which we were travelling rang hollow, and was evidently only roofed over by a thin layer of tough snow. Altogether we travelled about four miles to the north of our camp, but could see no sign from there of any sea to the north of us. Meanwhile, Mackay diverged somewhat to the west, climbed on the top of a high ice pyramid, but was unable to see any trace of the sea beyond. We now returned at a smart pace back to camp, arriving about 9 A.M. We were all pretty tired, and, as usual before entering the tent, we took off our spiked ski-boots so as to avoid puncturing the waterproof floorcloth, and put on finnesko. Hoosh was prepared, and we had a good meal of it, as well as of fried seal meat with blubber and seal oil.

It was evident now that even if we were to succeed

in crossing the Drygalski Glacier, the passage would occupy a good many days, even under the most favourable circumstances, and our provisions were running very short. After we finished our hoosh, Mawson, with the field-glasses, sighted a seal near to the big berg to the south of us, which we had passed some five days previously. We decided that we would go after this seal the next day; meanwhile, the seal disappeared. Fearing that a blizzard might spring up on the following day, I proposed to go that evening out to the berg in search of seals, but Mackay kindly volunteered instead and started off with his rück-sack and ice-axe and a small allowance of provisions, consisting of some cooked seal meat, biscuits and chocolate. He had a long journey before him. While he was gone I was chiefly occupied in dividing up our rations into half-ration lots. At 5 P.M. I was attracted by the notes of a penguin behind an ice mound at no great distance. Rousing Mawson, we both went in pursuit and after a long and severe chase, captured an Adelie penguin.

At about midnight, December 4–5, Mackay returned to camp after his fourteen miles' tramp over the sea ice. He brought back with him a most welcome addition to our larder in the shape of over thirty pounds of seal meat, liver and seal blubber. He reported that he had had great difficulty in crossing the large ice-crack where we had constructed the causeway for our sledge some five days previously, and he said that it would now have been impossible to have got the sledge over it. Mackay had been up over seventeen hours, and had been sledging, travelling over heavy ice, and carrying his heavy load of seal meat with only short intervals for meals. He had travelled a distance of about twenty-four miles, and of course under the circumstances was much exhausted and badly in need of a long rest. By securing the so much needed additional food-supply, he had rendered us an extremely important service. It now, of course, became necessary to give him the needed rest on the following day. Accordingly, the earlier part of December 6 we spent in the sleeping-bag.

Soon after midnight, December 5–6, we left our camp on the south side of the Drykalski Glacier, and struck across the high ridges of blue ice into the small valley in the glacier which we had prospected two days previously. As usual a keen wind was blowing off the plateau at this time of the morning, but the temperature soon rose to plus 23° Fahr. at 7.15 A.M. The sky was overcast with heavy stratus and cumulus clouds, especially in the direction of Cape Washington. We passed over a considerable number of crevasses without any serious accident. The day's sledging was heavy on account of the strongly undulating surface of the ice and the quantity of soft snow in the ice valley on the surface of the Drygalski Glacier.

The following day, December 7, was also dull and lowering with very dense cumulus clouds over Cape Washington and Mounts Melbourne and Nansen. We inferred that this dense cumulus was due to the presence of open water between the Drygalski Glacier and Cape Washington, and were not a little anxious as to whether, in the event of the sea ice having all drifted out on the north side of the Drygalski Glacier, it would be possible for us to travel shorewards on the surface of the glacier itself, when we got to the other side. We encountered many precipitous slopes from thirty to forty feet deep, often with overhanging cornices, barring our northward progress like those already met with in the part of the glacier from which we had retreated. These overhanging cliffs, however, were not quite as serious as those which we had left behind, and by making considerable *détours* we managed to circumvent them. At last we seemed to have got amongst an impassable belt of high crevassed ice ridges with precipitous chasms between. After a good deal of reconnoitring ahead a clue out of the labyrinth was discovered in the form of a series of high snow ridges which led backwards and forwards, in and out, amongst the high-pressure ridges, and eventually enabled us to land our sledges in a broad crevassed valley on the glacier surface.

After the hoosh at the end of our day's sledging, Mawson and I walked about two miles, looking out for a track for the next morning. The outlook was by no means encouraging, as the surface still bristled with huge ice undulations as far as the eye could reach. It was just as though a stormy sea had suddenly been frozen solid, with the troughs between its large waves here and there partly filled with snow, while the crests of the waves were raised by hard ridges of drift snow, terminating in overhanging cliffs, facing the north. It was obvious, too, that the glacier ice over which we would have to travel, was still very heavily crevassed. As we returned, a mild blizzard sprang up from the south-south-west, bringing low drift with it. The blizzard cleared off in about an hour and a half, and the sun came out strong and hot, and rapidly thawed the snow on our tent and on the food-bags stored on our sledges.

The following day, December 8, we dug away the drift snow piled by the blizzard against our sledges, and were pleased to find that the day was beautifully fine and sunny with a light breeze from the west-south-west. The sledging was very heavy up and down steep ice slopes with much soft snow between.

Mawson had a slight attack of snow blindness on December 9. The day was so warm that we even felt it oppressive, the temperature at midnight being as high as plus 19° Fahr. The glacier ice kept cracking from time to time with sharp reports. Possibly this may have been due to the expansion of the ice under the influence of the hot sun. At one spot the sledges had to be dragged up a grade of 1 in 3 over smooth blue glacier ice. This was exceedingly heavy work. At last, when we were near to our time for camping, Mackay, on going on a short distance ahead to reconnoitre with the field-glasses, sighted open water on the northern edge of the Drygalski Ice Barrier, about three to four miles distant. He announced his discovery with shouts of *θάλαττα, θάλαττα,* which thrilled us now as of old they thrilled the Ten Thousand. It was no sparkling waters of the Euxine that had met his gaze, but a Black Sea nevertheless, for so it appeared as its inky waves heaved under the leaden sky. But what a joy to have reached once more that friendly water world that went up by many a creek and river to our homes. It was now clear to us that we could not hope for sea ice over which to sledge westwards to the shore, where we proposed to make our final depôt before attempting the ascent of the great inland plateau in order to reach the Magnetic Pole.

During the day we were cheered by a visit from several snow petrels, which flew around our camp, as well as from three skua gulls. Mawson managed to snare one of the skuas with a fishing-line, but it got away when he was hauling in the line. Our sledging that day was not quite so heavy; the ice undulations were less formidable, and the belts of snow between became wider and firmer. Just after lunch a beautiful Wilson's petrel flew around us.

December 10 saw us still struggling to cross the Drygalski Glacier. We could see that we were now on a pretty high ridge, but the highest part of the glacier now lay to the south and therefore behind us. We were much rejoiced towards the end of the day's sledging to find ourselves at last off the true glacier type of surface, and on to what may be described as an undulating barrier type. This improvement of the surface to our west enabled us to do what we had been longing to do for the last six days, turn our sledge westwards. At first we had to incline somewhat to the north-west in order to skirt round some high ice ridges. Then, after making some northing, we were able to go nearly due west. The snow surface was largely of the pie-crust type; our ski-boots broke through it at every step and we sank in up to our ankles. At intervals we still crossed low ridges of solid glacier ice, traversed by crevasses. All the crevasses were more or less roofed over with tough snow lids. These lids sometimes were slightly in relief, or sometimes showed slight depressions in the general surface. In such areas the snow lids rung hollow as the sledges travelled over them. We found the snow lids always most treacherous close to either wall of the crevasse, and we frequently fell partially through at such spots, but had no very bad falls in this part of the glacier.

The following day, December 11, we had a fine view of "Terra Nova" Bay, and as far as could be judged the edge of the Drygalski Ice Barrier on the north was now scarcely a mile distant. We were much surprised at the general appearance of the outline of the Ice Barrier coastal ice and coast-line ahead of us. It did not agree, as far as we could judge, with the shape of this region as shown on the Admiralty chart, and we could see no certain indication whatever of what was called, on the chart, "the low, sloping shore." Accordingly we halted a little earlier than usual in order to reconnoitre. There was a conspicuous ice mound about half a mile to the north-west of this camp. Mackay started off with the field-glasses for a general look round from this point of vantage. Mawson started changing his plates in the sleeping-bag, while I prepared to go out with my sketch-book and get an outline panoramic view of the grand coast ranges now in sight. Crevasses of late had been so few and far between that I thought it was an unnecessary precaution to take my ice-axe with me, but I had scarcely gone more than six yards from the tent, when the lid of a crevasse suddenly collapsed under me at a point where there was absolutely no outward or visible sign of its existence, and let me down suddenly nearly up to my shoulders. I only saved myself from going right down by throwing my arms out and staying myself on the snow lid on either side. The lid was so rotten that I dared not make any move to extricate myself, or I might have been precipitated into the abyss. Fortunately Mawson was close at hand, and on my calling to him, he came out of our sleeping-bag, and bringing an ice-axe, chipped a hole in the firm ice on the edge of the crevasse nearest to me. He then inserted the chisel edge of the ice-axe in the hole and holding on to the pick point, swung the handle towards me: grasping this, I was able to extricate myself and climbed out on to the solid ice.

It was a beautiful day, the coast-line showing up

very finely, and I was able to get from the ice mound a sketch of the mountains. Mawson also took three photographs, making a panoramic view of this part of the coast. He was able, also, to get a valuable series of angles with the theodolite, which showed that the shape of the coast-line here necessitated serious modification of the existing chart.

Chapter Ten

PROFESSOR DAVID'S NARRATIVE (*Continued*)

Die schönste Jungfrau sitzet
Dort oben wunderbar,
Ihr goldnes Geschmeide blitzet,
Sie Kämmt ihr goldnes Haar.

Sie Kämmt es mit goldenem Kamme,
Und singt ein Lied dabei;
Das hat ein wundersame,
Gewaltige melodei.

HEINE.

FAR beyond the golden mountains to the north and west lay our goal, but as yet we knew not whether we were destined to fail or succeed. Meanwhile no time was to be lost in hurrying on and preparing for a dash on to the plateau, if we were to deserve success.

The following day, December 12, we sledged on for half a mile until we were a little to the west of the conspicuous ice mound previously described. We concluded that as this ice mound commanded such a general view of the surrounding country, it must itself be a conspicuous object to any one approaching the Drygalski Glacier by sea from the north; and so we decided that as there was still no trace of the "low, sloping shore" of the chart, and that as the spot at which we had now arrived was very near to the area so named on the chart, we would make our depôt here. We intended to leave at this depôt one of our sledges with any spare equipment, a little food, and all our geological specimens, and proceed thence shorewards and inland with one sledge only. We estimated that we still had fully 220 miles to travel from this depôt on the Drygalski Glacier to the Magnetic Pole. It was, therefore, necessary now to make preparations for a journey there and back of at least 440 miles. We thought that with *détours* the journey might possibly amount to 500 miles.

We could see, even from our distance of from twenty to thirty miles from the shore-line, that we had no light task before us in order to win a way on to the high inland plateau. Before we knew that the whole of the sea ice had gone out between us and Mount Melbourne, we had contemplated the possibility of travelling further northwards along the coast on sea ice, down to a spot marked on the chart as Gerlache Inlet. This inlet we now saw was situated amongst a wilderness of high sharp jagged mountain peaks rising to heights of from 6000 to 8000 ft. above sea-level, and as it could now be approached only from the land, it was now practically inaccessible. Nearer to us, and to the north-west of our position on the Drygalski Ice Barrier, was the giant form of Mount Nansen, one of the grandest and most imposing of all the mountains seen by us in the Antarctic. Further to the left and nearly due west of us was another fine dark mountain massif, Mount Larsen. Between Mount Larsen and Mount Nansen was a vast glacier with a rugged surface, steep ice falls and large crevasses. About midway between Mounts Larsen and Nansen was a huge nunatak of black rock, rising abruptly from the ice surface at a point several miles inland from the shore-line. Further to the left of Mount Larsen was another glacier less formidable in appearance and smaller in size than the Mount Nansen Glacier. This terminated near the coast in rather a steep slope, and gradually became confluent with the Drygalski Glacier. To the south of this glacier, which may be termed the Mount Larsen Glacier, was another great mountain massif with Mount Belligshausen on the north and Mount Neumaer on the south. The foot-hils of Mount Neumaer terminated in steep precipices forming the northern wall of the Drygalski Glacier.

Our first business was to lay in a stock of provisions sufficient to last us for our 500 miles of further journeying. Mackay started for a small inlet about a mile and a half distant from our camp, where he found a number of seals and Emperor and Adelie penguins. He killed some seals and Emperor penguins, and loaded a good supply of seal steak, blubber, liver and penguin steak and liver on to the sledge. In the course of his hunting, he fell through an ice bridge, at a tide-crack, up to his waist in the water. Mawson and I went out to meet him when the sledge was loaded, and helped to drag it back to camp. We found it very hot in the tent, the weather being fine and sunny. It was delightful to be able at last to rest our weary limbs after the many weeks of painful toil over the sea ice and the Drygalski Glacier.

We started cooking our meat for the sledging trip on the following day, December 13, our intention being to take with us provisions for seven weeks, in addition to equipment, including scientific instruments, &c. We estimated that the total weight would amount to about 670 lb. We were doubtful, in our then stale and weakened condition, whether we should be able to pull such a load over the deep loose snow ahead of us, and then drag it up the steep ice slopes of the great glaciers which guarded the route to the plateau.

The sun was so hot that it started melting the fat out of our pemmican bags, so that the fat actually oozed through not only the canvas of the bags themselves, but also through the thick brown canvas of the large fortnightly food-bags, which formed a sort of tank for containing the pemmican bags, and we found it necessary at once to shade the food-bags from the sun by piling our Burberry garments over them. Leather straps, tar rope, tins, sledge harness, lamp-black off the blubber cooker, warmed by the rays of the sun, all commenced to sink themselves more or less rapidly into the névé.

We unpacked and examined both sledges, and found that of the two, the runners of the Duff sledge were the less damaged. As the result of the rough treatment to which it had recently been subjected, one of the iron brackets of this sledge was broken, but we replaced it with a sound one from the discarded Christmas Tree sledge.

The following day, December 14, we were still busy preparing for the great trek on the morrow. Mackay was busy cooking Emperor penguin and seal meat for the plateau journey; Mawson was employed in transferring the scientific instrument boxes and the Venesta boxes in which our Primus lamp and other light gear were packed from the Christmas Tree sledge on to the Duff sledge. He also scraped the runners of the sledge with pieces of broken glass in order to make their surfaces as smooth as possible. I was busy fixing up depôt flags, writing letters to the Commander of the *Nimrod,* Lieutenant Shackleton, and my family, and fixing up a milk tin to serve as a post office on to the depôt flag-pole. When all our preparations were completed we drew the Christmas Tree sledge with some of our spare clothing, our blubber cooker, a biscuit tin with a few broken biscuits, and all our geological specimens to the top of the ice mound, about a quarter of a mile distant. On reaching the top of the mound we cut trenches with our ice-axes in which to embed the runners of the sledge, fixed the runners in these grooves, piled the chipped ice on top, then lashed to the sledge, very carefully, the flag-pole about six feet high, with the black flag displayed on the top of it. The wind blew keenly off the plateau before our labours were completed. We all felt quite sorry and downcast at parting with this sledge, which by this time seemed to us like a bit of home. We then returned to camp. Just previous to depôting this sledge, Mackay fixed another small depôt flag close to the open sea a few yards back from the edge of the ice cliff.

Soon after we had turned into our sleeping-bag, a gentle blizzard started to blow from west by south. This continued all night, increasing in intensity in the morning. We were able to see great whale-backed clouds, very much like those with which we had been familiar over Mount Erebus, forming over Mount Nansen. As this blizzard wind was blowing partly against us, we decided that we would wait until it had either slackened off or decreased in force.

The whole of the next day, unfortunately, the blizzard continued. The sun was very hot, and as the result of its heat we were to-day for the first time subject to a new trouble. The blizzard, of course, drifted snow all over our tent; and a strong thaw set in on the side of it which faced the sun. The wind, flapping the canvas of the tent against the tent-poles, brought the thaw water through on to the poles facing the sun. Inside the tent, however, the temperature was just below freezing-point, and as the water started to trickle down the poles it froze. With the flapping of the tent backwards and forwards against the tent-poles, the small ridge of ice on the upper surface of each tent-pole became drawn out into sharp, saw-like teeth, and these started cutting through the canvas. All through this day, consequently, we had to be continually getting out of the sleeping-bag and running our hands down the tent-poles so as to rub off the ice teeth.

The blizzard continued till midnight of December 15–16, when its force markedly decreased. We breakfasted accordingly just after midnight. I dug out the sledge from the snow which had drifted over it, and Mackay cached some seal meat in an adjoining ice mouth. At last, about 7 A.M., we made a start and were delighted to find that, chiefly as the result of the three days' rest in camp, we were able to pull our sledge—weighing about 670 lb.—with comparative ease. The snow, though soft, had become crusted over the surface through the thaw brought on by the blizzard, followed by freezing during the succeeding cold night. The sledging was certainly heavy, but not nearly so distressing as that which we had recently experienced in crossing the Drygalski Glacier. The "tablecloth was being laid" on the top of Mount Nansen in the form of a remarkable flattish thin white fleecy cloud. It looked as though a high-level blizzard was blowing over the summit. We steered towards the great black nunatak midway between Mount Nansen and Mount Larsen, as Mawson and Mackay both considered that in this direction lay our chief hope of finding a practical route to the high plateau.

On December 17 we had a very interesting day. The sledging was rather heavy, being chiefly over soft snow and pie-crust snow. It was difficult to decide sometimes whether we were on fresh-water ice or on sea ice. Here and there we crossed ice ridges, evidently pressure ridges of some kind. These would be traversed by crevasses which showed the ice in such places to be at least thirty to forty feet in thickness. Close to our final camping-ground for the day was a long shallow valley or barranca; it was from one hundred and twenty to one hundred and thirty yards in width. The near side was steep, though not too steep for us to have let our sledge down; but the far side was precipitous, being bounded by an overhanging cliff from twenty to thirty feet high. The floor of this valley was deeply and heavily crevassed. This sunken valley, therefore, formed a serious obstacle to our advance.

While Mackay was preparing the hoosh Mawson travelled to the right, and I to the left along this valley seeking for a possible crossing place. At last Mawson found a narrow spot where there had been an ice bridge over the valley, but this had become cracked through at the centre. It was nevertheless strong enough to bear our sledge. Near this ice bridge Mawson stated that he noticed muddy material containing what appeared to be foraminifera, squeezed up from below. The day had been calm and clear, and we were able to get detailed sketches of this part of the coast range.

The following day we made for the ice bridge with our sledge, and found that the crack crossing it had opened to a width of eighteen inches during the night. The far side had become too, somewhat higher than the near side. We had little difficulty in getting the sledge over, and after crossing several other cracks in the ice and névé without mishap reached once more a fairly level surface. A light plateau wind was now blowing from off the Mount Nansen glacier. The hard snow surface was furrowed by two very definite sets of sastrugi, one set coming from a south-westerly direction, and apparently caused by blizzard wind, the other from nearly north-west. The latter were evidently due to strong rushes of cold air from the high plateau down the broad valley occupied by the Mount Nansen glacier. That day we passed over a series of pressure ridges with their steeper sides directed towards the north-west. At the bottom of these steep slopes the ice was often crevassed, and sometimes we had some little difficulty in crossing them. They were probably due to pressure from the Drygalski Glacier.

At lunch time, soon after midnight, we reached some very interesting glacial moraines in the form of large to small blocks, mostly of eruptive rock, embedded in the ice. It was probable, from their general distribution, that they formed part of an old moraine of Mount Nansen, though now about fifteen miles in advance of the present glacier front. A conspicuous rock amongst the boulders was a greenish-grey to greenish-black diorite, very rich in sphene. The brown crystals of sphene were frequently intercrystallised with the felspars, and gave the rock a very pretty appearance. Small fragments of sandstone and clay shale were also represented in these moraines. The larger blocks were up to seven feet in diameter, and formed chiefly of reddish porphyritic granite. We collected a number of specimens from this moraine.

Fine rolls of cumulus clouds were gathering to our north-east. The day was calm with occasional gleams of sunshine. After the plateau wind had died down about 2 P.M. it commenced to snow a little, the snow coming from between south-west and west-south-west.

At midnight on December 19 we started sledging in the falling snow, guided partly by the direction of the wind, partly by that of the pressure ridges and crevasses, occasionally taking compass bearings. Before we had gone far we reached a tide-crack with open water three to four feet wide. There was also a width of about eighteen feet of recently formed thin ice at this tide-crack. We tasted the water in this crack and found that it was distinctly salt. It was clear then that at this part of our journey we were travelling over sea ice. About half a mile further on we reached another open tide-crack, and had to make a considerable *détour* in order to get over it. The surface of the ice was now thawing, and we trudged through a good deal of slushy snow, with here and there shallow pools of water as blue as the Blue Grotto of Capri. On the far side of this second tide-crack, and beyond the blue pools, we reached a large pressure ridge forming a high and steep scarped slope barring our progress. Its height was about eighty feet. There was nothing for it, if we were to go forward, but to drag our heavy sledge up this steep slope. It was extremely exhausting work, and we were forced to halt a few times, and had to take the sledge occasionally somewhat obliquely up the slope where it was very steep. In such cases the sledge frequently skidded. Our troubles were increased by the fact that this ice slope was traversed by numerous crevasses, which became longer and wider the further we advanced in this direction.

At last we got to the slope, only to see in the dim light that there were a succession of similar slopes ahead of us, becoming continually higher and steeper. The ice, too, became a perfect network of crevasses, some of which were partly open, but most of them covered over with snow lids. Suddenly, when crossing one of these snow lids, just as he was about to reach the firm ice on the other side, there was a slight crash and Mawson instantly disappeared from sight. Fortunately the toggle at the end of his sledge rope held, and he was left swinging in the empty space between the walls of the crevasse, being suspended by his harness attached to the sledge rope.

Mackay and I hung on to the rope in case it should part at the toggle, where it was somewhat worn. Meanwhile, Mawson called out from below to pass him down the Alpine rope. Leaving Mackay to keep hold of the toggle end of Mawson's harness rope, I hurried back to the sledge, which was about ten feet behind, and just as I was trying to disengage a coil of rope Mawson called out that he felt he was going. I ran back and helped Mackay to keep a strain on Mawson's harness rope. Mawson then said that he was all right. Probably at the time he felt he was going the rope had suddenly cut back through the lid of the crevasse and let him down for a distance of about a couple of feet. Altogether he was about eight feet down below the level of the snow lid. While I now held on to Mawson's harness rope Mackay hurried back to the sledge, and with his Swedish knife cut the lashing around the Alpine rope, and started uncoiling it, making a bowline at the end in which Mawson could put his foot. Meanwhile Mawson secured some ice crystals from the side of the crevasse, and threw them up for examination. The Alpine rope having been lowered, Mawson put his foot into the bowline and got Mackay to haul his leg up as high as his bent knee would allow it to go, then, calling to him to hold tight the rope, Mawson, throwing the whole weight of his body on to it, raised himself about eighteen inches by means of his arms so as to be able to straighten his right leg. Meanwhile, I took in the slack of his harness rope. He then called to me to hold tight the harness rope, as he was going to rest his whole weight on that, so as to take the strain off the Alpine rope. Mackay then was able to pull the Alpine rope up about eighteen inches, which had the effect of bending up Mawson's right leg as before. Mackay then held fast the Alpine rope, and Mawson again straightened himself up on it, resting his whole weight on that rope. Thus little by little he was hoisted up to the under surface of the snow lid, but as his harness rope had cut back a narrow groove in this snow lid several feet from where the snow gave away under him Mawson now found his head and shoulders pressing against the under side of the snow lid, and had some difficulty in breaking through this in order to get his head out. At last the top of his head emerged, a sight for which Mackay and I were truly thankful, and presently he was able to get his arms up, and soon his body followed, and he got safely out on the near side of the crevasse. After this episode we were extra cautious in crossing the crevasses, but the ice was simply seamed with them. Twice when our sledge was being dragged up ice-pressure ridges it rolled over sideways with one runner in a crevasse, and once the whole sledge all but disappeared into a crevasse, the snow lid of which had partly collapsed under its weight. Had it gone down completely it would certainly have dragged the three of us down with it, as it weighed nearly one-third of a ton. It was clear that these high-pressure ridges and numerous crevasses were caused now, not by the Drygalski, but by the Nansen Glacier.

It was now somewhat foggy, but occasionally the fog and mist lifted, and in the distance one caught glimpses of magnificent cliffs of reddish brown granite, with wisps and wreathes of white mist hanging around the summits. The view reminded me of the Grampians in Scotland near Ossian's Cave at the Pass of Glencoe. Later on in the day we saw in the dim light that we had before us a long steep descent into an ice valley, which appeared to be heavily crevassed at the bottom. As we were uncertain whether we could get across it at this spot we left our tent and sledge, and reconnoitred ahead, taking with us the Alpine rope and our ice-axes. We found a way of crossing this valley, but could see that the ice-surface ahead of us was apparently worse than ever. We returned to our tent and sledge, and put up the tent, and chopped lumps of ice off the glacier with which to load the skirt, as no snow was available at the time. It was just commencing to snow, and wind was freshening from the south-west. We were now in a perfect labyrinth of crevasses and pressure ridges. Snow continued falling heavily accompanied by a blizzard wind for the rest of that day and the whole of the succeeding night. Inside the tent we experienced some discomfort through the dripping of water caused by the thawing snow. As usual during a blizzard the temperature rose, and although the sun's heat rays were partly intercepted by the falling snow quite sufficient warmth reached the side of the tent nearest the sun to produce this thaw. Pools of water lodged on the foot of our sleeping-bag, but we were able to keep the head of it fairly dry by fixing up our Burberry blouses and trousers across the poles on the inside of the tent so as to make a temporary waterproof lining just above our heads. We were all thoroughly exhausted, and slept until about 7 A.M. the following day, December 20. By that time the snow cleared off.

About six inches of snow had fallen, and was lying deeply drifted in places. We dug away the drift snow from around the sledges, and after the morning hoosh held a council of war. The question was whether we should continue pulling on in the direction of the nunatak rising from the Mount Nansen Glacier, or whether we should retreat and try some other way which might lead us to the plateau. Mackay was in favour of hauling ahead over the Mount Nansen Glacier, while

Mawson and I favoured retreat, and trying a passage in some other direction.

At last we decided to retreat. Our fortunes now, so far as the possibility of reaching the Magnetic Pole were concerned, seemed at a low ebb. It was already December 20, and we knew that we had to be back at our depôt on the Drygalski Glacier not later than February 1 or 2, if there was to be a reasonable chance of our being picked up by the *Nimrod.* We had not yet climbed more than 100 ft. or so above sea-level, and even this little altitude was due to our having climbed ice-pressure ridges, which from time to time dipped down again to sea-level. We knew that we had to travel at least 480 to 500 miles before we could hope to get to the Magnetic Pole and back to our depôt, and there remained only six weeks in which to accomplish this journey, and at the same time we would have yet to pioneer a road up to the high plateau. Now that everything was buried under soft snow it was clear that sledging would be far slower and more laborious than ever. We soon proved that this was the case, for after starting the sledge it gathered masses of soft snow around it and under it as it went, and at the end of 200 yards we had to halt for a temporary rest, hoist the sledge up on one side and knock away the masses of clogged snow from underneath it. This had to be repeated every few hundred yards, and after we had gone half a mile we decided to leave the sledge and go ahead with ice-axes and Alpine rope to reconnoitre.

We started off in a south-westerly direction with the intention of seeing whether the Mount Bellingshausen Glacier slope would be more practicable for our sledges than the Mount Nansen Glacier. We trudged through soft thawing snow with here and there shallow pools of water on the surface of the ice. This, of course, saturated our socks, which froze as the temperature fell during the night. After proceeding about two and a half miles we observed with the field-glasses that the foot of the Mount Bellingshausen Glacier was not only steep but broken and rugged. We decided to examine what appeared to be a narrow stretch of snow mantling around the base of a granite mountain, one of the offshoots from the Mount Larsen massif. After crossing a good deal of pressure ice and crevasses, and floundering amongst the boulders of old moraines we reached some shallow lakes of thawed snow near the junction between the sea ice and the foot of the snow slope for which we had been steering. In the neighbourhood of the moraines, which here consisted of great blocks of eruptive rock partly or wholly imbedded in ice, the blocks became so warmed up by the sun's heat that they partially thawed the ice around, and in some cases above them: and so when one stepped near one of these blocks, or over a concealed block, the ice gave way with a crash letting one down a depth of from one to three feet. At one place, before reaching the shallow lakes, we found quite a strong stream of water flowing just under the surface of the ice. This was evidently supplied from thaw water from the slopes near the shore-line.

After paddling, unwillingly, in the shallow lakes we reached the foot of what proved now to be not a snow slope but a small branch glacier. This was covered with a considerable depth of soft newly drifted snow, and we found the ascent in consequence very tiring as we sunk at each step in the soft snow over our knees. At last we attained an altitude of 1200 ft. above sea-level, and were then high enough to see that the upper part of this branch glacier joined the Mount Bellingshausen Glacier at about 800 ft. higher and some half

Depot on the Drygalski Barrier

"Backstair's Passage" on the Ascent from the Sea-Ice to the Plateau.
Mount Larsen on the Left

mile further on. We were well pleased with this discovery, but as the glacier front ascended about 1500 ft. in less than a mile we did not look forward to the task of getting our heavy sledge up this steep slope, encumbered as it was with soft deep thawing snow.

On our return to the shore-line down the glacier slope we discovered that it was slightly crevassed in places, though not heavily so. At the foot of the glacier, and a short distance towards our camp, we found a moraine gravel. This was intermixed with a dark marine clay containing numerous remains of serpulæ, pecten shells, bryozoa, foraminifera, &c. Mackay also found a perfect specimen of a solitary coral, allied to Delto-cyathus, and also a Waldheimia. All these specimens were carefully preserved and brought into camp. While we were collecting these specimens we could hear the roar of many mountain torrents descending the steep granite slopes of the great mountain mass to the south of our branch glacier. Occasionally, too, we heard the boom and crash of an avalanche descending from the high mountain top. Such sounds were strange to our ears, accustomed so long to the almost perfect solitude and silence of the Antarctic, hitherto broken only by the bleating of baby seals and the call of the penguins.

Mawson discovered in another part of the moraine, nearer to our camp, a bright green mineral forming thin crusts on a very pretty quartz and felspar porphyry. These we decided to examine more carefully on the morrow. We were all thoroughly exhausted after the day's work, and Mackay had a rather bad attack of snow-blindness. For some time after we got into the sleeping-bag, and before we dozed off, we could still hear the intermittent roar of avalanches like the booming of distant artillery.

The following day, December 22, we picked our way with our sledge cautiously amongst the crevasses and over the pressure mounds, the traversing of which gave us some trouble in places, and eventually reached a fairly good track along the ice parallel to the moraine from which we had been collecting the day previous. We found a large pool of thaw water on the surface of the ice. This was fed by a sub-glacial stream coming from an old rock moraine. We could hear this stream rolling the pebbles along in its channel. At another point the moraine showed a remarkable cone, which at first sight we took for a typical esker, but a nearer examination revealed the fact that the whole cone, with the exception of the exterior, was formed of solid ice with only an outer coating of sand, mud and gravel associated with abundant marine organisms similar to those collected by us the previous day. We halted when we arrived opposite the green mineral observed by Mawson the previous day. We collected a good deal of this. At first sight we thought it was the common mineral epidote, but its hardness and the fact that it had turned yellow, where it was weathered, made this hypothesis untenable. The green crusts formed by it were about one fourteenth to one-sixteenth of an inch in thickness, and it was evidently fairly widely distributed in that locality, as numerous large joint faces of the quartz and felspar porphyry were completely coated with it. A little further on we came upon an enormous silicious sponge, eighteen inches by two feet in diameter, adhering firmly to one of the moraine boulders. We secured specimens of this.

Altogether the locality was most fascinating, and we longed to have been able to spend more time there. Amongst other interesting problems was the question as to how the material of the sea floor came to be uplifted here to a height of twenty to thirty feet or more above sea-level, and as to how the marine sediments came to be resting on an old conical surface of dense ice. We tested the latter to see whether it was of salt- or of fresh-water origin; it was not distinctly saline, though slightly so—much as glacier ice would be if it were sprayed by the sea. None of us could account for this curious phenomenon. It seemed as though the marine muds had been subjected to considerable pressure, as numbers of the fossils in it were triturated and shattered. It is of course just possible that in the forward movement of the Mount Nansen Glacier it may have pushed up some of the sea bottom above sea-level, and still there remains the question as to how masses of ice came to find their way under the moraine sediments. It is possible that after an extensive glaciation of this region the glacier ice from inland spread over the spot where this moraine is now situated, but on the retreat of the ice inland, while still a small thickness of ice was left in this bay, a submergence ensued, and during that submergence a marine mud was deposited over the ice together with the larger organisms found in association with the mud. Then there was an advance of the ice once more, and moraines of large blocks of rock were laid down over the top of the moraine muds and the relics of the ancient glacier ice. Then once more the ice retreated to its present position leaving the moraine blocks and moraine muds of the old ice in the relative situations mentioned.

As we skirted the foot of the small branch glacier we noticed several small puffs of snow near the top angle of the snow slope which we proposed to escalade. Just as we were pulling our sledge to the foot of this slope the puff of wind with drift snow developed suddenly into a strong blizzard. We pulled in against this with great difficulty for half an hour, then camped at the foot of the slope. The blizzard with its heavy drift snow and the occasional gleams of warm sunshine cast much drift over our tent with accompanying thaw. Consequently inside the tent water dripped heavily all over our clothing and sleeping-bag. Fortunately we were just above the level of the thaw water of the small lakes, but we could hear water trickling close underneath our tent amongst the granite boulders of the moraines just under the ice.

We were able now to economise fuel, as we could bale the water out of these rock pools and streams for making our hoosh, tea and cocoa. All that night the blizzard raged, and we thought any moment that the tent would be ripped up from top to bottom. It was getting very thin by this time and had already been frequently repaired by Mackay and Mawson. On this occasion several new rents started from near the top of the tent and spread downwards. Moreover, the canvas cap of our tent was broken by the force of the wind and the pressure of the drift snow.

The following day, about 7 A.M., I got up and dug away the drift snow from the lee side of the tent, which was cramping our feet and legs, and found that it was still snowing heavily outside, and blowing hard as well. In the afternoon the blizzard slacked off somewhat, and the drift nearly ceased. We got up accordingly and had a meal. We halved our sledge load, repacked the sledge, and by dint of great exertions dragged it up the steep snow and ice slope to a height of 800 ft. above the sea. This was done in the teeth of a mild but freshening blizzard. The blizzard at last got too strong for us, so that we left the load at the altitude mentioned and returned back to our tent with the empty sledge.

A PARSELENE

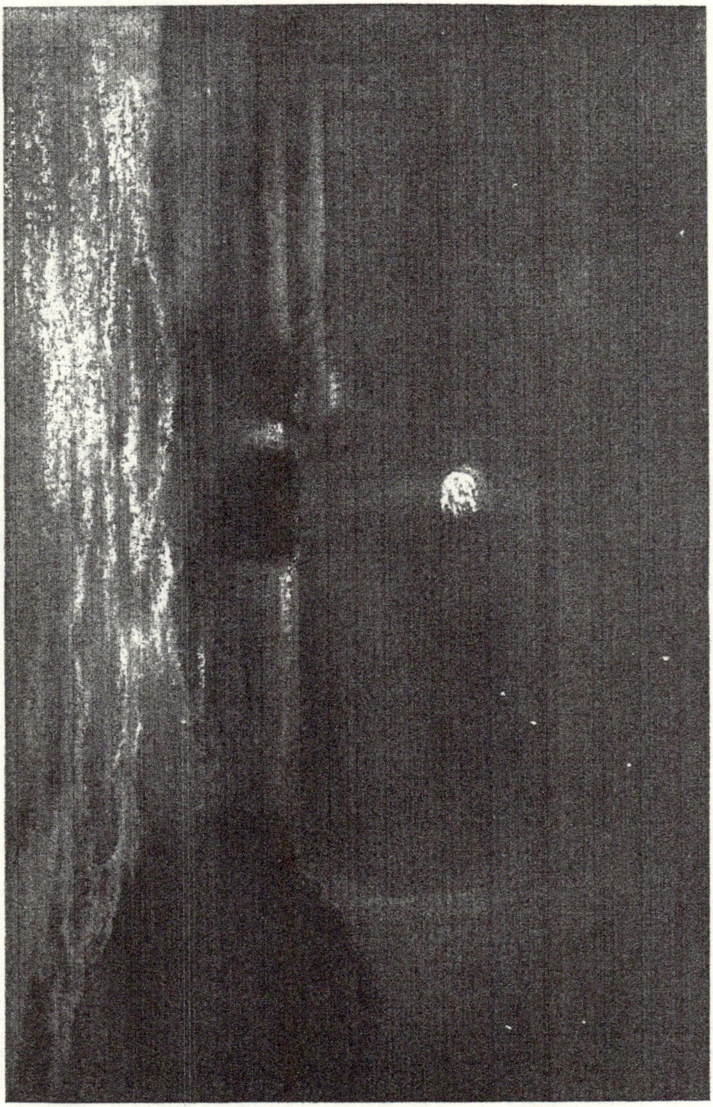

We had been pleased to find that the blizzard, although it had delayed us and damaged our tent, had proved a blessing in disguise. It had not brought with it much fresh snow, but had blown away most of the loose snow left by the preceding blizzard, leaving behind it now a fairly hard snow surface suitable for sledging.

Mackay's eyes, still suffering from the effects of snow-blindness, were treated with a solution of thin tabloids (laminæ) of sulphate of zinc and cocaine, with the result that his eyes were much better the following day, December 25. This day there was still a strong breeze coming off the plateau, and sweeping over our tent. A little later in the morning the weather became calm, and a glorious sunny day smiled upon us. Mawson and Mackay repaired the rents in the tent, while I saw to repacking of the sledge with the remaining half load, and collected some geological specimens.

We started shortly before noon and commenced dragging up the second part of our load to the accompaniment of the music of murmuring streams. During our interval for lunch, Mawson was able to get some theodolite angles. We had the great satisfaction, when we turned in at 10 P.M. on Christmas Eve, to find that we were above the uncomfortable zone of thaw, and everything around us was once more crisp and dry, though cold. Our spirits, too, mounted with the altitude. We were now over 1200 ft. above sea-level.

Chapter Eleven

PROFESSOR DAVID'S NARRATIVE (*Continued*)

THE following day, December 25, was Christmas Day. When I awoke, I noticed a pile of snow on top of the sleeping-bag close to my head. At first, before I was fully awake, I imagined that it was the moisture condensed from Mawson's breath. Then I heard the gentle patter of snow-flakes, and, on turning my head in the direction in which the rustling proceeded, saw that the wind had undermined the skirt of our tent, and was blowing the snow in through a small opening it had made. Accordingly, I slipped out and snowed up the skirt again, trampling the snow down firmly. A plateau wind was now blowing with almost blizzard force.

About two hours later we got up, and, after some trouble with the Primus lamp on account of the wind, had our breakfast, but, as the wind was blowing dead against us, we turned into the sleeping-bag for a short time. It was nearly noon before the wind died down, and we started off with our sledge, still relaying with half loads, the day being now beautifully clear and sunny. At the 1300 ft. level we started our sledge meter again, having lifted it off the ice while we were going up the steep slope. A little further on we were able to put the whole of our load again on to the sledge and so dispense with further relay work. This, too, was a great blessing.

When we arrived at our spot for camping that night we had the satisfaction of finding that we were over 2000 ft. above sea-level, and that we had, in addition to the climbing, travelled that day about four miles. The plateau wind had almost gone, and once more we revelled in being not only high, but dry. Having no other kind of Christmas gift to offer, Mawson and I presented Mackay with some sennegrass for his pipe, his tobacco having long ago given out. We slept soundly that Christmas night.

On December 26 we observed dense dark snow clouds to the north-east, and a little light snow commenced to fall, but fortunately the weather cleared towards the afternoon. Mawson lost one of his blue sweaters off the sledge, but he and Mackay went back some distance and recovered it. Towards the afternoon we found it necessary to cross a number of fairly large crevasses. These were completed snowed over, and although we frequently fell through up to our knees, we had no serious trouble from them on this occasion. Some of them were from twenty to thirty feet in width, and it was fortunate for us that the snow lids were strong enough to carry safely the sledge and ourselves. Mackay suggested, for greater security, fastening the Alpine rope around Mawson, who was in the lead, and securing the other end of it to the sledge. The rope was left just slack enough to admit of the strain of hauling being taken by the harness rope, hence Mawson had two strings to his bow in case of being suddenly precipitated into a crevasse. This was a good system, which we always adopted afterwards in crossing heavily crevassed ice.

The following day, December 27, we decided to make a small depôt of our ski-boots (as by this time it appeared we were getting off the glacier ice on to hard snow and névé where we should not require them) and also of all our geological specimens, and about one day's food-supply, together with a small quantity of oil—a supply for about two days in one of our oil-cans. The following is a list of the provisions:

Powdered cheese (enough for two meals).
Tea (for four meals).
Twenty-five lumps of sugar.
Hoosh for one meal.
Chocolate (for one and a half meals).
Twelve biscuits.

We also left an empty biscuit tin into which we crammed our ski-boots, and our three ice-axes, using one of them stuck upright as a staff for a small blue flag to mark the depôt. Mawson took some good bearings with the prismatic compass, and we then proceeded on our way. This depôt we called the Larsen Depot, as it was close to one of the southern spurs of Mount Larsen.

All eyes were now strained, as we advanced with our sledge, to see whether there was still any formidable range of mountains ahead of us barring our path to the plateau. At one time it seemed as though there was a high range in the dim distance, but a careful examination with the field-glasses showed that this appearance was due only to clouds. Our joy and thankfulness was unbounded when we at last realised that apparently there was now a fairly easy ascent of hard névé and snow on to the plateau. That day we sledged a little over ten miles. During the night there was a very strong radiant in the sky from about south-west to north-east, with a movement of altro-stratus cloud from north-west to south-east. Therefore, probably, this radiant was due to formation of great rolls of cloud curled over by the anti-trade wind as it pressed forward in a south-easterly direction. The rolls of clouds were distinctly curved convexly towards the south-east.

The following day, December 28, we travelled on north-westwards in thick cloudy weather, at first quite calm. At about 10 A.M. a breeze set in from the sea, spreading westwards over the top of Mount Nansen over 8000 ft. above sea-level. Above Nansen it met the upper current wind and was obviously deflected by it in a south-easterly direction. Meanwhile, in the direction of the coast the sky was very dark and lowering, and probably snow was falling there. Remarkable pillars of cloud formed over the Mount Larsen group. These were photographed by Mawson. We passed over occasional patches of nearly bare glacier ice, alternating with stretches of hard névé. When we camped that evening we had sledged a little over ten miles, and a keen, cold wind was blowing gently off the high plateau to our west.

The following day, December 29, was clear, calm and cold. At noon a pretty strong wind was blowing off the plateau. The surface of the snow was fairly strongly ridged with sastrugi. One set was made by winds coming from between west-south-west and west by north, the other by winds nearly north-westerly, or between west-30°-north and west-40°-north. As this latter bearing was not far off the direction in which we were travelling, we were able from time to time to follow these minor sastrugi, which were thus of considerable help to us in bringing over the sastrugi more oblique to the direction in which we were travelling.

The following day, December 30, Mounts Larsen and Bellingshausen were disappearing below the horizon, and several mountains were showing up clearly and sharply to the north of us, the principal peaks of which were at first identified by us as Mount New Zealand

and Mount Queensland of Captain Scott's chart. Later Mawson concluded that the western of the two at any rate was new and unnamed.

There was still a strong plateau wind. We were now at an altitude of about 4500 ft. Once more, as in winter time, our breath froze into lumps of ice, cementing our Burberry helmets to our beards and moustaches. In putting up the tent for lunch in the strong plateau wind, it became badly torn near the cap piece. This wind had started before midnight on the previous night, and was blowing strongly until the afternoon, at from twelve to about fifteen miles an hour. It carried along with it a little low drift. The plateau wind did not die down until the evening. Our distance travelled was eleven miles, and we were still travelling on an up grade, being now nearly 5000 ft. above sea-level.

December 31 passed off without any special event other than that after Mackay had repaired the tent in the morning it became torn again at lunch-time when we were fitting it over the tent-poles. Mawson took a fresh set of magnetic observations. We camped for this purpose at the bottom of a wide undulation in the névé surface. We were disappointed at his announcement that he made out that the Magnetic Pole was further inland than had been originally estimated. What with the observations with the Lloyd-Creak dip circle, and the time occupied in repairing the rents in the tent, we ran ourselves somewhat short of time for our sledging that day, and did not camp until a little before midnight. We were still dragging the sledge on an up grade; the surface was softer and more powdery than before, and the sastrugi heavier. Also since the previous Tuesday we had been obliged to put ourselves on somewhat shorter rations than before, as we had to take one-eighth of our rations out in order to form an emergency food-supply in the event of our journey to and from the Magnetic Pole proving longer than we originally anticipated.

That night, about a mile before reaching camp, we sighted to the west of us, much to our surprise, some distinct ice falls. This showed us that the snow desert over which we were travelling had still some kind of creeping movement in it. A skua gull came to visit us this New Year's Eve. He had been following us up for some time in the distance, mistaking us, perhaps, for seals crawling inland to die, as is not infrequently the habit of these animals. We were now about eighty miles inland from the nearest open water. Being disappointed of his high hopes, he left us after that day and we saw him again no more. The run for the day was about ten miles. We felt very much exhausted when we turned into our sleeping-bag that night.

January 1, 1909 (New Year's Day), was a beautiful calm day with a very light gentle plateau wind, with fairly high temperature. The sky was festooned in the direction of Mount Nansen with delicate wispy cirrus clouds converging in a north-east direction. Later on, towards the evening, it was evident that these cirrus clouds were strongly bent round from south-west in a northerly direction. Possibly this bending with the concave side to the west-north-west was due to the pressure at a high level of the anti-trade wind blowing towards the east-south-east. Mawson took observations for latitude and for magnetic deviation at noon. He made our latitude at noon to be 74° 18′. That night Mawson gave us a grand hoosh and a rich pot of cocoa in celebration of New Year's Day.

We all thoroughly enjoyed this meal after our exhausting march.

On January 2 we noticed that the sastrugi were gradually swinging round into a direction a little north of west. The snow was frequently soft in large patches, which made sledging very heavy. We ascended altogether about 290 ft., but we crossed a large number of broad undulations, the troughs of which were from thirty to forty feet below their crests. These undulations considerably increased the work of sledging, and the loose patches of snow were so very soft and powdery that the runners of our sledge sunk deeply into them, so that it was only with our utmost efforts that we were able that day to finish our usual ten miles. Again we were much exhausted when the time came for camping. We were beginning to suffer, too, from hunger, and would have liked more to drink if we could have afforded it. We talked of what we would have drunk if we had had the chance. Mackay said he would have liked to drink a gallon of buttermilk straight off; Mawson would have preferred a big basin of cream; while I would have chosen several pots of the best coffee with plenty of hot milk.

We were still climbing on January 3, having ascended another 500 ft. It proved the heaviest day's sledging since we reached the plateau. The snow was still softer than on the previous day, and the surface was more undulating than ever, the troughs of the undulations being about fifty feet below the crests. The sastrugi themselves were from two to three feet in height. The crests of the large undulations were usually formed of hard snow, the strong winds having blown any loose material off them. This loose material had accumulated to some depth in the troughs, and hence made the wide patches of soft snow which made our sledge drag so heavily as we crossed them. By dint of great efforts we managed to finish our ten miles for that day.

The next day, January 4, we were pleased to find that there was less up grade than on the previous day. We were now at an altitude of over 6000 ft., and found respiration in the cold, rarified air distinctly trying. It was not that we suffered definitely from mountain sickness, but we felt weaker than usual as the result, no doubt, of the altitude combined with the cold. Towards evening, large clouds developed, much like the whaleback clouds which we had often observed forming over Erebus about the time of blizzards. Great rolls of cumulus spread rapidly from the north-west towards the south-east, and we feared that a blizzard was impending. On the whole the sledging was a little easier to-day than the preceding day, and again we managed to do our ten miles.

On the morning of January 5 we found the sky thickly overcast, except to the south and the south-east, where clear strips of blue were showing. We thought that snow was coming. The weather was perfectly calm, comparatively warm, but the light dull. We could still see the new inland mountain and Mount New Zealand distinctly. The sun was so oppressively hot when it peeped out from behind the clouds that one could feel it burning the skin on one's hands.

The surface was more marked by sastrugi than ever, but on the whole firm. We sledged ten miles. I will quote from my diary the notes regarding some succeeding days.

January 6.—To-day the weather was gloriously fine. Bright, warm sunshine with a crisp, cold air in the early morning and the weather almost calm. The pulling was rather heavy during the afternoon; possibly the hot sun may have somewhat softened the surface of the snow. This morning I left off my crampons and put on a new pair of finnesko. These latter proved somewhat slippery, and in falling heavily this afternoon over one of the sastrugi I slightly strained some muscles on the inner side of my left leg, just below the knee. This gave me a considerable amount of pain for the rest of the journey. Mackay lost all his stockings and socks off the bamboo pole of the sledge, but was fortunate enough to recover them after walking back over a mile on our tracks.

January 7.—We were up at 5 A.M., when the temperature was minus 13° Fahr. We were anxious to arrive at the end of our first five miles in good time for Mawson to get a meridian altitude, and take theodolite angles to the new mountain and Mount New Zealand, which were now almost disappearing from view below the horizon. Mawson made our latitude to-day 73° 43′. This was one of the coldest days we had as yet experienced on the plateau, the wind blowing from west by north. We all felt the pulling very much to-day, possibly because it was still slightly uphill, and probably partly on account of mountain lassitude. The distance travelled was ten miles.

Friday, January 8.—To-day, also, was bitterly cold. The wind blew very fresh for some little time before noon from a direction of about west by north, raising much low drift. Our hands were frost-bitten several times when packing up the sledge. The cold blizzard continued for the whole day. At lunch time we had great difficulty in getting up the tent, which became again seriously torn in the process. Our beards were frozen to our Burberry helmets and Balaclavas, and we had to tear away our hair by the roots in order to get them off. We continued travelling in the blizzard after lunch. Mawson's right cheek was frost-bitten, and also the tip of my nose. The wind was blowing all the time at an angle of about 45° on the port bow of our sledge. We just managed to do our ten miles and were very thankful when the time came for camping.

The following day, January 9, a very cold plateau wind was still blowing, the horizon being hazy with low drift. We were now completely out of sight of any mountain ranges, and were toiling up and down amongst the huge billows of a snow sea. The silence and solitude were most impressive. About 10.30 A.M. a well-marked parphelion, or mock sun, due to floating ice crystals in the air, made its appearance. It had the form of a wide halo with two mock suns at either extremity of the equator of the halo parallel to the horizon and passing through the real sun. Mawson was able to make his magnetic deviation observation with more comfort, as towards noon the wind slackened and the day became gloriously bright and clear. In the afternoon it fell calm.

We were feeling the pinch of hunger somewhat, and as usual our talk, under these circumstances, turned chiefly on restaurants, and the wonderfully elaborate dinners we would have when we returned to civilisation. Again we accomplished our ten miles, and were now at an altitude of over 7000 ft.

January 10 was also a lovely day, warm and clear; the snow surface was good and we travelled quickly. There was a strong "Noah's Ark" structure in the high-level cirrus clouds, there being a strong radiant point respectively in the north-west and south-east, and this made us somewhat apprehensive that we were in for another blizzard. These cirrus clouds were also strongly curved with the concave side of the curve facing the north-east. We thought this curve was, perhaps, due to the anti-trade wind bending round in a direction following that of the curve in the wisps of cirrus.

January 11.—We were up about 7 A.M., the temperature at that time being minus 12° Fahr. It was a cold day to-day, and we had a light wind nearly southerly. At first it blew from between south and south-south-east; this gradually freshened at lunch time and veered towards the west. It then returned again more towards the south-south-east. Mawson had a touch of snow-blindness in his right eye. Both he and Mackay suffered much through the skin of their lips peeling off, leaving the raw flesh exposed. Mawson, particularly, experienced great difficulty every morning in getting his mouth opened, as his lips were firmly glued together by congealed blood.

That day we did eleven miles, the surface being fairly firm, and there being no appreciable general up grade now, but only long-ridged undulations, with sastrugi. We noticed that these sastrugi had now changed direction, and instead of trending from nearly west, or north of west, eastwards, now came more from the south-east directed towards the north-west. This warned us that we might anticipate possibly strong head winds on our return journey, as our course at the time was being directed almost north-west, following from time to time the exact bearing of the horizontal magnetic compass. The compass was now very sluggish, in fact the theodolite compass would scarcely work at all. This pleased us a good deal, and at first we all wished more power to it; then amended the sentiment and wished less power to it. The sky was clear, and Mawson got good magnetic meridian observations by means of his very delicately balanced horizontal moving needle in his Brunton transit instrument.

January 12.—The sky to-day was overcast, the night having been calm and cloudy. A few snow-flakes and fine ice crystals were falling. The sun was very hot and it somewhat softened the snow surface, thereby increasing of course the difficulty of sledging. We sledged to-day ten and three-quarter miles.

The evening, after hoosh, Mawson, on carefully analysing the results set forth in the advance copy of the Discovery Expedition Magnetic Report, decided that although the matter was not expressly so stated, the Magnetic Pole, instead of moving easterly, as it had done in the interval between Sabine's observations in 1841 and the time of the *Discovery* expedition in 1902. was likely now to be travelling somewhat to the north-west. The results of dip readings taken at intervals earlier in the journey also agreed with this decision. It would be necessary therefore to travel further in that direction than we had anticipated in order to reach our goal. This was extremely disquieting news, for all of us as we had come almost to the limit of our provisions, after making allowance for enough to take us back on short rations to the coast. In spite of the anxiety of the situation extreme weariness after sledging enabled us to catch some sleep.

The following morning, January 13, we were up about 6 A.M. A light snow was falling, and fine ice crystals made the sky hazy. There was a light wind blowing from about south-south-east. About 8 A.M. the sun peeped through with promise of a fine day. We had had much discussion during and after breakfast as to our future movements. The change in the position of the Pole necessitated of course a change in our plans. Mawson carefully reviewed his observations as to the position of the Magnetic Pole, and decided that in order to reach it we would need to travel for another four days. The horizontally moving needle had now almost ceased to work. We decided to go on for another four days and started our sledging. It was a cold day with a light wind. The temperature at about 10.30 A.M. being minus 6° Fahr. At noon Mawson took a magnetic reading with the Lloyd-Creak dip circle, which was now fifty minutes off the vertical, that is, 89° 10′. At noon the latitude was just about 73° South. The sastrugi were now longer and higher than usual, and there were two distinct sets. The strongest sastrugi trended from south to north; a subordinate set from south-east to north-west. That day we sledged thirteen miles.

January 14.—The day was gloriously clear and bright with a warm sun. A gentle wind was blowing from about south-south-east, and there was a little cumulus cloud far ahead of us over the horizon. The surface of the snow over which we were sledging was sparkling with large reconstructed ice crystals, about half an inch in width and one-sixteenth of an inch in thickness. These crystals form on this plateau during warm days when the sun's heat leads to a gentle upward streaming of the cold air with a small amount of moisture in it from beneath. Under these influences combined with the thawing of the surface snow, these large and beautiful ice crystals form rapidly in a single day. We observe that after every still sunny day a crop of these crystals develops on the surface of the névé, and remains there until the next wind blows them off. They form a layer about half an inch in thickness over the top of the névé. In the bright sunlight the névé, covered with these sheets of bright reflecting ice crystals, glittered like a sea of diamonds.

The heavy runners of our sledge rustled gently as they crushed the crystals by the thousand. It seemed a sacrilege. The sastrugi were large and high, and our sledge bumped very heavily over them with a prodigious rattling of our aluminium cooking-gear. It was clear that the blizzard winds blow over this part of the plateau at times with great violence. Apparently all the winds in this quarter, strong enough to form sastrugi, blow from south or west of south or from the south-east. Our run to-day was twelve miles one hundred and fifty yards.

January 15.—We were up to-day at 6 A.M. and found a cold southerly breeze blowing, the temperature being minus 19° Fahr. at 6.30 A.M. Mawson got a good latitude determination to-day, 72° 42′.

At about twenty minutes before true noon Mawson took magnetic observations with the dip circle, and found the angle now only fifteen minutes off the vertical, the dip being 89° 45′. We were very much rejoiced to find that we were now so close to the Magnetic Pole. The observations made by Bernacchi, during the two years of the *Discovery* expedition's sojourn at their winter quarters on Ross Island, showed that the amptitude of daily swing of the magnet was sometimes considerable. The compass, at a distance from the Pole, pointing in a slightly varying direction at different times of the day, indicates that the polar centre executes a daily round of wanderings about its mean position. Mawson considered that we were now practically at the Magnetic Pole, and that if we were to wait for twenty-four hours taking constant observations at this spot the Pole would, probably, during that time, come vertically beneath us. We decided, however, to go on to the spot where he concluded the approximate mean position of the Magnetic Pole would lie. That evening the dip was 89° 48′. The run for the day was fourteen miles.

From the rapid rate at which the dip had been increasing recently, as well as from a comparison of Bernacchi's magnetic observations, Mawson estimated that we were now about thirteen miles distant from the probable mean position of the South Magnetic Pole. He stated that in order to accurately locate the mean position possibly a month of continuous observation would be needed, but that the position he indicated was now as close as we could locate it. We decided accordingly, after discussing the matter fully that night, to make a forced march of thirteen miles to the approximate mean position of the Pole on the following day, put up the flag there, and return eleven miles back on our tracks the same day. Our method of procedure on this journey of twenty-four miles is described in the journal of the following day.

Saturday, January 16.—We were up at about 6 A.M. and after breakfast we pulled on our sledge for two miles. We then depôted all our heavy gear and equipment with the exception of the tent, sleeping-bag, Primus stove and cooker, and a small quantity of food, all of which we placed on the sledge together with the legs of the dip circle and those of the theodolite to serve as marks. We pulled on for two miles and fixed up the legs of the dip circle to guide us back on our track, the compass moving in a horizontal plane being now useless for keeping us on our course. At two miles further we fixed up the legs of the theodolite, and two miles further put up our tent, and had a light lunch. We then walked five miles in the direction of the Magnetic Pole so as to place us in the mean position calculated for it by Mawson, 72° 25′ South latitude, 155° 16′ East longitude. Mawson placed his

THE NORTHERN PARTY ON THE PLATEAU.

* POOL OF THAW WATER FORMED BY THE EMERGENCE OF A SUB-GLACIAL STREAM SOUTH-EAST OF MOUNT LARSEN

camera so as to focus the whole group, and arranged a trigger which would be released by means of a string held in our hands so as to make the exposure by means of the focal plane shutter. Meanwhile, Mackay and I fixed up the flag-pole. We then bared our heads and hoisted the Union Jack at 3.30 P.M. with the words uttered by myself in conformity with Lieutenant Shackleton's instructions, "I hereby take possession of this area now containing the Magnetic Pole for the British Empire." At the same time I fired the trigger of the camera by pulling the string. Thus the group were photographed in the manner shown on the plate. The blurred line connected with my right hand represents the part of the string in focus blown from side to side by the wind. Then we gave three cheers for his Majesty the King.

There was a pretty sky at the time to the north of us with low cumulus clouds, and we speculated at the time as to whether it was possible that an arm of the sea, such as would produce the moisture to form the cumulus, might not be very far distant. In view of our subsequent discovery of a deep indent in the coast-line in a southerly direction beyond Cape North, it is possible that the sea at this point is at no very considerable distance.

The temperature at the time we hoisted the flag was exactly 0° Fahr. It was an intense satisfaction and relief to all of us to feel that at last after so many days of toil, hardship and danger we had been able to carry out our leader's instructions, and to fulfil the wish of Sir James Clarke Ross that the South Magnetic Pole should be actually reached, as he had already in 1831 reached the North Magnetic Pole. At the same time we were too utterly weary to be capable of any great amount of exultation. I am sure the feeling the was uppermost in all of us was one of devout and heartfelt thankfulness to the kind Providence which had so far guided our footsteps in safety to that goal. With a fervent "Thank God" we all did a right-about turn, and as quick a march as tired limbs would allow back in the direction of our little green tent in the wilderness of snow.

It was a weary tramp back over the hard and high sastrugi and we were very thankful when at last we saw a small dark cone, which we knew was our tent, rising from above the distant snow ridges. On reaching the tent we each had a little cocoa, a biscuit and a small lump of chocolate. We then sledged slowly and wearily back, picking up first the legs of the theodolite, then those of the dip circle, and finally reached our depôt a little before 10 P.M.

In honour of the event we treated ourselves that night to a hoosh, which though modest was larger in volume than usual, and was immensely enjoyed. Mawson repacked the sledge after hoosh time, and we turned into the sleeping-bag faint and weary, but happy with the great load of apprehension of possible failure, that had been hanging over us for so many weeks, at last removed from our minds. We all slept soundly after twenty-four miles of travel.

Chapter Twelve

PROFESSOR DAVID'S NARRATIVE (*Continued*)

Coldly thy rosy shadows bathe me, cold
Are all thy lights, and cold my wrinkled feet.
TENNYSON.

I CALLED the camp at a little before 10 A.M. the following morning. We now discussed the situation and our chances of catching the *Nimrod,* if she came in search of us along the coast in the direction of our depôt on the Drygalski Glacier. We had agreed, before we decided to do the extra four days' march to the shifted position of the Magnetic Pole, that on our return journey we would do not less than thirteen miles a day. At the Magnetic Pole we were fully 260 statute miles distant, as the skua flies, from our depôt on the Drygalski Glacier. As we had returned eleven of these miles on the day previous we still had 249 miles to cover. It was now January 17, and the *Nimrod* was due to start to search for us on February 1. As there was of course plenty of sunlight day and night, we thought it quite possible that she might be up to the Drygalski Glacier on February 2—possibly on the morning of that day. We accordingly decided to try and make back to our Drygalski depôt by February 1. This gave us fifteen days. Consequently we would have to average sixteen and two-thirds miles a day in order to reach the coast in the time specified. This of course did not allow of any delay on account of blizzards, and we had seen from the evidence of the large sastrugi that blizzards of great violence must occasionally blow in these quarters, and from the direction of the sastrugi during our last few days' march it was clear that the dominant direction of the blizzard would be exactly in our teeth. The prospect, therefore, of reaching our depôt in the specified time did not appear bright. Providentially we had most beautiful and glorious weather for our start on January 17. It remained fine for the whole day, and we were greatly favoured by a light wind which now blew from between north-west and west-north-west—a perfectly fair wind for our journey. In fact the wind changed direction with us. It had helped us by blowing from the south-east, just before we reached the Magnetic Pole, and now it was blowing in the opposite direction, helping us home. That day, in spite of the late start, we sledged sixteen miles.

On January 18 the weather again was fine, and we had a hard day's sledging. Unfortunately Mawson's left leg became very lame and pained him a good deal. There was a strong radiant in the clouds towards the north-west, and we were a little apprehensive of the wind in consequence. Our run for the day was sixteen miles two hundred yards. This was the end of my week's cooking, and we were able to indulge that night in a fairly abundant hoosh, also in very milky and sweet cocoa, and Mackay admitted that he actually felt moderately full after it for the first time since we had left the Drygalski Depôt.

The following day, January 19, we boiled the hypsometer at our camp, and found the level to be about 7350 ft. above the sea. The boiling-point was 196.75° Fahr. There was a cold fresh wind blowing from the south-east, a head wind for us, the temperature at the time being minus 11° Fahr. There were still low cumulus clouds to the north of us. The wind freshened in the afternoon to a mild blizzard, and we found pulling against it very severe work. That morning we had quite an unusual diversion. Mawson, who is a bold culinary experimenter, being messman for the week, tried the experiment of surreptitiously introducing a lump of sugar into the pemmican. Mackay detected an unusual flavour in the hoosh, and cross-questioned Mawson severely on the subject. Mawson admitted a lump of sugar. Mackay was thereupon roused to a high pitch of indignation, and stated that this awful state of affairs was the result of going out sledging with "two foreigners." This mild blizzard partly obliterated our old sledge tracks by piling over them new sastrugi of fine hard snow. We had a great struggle that day to make our sixteen miles, but fortunately the blizzard slackened off towards 9 P.M., and we just managed it.

Owing to some miscalculation, for which I was responsible, we discovered that we had no tea for this week, our sixth week out, unless we took it out of the tea-bag for the seventh week. Accordingly we halved the tea in the seventh week bag, and determined to collect our old tea-bags at each of our old camps as we passed them, and boil these bags together with the small pittance of fresh tea. And here I may mention the tastes of the party in the matter of tea somewhat differed. Mackay liked his tea thoroughly well and long boiled, whereas Mawson and I liked it made by just bringing the water to the boil; as soon as we smelt the aroma of tea coming from underneath the outer lid of the cooker we used to shut off the Primus lamp immediately and decant the tea into the pannikins. Mackay had always objected to this procedure when we were sledging along the sea ice where water boils at about 212° Fahr.; now, however, he had a strong scientific argument in his favour for keeping the pot boiling for a few minutes after the tea had been put in. He pointed out that at our present altitude water boiled at just over 196° Fahr., a temperature which he maintained was insufficient to extract the proper juices and flavour from the tea, unless the boiling was very much prolonged. Mawson, however, averred—on chemical and physical grounds—that with the diminished atmospheric pressure certain virtuous constituents of the tea could be extracted at a lower temperature. The discussion was highly scientific and exhilarating, though not very finite. It was agreed as a compromise to allow the boiling to continue for three or four minutes after the water had come to the boil before the tea was poured out. As in our progress coastwards we were continually coming upon more old tea-bags at our old camps, and always collected these and did not throw away any that had been used before we soon had quite an imposing collection of muslin bags with old tea leaves, and with the thorough boiling that they now got there was a strong flavour of muslin superadded to that of old tea. Nevertheless this drink was nectar.

January 20.—We were still able to-day to follow our sledge tracks, which was a great blessing, the magnetic needle being of so little use to us. We had the wind slightly against us bringing up a little low drift. Again we made our sixteen-mile run, though with great difficulty, for the wind had been blowing freshly all day on our starboard bow.

In view of the good progress that we had made, and after carefully calculating out the provisions left over, Mawson, who was at this time messman, proposed that we should return to nearly full rations, as we were becoming much exhausted through insufficient food. This proposal was, of course, hailed with delight.

On *January* 21 there was a light wind with low temperature, clear sky and hot sun, which combined to consolidate the surface over which we were sledging. By this time Mackay and Mawson's raw lips, which had been cracked and bleeding for about a fortnight previously, were now much better. Mawson's lame leg had also improved. Again we did our sixteen-mile run.

January 22.—We were up soon after 7 A.M. It was a clear day with bright sunshine. The wind started soon after 5 A.M., constantly freshening, as it usually did in this part of the plateau, till about 3 P.M. Then it gradually died down by about 10 P.M. The temperature at 7.15 A.M. was minus 20° Fahr., and at this altitude we found the wind at this temperature very trying. To-day we had to sledge over a great deal of pie-crust snow, which was very fatiguing. Again we did sixteen miles. We had since the day before yesterday lost our old sledge tracks. Mackay earned a pound of tobacco, to be given him when we returned to civilisation, by being the first to make the " land fall "—new mountain, west of Mount New Zealand—which showed out now in the far distance very faintly a little to the left of our course. It was a welcome sight to all of us. To-day we sledged fifteen miles.

January 23.—The weather was bright and cold with a light southerly wind. This day was very fatiguing, the sledging being over patches of soft snow and pie-crust snow. At the same time we were conscious now that although we were sledging up and down wide undulations we were on the whole going down hill, and the new mountain was already showing up as an impressive massif. The air was cold and piercing. Mawson's right leg was still painful. That night we were all very much exhausted, and were obliged to allow ourselves fully eight hours sleep. Our run was sixteen miles.

January 24.—To-day we had more heavy sledging over a lot of pie-crust snow and soft snow. The wind was blowing somewhat against us at about twelve miles an hour, the temperature being minus 4° Fahr. in the afternoon. A low drift was sweeping in waves over the snow desert; it was a desolate scene. Later in the day we were cheered by the sight of Mount Baxter.

Towards evening we had some discussion as to whether we were following approximately our old outgoing tracks. Mackay thought we were nearer to the new mountain than before, I thought we were further to the south-west, Mawson, who was leading, contended that we were pretty well on our old course. Just then I discovered that we were actually on our old sledge tracks, which showed up plainly for a short distance between the newly formed sastrugi. This spoke volumes for Mawson's skill as a navigator. Distance sledged sixteen miles.

January 25.—It was blowing a mild blizzard. We estimated at lunch time that we were about eighty and a half miles distant now from our Mount Larsen Depôt. The temperature during the afternoon was minus 3° Fahr. We all felt, as usual, much fatigued after the day's sledging. For the past four or five days we each took an Easton syrup tabloid for the last stage but one before reaching camp, and this certainly helped to keep us going. This evening the blizzard died down about 8 P.M., and Mount Nansen was sighted just before we camped.

January 26.—We lost our old sledge tracks again to-day. The weather turned cloudy in the afternoon, and the light was very bad. We now reached a surface of hard marble-like névé, which descended by short steep slopes. We did not at first realise that we were about to descend what we had termed the Ice Falls on the outward journey. Every now and then the sledge would take charge and rush down this marble staircase, bumping very heavily over the steps. Mawson and I frequently came heavy croppers. Mawson put on crampons outside his finnesko to enable him to get a grip of the slippery surface, but my crampons were frozen so hard and so out of shape that I was unable to get them on, so I followed behind and steadied the sledge as it continued bumping its way down the marble steps. At last we reached once more a flattened surface and camped. Our run for the day was fourteen and a half miles.

January 27.—This morning we all felt very slack after the night spent in the closely covered sleeping bag, the sky at the time being cloudy. Under these circumstances, as we now had come down from our highest altitude by about 4000 ft., and the temperature of course, had somewhat risen, we felt stifled and depressed. During the morning fine snow fell and the weather was quite thick to the south and east of us. Mawson steered us by the trend of the sastrugi. As the day wore on, the weather cleared up and we had a good view of the new mountain, Mount New Zealand, and Mount Baxter. The pulling at first was very hard work, being up-hill, but later we had a good run down hill to the spot where we camped for lunch. After lunch we sledged down a still steeper slope, the sledge occasionally take charge. At this spot Mackay partially fell into a crevasse. To-day we were much cheered by the sight at last of Mount Larsen. By the time we reached the spot where we camped that night we had a good clear view of Larsen. The distance travelled was sixteen miles. We were now only about forty miles from our Mount Larsen Depôt.

January 28.—We turned out of the sleeping-bag to-day at about 6.30 A.M. A blizzard was blowing, and after breakfast we had much difficulty in the cold wind in getting up the mast and sail. Mackay, who usually did the greater part of this work, got his hands rather badly frost-bitten before our preparations were completed. We used the thick green canvas floorcloth as a sail; the tent-poles served us for a mast, and a piece of bamboo did duty as a yard.

The wind was blowing at, perhaps, about twenty-five miles an hour, and as soon as we started the sledge, it began to travel at such a hot pace that Mackay and Mawson, with their long legs, were kept walking at the top of their speed, while I, with my shorter ones, was kept on a jog trot. Occasionally, in an extra strong puff of wind, the sledge took charge. On one of these occasions it suddenly charged into me from behind, knocked my legs from under me, and nearly juggernauted me. I was quickly rescued from this undignified position under the sledge runners by Mawson and Mackay. We had now arrived at a part of the plateau where the monotonous level or gently undulating surface gave place to sharp descents. It was necessary in these cases for one of us to untoggle from the front of the sledge and to toggle on behind, so as to steer and steady it. About noon, when we were in full career, the bow of the sledge struck one of the high sastrugi obliquely and the sledge was capsized heavily, but fortunately nothing was broken. After righting the sledge we camped for lunch.

At lunch, with a faint hope of softening the stern heart of our messman for the week—Mackay—and inducing him to give us an extra ration of food, I mildly informed him that it was my birthday. He took the hint and we all fared sumptuously at lunch and dinner that day. The day's run was twenty miles. It had been one of the most fatiguing days that we had as yet experienced, and we were all utterly exhausted when we turned into our sleeping-bag at 8.30 P.M.

January 29.—We were up at about 8 A.M., and found that the plateau wind was still blowing at a speed of about fifteen miles an hour. After our experience of the preceding day we decided that we would not make sail on the sledge, and as a matter of fact, found that pulling the sledge in the ordinary way was far less wearying than the sailing had proved the preceding day. We pulled on steadily hour after hour, and Mounts Nansen and Larsen grew every moment clearer and larger, and we began to hope that we might be able to reach our depôt at Mount Larsen that night. After we had sledged about ten miles, descending at a gentle grade all the way, we found that there was a slight up grade in the snow surface towards the foot of Mount Larsen, but it was not steep enough to cause us any trouble. But later in the day Mawson's sprained leg caused him a good deal of pain, and we had almost decided to camp at a point nearly twenty miles from our preceding camp, when Mackay's sharp eyes sighted, at a distance of about a mile, our little blue flag, tied to the ice-axe at our depôt. We soon reached the depôt, fixed up the tent, had a good hoosh, and turned into the sleeping-bag past midnight.

We were up at 9 A.M. on January 30. The day was sunny, but ominous clouds were gathering overhead as well as to the south. After breakfast we collected the material at our depôt, chiefly ski-boots, ice-axes, oil, a little food, and geological specimens, and loaded these on to our sledge. We found that, owing to the alternate thawing and freezing of the snow at our depôt, our ski-boots were almost filled with solid ice. The work of chipping out this ice proved a slow and tedious job, and we did not get started until about 11 A.M. Soon after we got going we found ourselves for a time in a mesh-work of crevasses. These were from a foot up to about twenty feet in width. Nearly all of them were roofed over with a hard layer of snow. The only visible evidence of the existence of a crevasse was a slight depression in the snow surface at the inner edges of the two walls bounding the crevasses, the whole of the snow roof or lid being slightly counter-sunk below the general level of the surrounding snow surface.

This, however, was not always the case, and crevasses not infrequently existed entirely concealed from view under a perfectly smooth hard snow surface. On account of the fact, as already explained, that the snow lids were thinner next to the walls of the crevass, and thicker towards a position central between the walls, we always used to take care, if we could see the little depression in the snow surface—a sure indication of a crevasse—not to put our foot down near the edge of the depression, but to alight on the snow lid some feet away from the crevasse wall.

On stepping out on to one of these snow lids a large piece suddenly gave way under me, and I was instantly precipitated into the chasm below, but fortunately caught the Alpine rope under my arm as I was falling; this broke the force of the jerk on my sledge harness. I was down about six feet below the snow lid, and Mawson and Mackay holding on to the harness and Alpine ropes which were supporting me, I was able to climb out quickly, and we resumed our journey. Shortly, after this, and after crossing a number of other crevasses, we discovered that the wheel of our sledge meter had disappeared. Probably it had got into one of the crevasses, and gone to the bottom. As we were now so close to the end of our journey, the loss of this, which earlier in our travels would have been a serious disaster, was not of much importance. We had run about eight miles before this lunch, previous to the loss of our sledge meter wheel. At lunch-time Mawson compounded a wonderful new hoosh made out of seal liver, pounded up with a geological hammer, and mixed with crushed biscuit.

We had some discussion as to whether it would be better to descend on to the sea ice by the old track up which we had come, which we termed Backstairs Passage, or make down the main Larsen Glacier to the point where it junctioned with the Drygalski Glacier. Mackay was in favour of the former, Mawson and I of the latter. Mackay thought the devil one knew was better than the devil one didn't know, while Mawson and I feared that during the thaw, which was rapidly breaking up the sea ice at the time when we were ascending the plateau, the ice might have gone away from the base of Backstairs Passage right up to the steep granite cliffs of the coast. Had this been the case, and had we descended by our old route, we should have had to retrace our steps and become involved in a very arduous uphill piece of sledging necessitating an ascent of at least 1000 to 1500 ft. in a distance of a little over a mile. As subsequent events proved, Mackay was right and we were wrong.

We held on down the main glacier with the imposing cliffs and slopes of dark-red granite and blackish eruptive rock intermixed with it close on our left. Mawson's leg was now so bad that it was only with considerable pain and difficulty that he could proceed, and both

Mackay's and my eyes were affected a good deal by snow-blindness and were painful. We found as we advanced that at about six miles easterly from our lunch camp, the surface of the Mount Larsen Glacier descended at a very steep angle. Somewhat ahead to the right it was clear that, where it junctioned with the Drygalski Glacier, it was seamed by enormous crevasses and traversed by strong pressure ridges. We held on with our sledge on a course which took us close to the north side of the glacier. At last the descent became so steep that it was with the utmost difficulty that we could hold the sledge back and prevent its charging down the slope. We halted here and Mackay went ahead to reconnoitre. Presently he came back and said that the narrow strip of snow covering the glacier ice, near its contact with the rocky cliffs on our left, was continuous right down to the bottom of the slope, and he thought it was practicable, if we made rope brakes for the runners on our sledge, to lower it down this steep slope in safety. He fixed on some brakes of brown tarred rope by just twisting the rope spirally around the sledge runners. We then cautiously started the sledge down the steepest bit of the slope, all of us ready to let go in case the sledge took charge. The rope brake worked wonders, and it was even necessary to put a slight pull on the sledge in places in order to get it down the steep snow surface. We had left the great crevasses and ice falls near the junction of the Mount Larsen and Drygalski Glaciers a little to our right.

We now found ourselves on an ice-surface quite unlike anything which we had hitherto experienced. In the foreground were some small frozen lakes close to the foot of the granite hills; on the far side of the lakes were beautiful glacial moraines. All around the lakes, and for a considerable distance up the ice slopes descending towards them, the surface of the ice was formed of a series of large thin anastomosing curved plates of ice. These were pieced together in such a way as to form a pattern on a large scale resembling the cups of some of the recent compound corals, or the ancient extinct form known to geologists as *Alveolites*. These curved ice plates or tiles sloped at an angle of about 45°, and formed, of course, an immense obstruction to sledging, as their sharp edges caught and held our sledge runners. We found, too, that it was very distressing travelling over this extraordinary surface, which, from a scenic point of view, was exquisitely beautiful. As we stepped forwards, out feet usually crashed through the ice tiles, and our legs were imbedded in the formation up to our knees. Frequently, under these circumstances, we would stumble forwards, and had some difficulty in dragging our legs out. It was like sledging over a wilderness of glass cucumber-frames set up at an angle of 45°. Another moment one would find the tiles thick enough and strong enough to support one, but their surfaces being at an angle of 45° to the horizontal, our feet would slip down them sideways and we ran an imminent risk of spraining our ankles. At every step we took we did not know until after the event which of the above two experiences would follow.

After sledging for a short distance over surfaces of this kind, sloping somewhat steeply to the small lakes, we decided to camp on the pale green ice of one of these lakes. Mawson tested this ice and found that it was strong enough to hold, though evidently of no great thickness. We sledged along this lake for a few hundred yards to its north-east end. There was a little snow here which would do for loading the skirt of our tent. By this time the sky was thickly overcast. We fixed up the tent, chopping little holes in the surface of the smooth ice, in which to socket the ends of the tent-poles, and while Mackay cooked, Mawson and I snowed the skirt. This was subsequent to a little reconnoitring which we each did. It was 2 A.M. before we camped on the lake ice, and 4 A.M. before we turned into our sleeping-bag.

Close to our tent was the most beautiful lateral moraine which we had yet discovered. It was formed of blocks of bright red granite, together with quartz porphyries with much rusty stain due to oxidation of iron pyrites, and masses of dark brown rocks, more basic, perhaps of an intermediate character between granite and diorite. We found that immediately to the right of us, in an easterly direction—that is, directly between us and our depôt on the Drygalski Glacier—were great pressure ridges of ice, and a vast entanglement of crevasses. In fact, in that direction the glacier seemed impassable. The only possible outlet for us with our sledge appeared to be close alongside of the lateral moraine at the point where the glacier ice joined it. Even this route was obviously a very difficult one, and we decided before we turned in that on the morrow we should have to unload our sledge and make a portage, or a plurality of portages. The ice on the small lake on which we were camped was only between two and three inches thick, and had obviously formed quite recently after the thaw. It commenced to thaw now under the influence of the warmth of our sleeping-bag, as we lay in it, and we found shallow pools of water all around us when we awoke the next morning.

January 31.—We were up about 11 A.M., having slept soundly after the very exhausting work of our previous day's sledging. During the night it had snowed heavily, there being fully from three to four inches of newly fallen snow covering everything around us, and it was still snowing while we were having breakfast. After breakfast the snow nearly ceased, and we took half the load off our sledge and started with the remainder to try and work a passage out of the ice-pressure ridges of the combined Drygalski and Larsen Glaciers on to the smoother sea ice, and eventually on to the Drygalski Ice Barrier. While Mawson and Mackay pulled, I steadied the sledge on the lower side in rounding the steep sidelings. We were still sledging over the leafy or tile ice, which mostly crunched underfoot with a sharp tinkling sound. We skirted the lateral moraine for a distance of over half a mile, following a depression in the ice-surface apparently produced by a stream, the outlet of the waters of the small lakes. At one spot Mawson crashed right through into the water beneath, and got wet up to his thighs. In spite of my efforts to keep it on even keel, the sledge frequently capsized on these steep sidelings. At last, after struggling up and down heavy slopes, and over low-lying areas of rotten ice, which every here and there let us through into the water beneath, we arrived at the foot of an immense ice-pressure ridge. It was a romantic-looking spot, though at the time we did not exactly appreciate its beauties. To our left was a huge cliff of massive granite rising up steeply to heights of about 2000 ft. The combined pressure of the Drygalski and Mount Larsen Glaciers had forced the glacier ice up into great ridges, trending somewhat obliquely to the coast cliff.

We went back to the tent where we got some hot tea, of which Mawson, particularly, was very glad, as he was somewhat cooled down as the result of his wetting. Then we packed up the remainder of our belongings on the sledge and dragged it down to where we had dumped the half load on the near side of the pressure ridge. Mackay reconnoitred ahead, and found that the large-pressure ridge, which appeared to bar our progress towards our depôt, gradually came nearer and nearer in to the granite cliff, until it pressed hard against the cliff face. Obviously, then, we were impounded by this huge pressure ridge, and would have to devise some mean of getting over it. Taking our ice-axes we smoothed a passage across part of the ridge. This proved a very tough piece of work. We then unloaded the sledge and passed each one of our packages over by hand. Finally we dragged the sledge up and hoisted it over and lowered it down safely on the other side. After this we reloaded the sledge and dragged it for some considerable distance over more of the leafy ice-surface alternating with flattish depressions of rotten ice and snow, with water just beneath. We were now troubled, not only by the tile-ice surface, but also by small channels with steep banks, apparently eroded by glacial streams which had been flowing as the result of the thaw while we were on the Magnetic Pole plateau. We were also worried from time to time as to how to get over the vast number of intersecting crevasses which lay in our path.

Little by little the surface improved as we sledged towards our depôt. The platy structure on the ice became less and less pronounced, giving place to a surface like that of innumerable frozen wavelets with sharp crests. By lunch-time we arrived at a grand old glacial moraine. Amongst its boulders was a handsome coarsely crystalline red granite of which Mackay secured a good specimen. Numbers of boulders projected a few feet above the surface of the ice, but most of them were wholly encased in ice. After lunch, the sledging surface, though still heavy, owing to the newly fallen snow, improved a little, but we soon found our progress barred by what may be termed an ice donga, apparently an old channel formed by a river of thaw water. We encountered three such dongas that afternoon. They were from a few feet up to fifty or a hundred feet or more in width, and from ten to twenty feet deep, and bounded by precipitous or overhanging sides.

After a considerable amount of reconnoitring by Mackay and Mawson, and often making considerable *détours* with our sledge, we managed to cross them. Our difficulties were increased by the innumerable crevasses and steep ice ridges. Some of these crevasses were open, while others were roofed over with tough snow. We fell into these crevasses from time to time, and on one occasion, Mackay and I fell into the same crevasse simultaneously, he up to his shoulders and I up to my waist. Fortunately we were able, by throwing out our arms, to prevent ourselves from falling right through the snow lid. While we were sledging on through the night amongst this network of crevasses, the sky became heavily overcast, and it commenced to snow. At last we succeeded in getting within less than a mile of the moraine containing the boulders of remarkable sphene-diorite, specimens of which we had collected at that spot on our outward journey. Here we camped and turned into our sleeping-bag at 7 A.M. on February 1.

It continued snowing heavily during the day, the fall being about six inches in depth. The snow on the side of the tent facing the sun thawed rapidly, and the thaw water dripped through and formed pools on top of our sleeping-bag. Mawson's sprained leg pained him a great deal. We estimated that we were now only about sixteen miles, as the skua flies, from our depôt on the Drygalski Glacier, but as we had only two days' food left, it became imperative to push on without delay. We started sledging in the thick driving snow on the evening of February 1. The surface was covered with a layer of soft snow, nine inches in thickness, but in the drifts it was, of course, deeper. The work of sledging under these circumstances was excessively laborious and exhausting, and besides it was impossible to keep our proper course while the blizzard lasted. Acoordingly, we camped at 8 P.M., and after our evening meal we rolled into our sleeping-bag and slid into the dreamless sleep that comes to the worn and weary wanderer.

At 8 A.M. on February 2 we were rejoiced to find the sun shining in a clear sky. We intended making a desperate attempt this day to reach our depôt, as we knew that the *Nimrod* would be due—perhaps overdue—by the night. We saw as we looked back that our track of yesterday was about as straight as a corkscrew. Once more we pulled out over the soft snow, and although refreshed somewhat by our good sleep we found the work extremely trying and toilsome. We crossed an ice donga, and about four miles out reached the edge of a second donga. Here we decided to leave everything but our sledge, tent, sleeping-bag, cooking-apparatus, oil and food, and make a forced march right on to the Drygalski Depôt. Accordingly we camped, had tea and two biscuits each, and fixed up our depôt, including the Lloyd-Creak dip circle, theodolite and legs, geological collections, &c., and marked the spot with a little blue flag tied on to an ice-axe.

We now found the sledge, thus lightened, distinctly easier to pull, and after making a slight *détour,* crossed the donga by a snow bridge. Soon we reached another donga, and successfully crossed it. At three and a half miles further at 8 P.M. we camped again and had

Watching for Seals at the Ice-edge

a little cheese and biscuit. After this short halt we pulled on again, steering north-8°-east magnetic. Mawson occasionally swept the horizon with our excellent field-glasses in hopes of sighting our depôt. Suddenly he exclaimed that he saw the depôt flag distinctly on its ice mound, apparently about seven miles distant, but it was well round on the starboard bow of our sledge on a bearing of south-38°-west magnetic. Mackay and I were much excited at Mawson's discovery. Mackay seized the field-glasses as soon as Mawson put them down and directed them to the spot indicated, but could see no trace of the flag; then I looked through the glasses with equally negative results. Mawson opined that we must both be snow-blind. Then he looked through them again, and at once exclaimed that he could see no trace of the flag now. The horizon seemed to be walloping up and down, just as though it was boiling, evidently the result of a mirage. Mawson, however, was so confident that he had seen the flag when he first looked, that we altered course to south-38°-west magnetic, and after we had gone a little over a mile, and reached the top of a slight eminence in the ice-surface, we were rejoiced to hear the announcement that he could now see the depôt flag distinctly. We kept on sledging for several miles further. At midnight, when the temperature had fallen to zero, I felt that the big toe of my right foot was getting frostbitten. My ski-boots had all day been filled with the soft snow and the warmth of my foot had thawed the snow, so that my socks were wet through; and now, since the springing up of the wind and the sudden fall in temperature, the water in the socks had turned to ice. So we halted, got up the tent, started the Primus and prepared for a midnight meal, while, with Mawson's assistance, I got off my frozen ski-boots and socks and restored the circulation in my toe, and put on some socks less icy than those I had just taken off.

We were much refreshed by our supper, and then started off again, thinking that at last we should reach our depôt, or at all events, the small inlet a little over a mile distant from it, but "the best laid schemes of mice and men gang aft agley." There was an ominous white streak ahead of us with a dark streak just behind it, and we soon saw that this was due to a ravine or barranca in the snow- and ice-surface interposing itself between ourselves and our depôt. We soon reached the near cliff of the barranca.

The barranca was about two hundred yards in width, and from thirty to forty feet deep. It was bounded by a vertical cliff or very steeply inclined slope on the near side, the north-west side, and by an overhanging cliff festooned with stalactites on the south-east side. To the north-east a strip of dark sea-water was visible between the walls of the barranca, which evidently communicated by a long narrow channel with the ocean outside, some three miles distant. Inland, the barranca extended for many miles as far as the eye could reach. The bottom of the barranca immediately beneath us was floored with sea ice covered with a few inches of snow. This ice was traversed by large tide-cracks, and we were much excited to see that there were a number of seals and Emperor penguins dotted over the ice floor. We determined to try and cross the barranca. We looked up and down the near cliff for a practicable spot where we could let down our sledge, and soon found a suitable slope, a little to the north-east of us, formed by a steep snow drift. We sledged on to this spot, and making fast the Alpine rope to the bow of the sledge, lowered it cautiously, stern first, to the bottom. The oil-cans in the rear of the sledge were rattled up somewhat when it struck bottom, but no harm was done. At the bottom we had some trouble in getting the sledge over the gaping tide-cracks, some ten to fifteen feet deep and three to five feet wide.

Arrived at the middle of the floor of the barranca, Mackay killed two Emperor penguins, and took their breasts and livers to replenish our exhausted larder. Meanwhile, Mawson crossed to the far side of the floor of the barranca on the look-out for a possible spot where we might swarm up. I joined him a few minutes later, and as I was feeling much exhausted after the continuous forced marches back from the Magnetic Pole, asked him to take over the leadership of the expedition. I considered that under the circumstances I was justified in taking this step. We had accomplished the work assigned to us by our leader, having reached the Magnetic Pole. We were within two or three miles of our Drygalski Depôt, and although the only food left there was two days' supply of broken biscuits with a little cheese, we had a good prospect of meat-supply, as the barranca abounded in seals and penguins, so that for the present we had no reason to apprehend the danger of starvation. On the other hand, as regards our ultimate personal safety, our position was somewhat critical. We were not even certain that the *Nimrod* had arrived at all in Ross Sea that season, though we thought it, of course, very probable that she had. In the next case, on the assumption that she had arrived, it was very possible that in view of the great difficulties of making a thorough search along the two hundred miles of coast, at any part of which we might have been camped—difficulties arising from heavy belts of pack-ice and icebergs, as well as from the deeply indented character of that bold and rugged coast—it was quite possible that the *Nimrod* would miss sighting our depôt flags altogether. In the event of the *Nimrod* not appearing within a few days, it would be necessary to take immediate and strenuous action with a view either to wintering at the spot, or with a view to an attempt to sledge back around the great mountain massifs and over the many steeply crevassed glaciers for over two hundred miles to our winter quarters at Cape Royds. Even now, in the event of some immediate strenuous action being necessary, if the *Nimrod* were to suddenly appear at some point along the coast, I thought it would be best for Mawson, who was less physically exhausted than myself, to be in charge. He had, throughout the whole journey, shown excellent capacity for leadership, fully justifying the opinion held of him by Lieutenant Shackleton when providing in my instructions that in the event of anything happening to myself Mawson was to assume the leadership. When I spoke to him on the subject, he at first demurred, but finally said that he would act for a time, and would think the matter over at his leisure before definitely deciding to become permanently the leader. I offered to give him authority in writing as leader, but this he declined to receive.

Meanwhile, the examination of the cliff face on the south-east side of the barranca showed that there was one very difficult but apparently possible means of ascent. We returned to where we had left Mackay, and then we three dragged the sledge around to the edge of a rather formidable tide-crack, behind which lay the mound of snow up which we hoped to climb; our idea being to unpack our sledge, drag it to the top of this steep mound, and rearing it on end at the top of the mound, use it as a ladder for scaling the overhanging cliff above. Mackay managed to cross the tide-crack, using the bamboo poles of our tent as a bridge, and after some difficulty, reached the top of the snow mound under the overhanging cliff. Much to our disappointment, however, he discovered that the mound was formed of very soft snow, his ice-axe sinking in to the whole depth of the handle directly he placed it on top of the mound. It was obvious that as our sledge would sink in to at least an equal depth, the top of it would then be too short to enable any of us to scale the overhanging cliff by its means. We were, therefore, reluctantly compelled to drag our sledge back again over the tide-cracks to the north-west side of the barranca down which we had previously lowered our sledge. We then discovered that, as in classical times, while the descent to Avernus was easy, it was difficult and toilsome to retrace one's steps. With Mawson ahead with the ice-axe and towing rope, and Mackay and I on either side of the seldge in the rear, we managed by pulling and pushing together to force the sledge up a few inches at a time. At each short halt, Mawson would stick in the ice-axe, take a turn of the leading rope around it, and support the sledge in this way for a brief interval while we all got our breath. At last the forty feet of steep slope was successfully negotiated, and we found ourselves once more on the level plain at the top of the barranca, but of course, on the wrong side in reference to our depôt. As we were within three miles of the open sea we thought it would be safe to camp here, as had the *Nimrod* sighted our depôt flag and stood in to the coast, we could easily have hurried down to the entrance of the inlet and made signals to her.

We had now been up since 8 A.M. on the previous day, and were very thankful to be able to enter our tent, and have a meal off a stew of minced penguin liver. We then turned into the sleeping-bag at about

7 A.M. Just about a quarter of an hour after we had turned in, as we learnt later, the *Nimrod* must have passed, bound north towards Mount Melbourne, within three miles of the ice cliff on which our tent was now situated. Owing, however, to a light wind with snow drift she was unable to sight either our depôt flag or tent.

Chapter Thirteen

PROFESSOR DAVID'S NARRATIVE (*Concluded*)

FEBRUARY 3.—After sleeping in the bag from 7 A.M. until 11 A.M. we got up and had breakfast, packed our sledge, and started along the north bank of the snow cañon. The snow and ice at the bottom were dotted with basking seals and moulting Emperor penguins. Fully a hundred seals could be counted in places in a distance of as many yards along the cañon. At about one mile from the camp we reached a small branch cañon, which we had to head off by turning to our right. We now proceeded about one and a half miles further along the edge of the main cañon, and in our then tired and weak state were much dispirited to find that it still trended inland for a considerable distance. We now halted by the sledge while Mackay went ahead to try and find a crossing, and presently Mawson and I were rejoiced to hear him shout that he had discovered a snow bridge across the cañon. Presently he rejoined us, and together we pulled the sledge to the head of the snow bridge. It was a romantic spot. A large slice of the snow or névé cliff had fallen obliquely across the cañon, and its surface had then been raised and partially levelled up with soft drift snow. There was a crevasse at both the near and far ends of the bridge, and the middle was sunk a good deal below the abutments. Stepping over the crevasse at the near end we launched the sledge with a run down to the centre of the bridge, then struggled up the steep slope facing us, Mackay steadying the sledge from falling off the narrow causeway, while we all three pulled for all we were worth. In another minute or two we were safely across with our sledge, thankful that we had now surmounted the last obstacle that intervened between us and our depôt.

While heading for the depôt we sighted an Emperor penguin close to our track. Mackay quickly slew him, and took his flesh and liver for our cooking-pot. Two miles further on we camped. Mawson minced the Emperor's flesh and liver, and after adding a little snow, I boiled it over our Primus so as to make one and a half pots of soupy mincemeat for each of us. This was the most satisfying meal we had had for many a long day. After lunch we sledged on for over one and a half miles further towards the depôt, and at about 10.30 P.M. reached an ice mound on the south side of the inlet in which the snow cañon terminated seawards. This camping spot was a little over a mile distant from our depôt. We were now all thoroughly exhausted and decided to camp. The spot we had selected seemed specially suitable, as from the adjacent ice mound we could get a good view of the ocean beyond the Drygalski Barrier. While Mawson and I got up the tent, Mackay went to kill a seal at the shore of the inlet. He soon returned with plenty of seal meat and liver. He said that he had found two young seals, and had killed one of them; that they had both behaved in a most unusual manner, scuttling away quickly and actively at his approach, instead of waiting without moving, as did most of the Weddell seals of which we had hitherto had experience. We discovered later that these two seals belonged to the comparatively rare variety known as Ross seal. After a delicious meal of seal blubber, blood and oil, with fried meat and liver, cooked by Mawson, Mawson and I turned into

the sleeping-bag, leaving Mackay to take the first of our four hour watches on the look-out for the *Nimrod*. During his watch he walked up to our depôt and dug out our biscuit tin, which had served us as a blubber lamp and cooker, together with the cut-down paraffin tin which we had used as a frying-pan. Both these he carried down to our tent. There he lit the blubber lamp just outside the tent and cooked some penguin meat, regaling himself at intervals during his four hour's watch with dainty morsels from the savoury dish. When he called me up at 4 A.M. I found that he had thoughtfully put into the frying-pan a junk of Emperor's breast, weighing about two pounds, for me to toy with during my watch. A chilly wind was blowing off the plateau, and I was truly thankful for an occasional nibble at the hot penguin meat. After cooking some more penguin meat I called up Mawson soon after 8 A.M. on February 4, and immediately afterwards turned into the bag, and at once dropped off sound asleep.

Mawson did not call Mackay and myself until after 2 P.M. We at once rolled up the sleeping-bag, and Mawson cooked a generous meal of seal and penguin meat and blubber, while Mackay made a thin soupy broth on the Primus. Meanwhile, I went on to the ice mound with the field-glasses, but could see nothing in the way of a ship to seaward and returned to the tent. We all thoroughly enjoyed our liberal repast, and particularly relished the seal's blood, gravy and seal oil.

After the meal we discussed our future plans. We decided that we had better move the tent that afternoon up to our old depôt, where it would be a conspicuous object from the sea, and where, too, we could command a more extensive view of the ocean. We also talked over what we had best do in the event of the *Nimrod* not turning up, and decided that we ought to attempt to sledge overland to Hut Point, keeping ourselves alive on the way, as best we might, with seal meat. It must be admitted that the prospect of tackling two hundred miles of coast, formed largely of steep rocky foreshores, alternating with heavily crevassed glacier ice, was not a very bright one. We also discussed the date at which we ought to start trekking southwards. Mackay thought we ought to commence making our preparations at once, and that unless the *Nimrod* arrived within a few days we ought to start down the coast with our sledge, tent, sleeping-bag, cooker and seal meat, leaving a note at the depôt for the *Nimrod,* in case she should arrive later asking her to look out for us along the coast, and if she couldn't sight us, to lay depôts of food and oil for us at certain specified spots. He considered that by this method we could make sure of beginning the long journey in a sound state of health, and, if fortunate, might reach Hut Point before the beginning of the equinoctial gales in March. Mawson and I, on the other hand, thought that we ought to wait on at our present camp until late in February.

From whatever point of view we looked at it, our present lot was not a happy one. The possibility of a long wait in the gloomy region of the Drygalski Glacier, with its frequent heavy snows at this season of the year, and leaden sky vaulted over the dark sea, was not pleasing to contemplate. Still less cheerful was the prospect of a long, tedious and dangerous sledge journey towards Hut Point. Even the diet of seal and penguin, just for the moment so nice, largely because novel, would soon savour of *toujours perdrix.*

Dispirited by forebodings of much toil and trouble, we were just preparing to set our weary limbs in motion to pack up our belongings for the short trek up to the depôt, when Bang! went something, seemingly close to the door of our tent; the sound thrilled us: in another instant the air reverberated with a big boom! much louder than the first sound. Mawson gave tongue first, roaring out, "A gun from the ship!" and dived for the tent door. As the latter was narrow and funnel-shaped there was for the moment some congestion of traffic. I dashed my head forwards to where I saw a small opening, only in time to receive a few kicks from the departing Mawson. Just as I was recovering my equilibrium, Mackay made a wild charge, rode me down and trampled over my prostrate body. When at length I struggled to my feet, Mawson had got a lead of a hundred yards, and Mackay of about fifty. "Bring something to wave," shouted Mawson, and I rushed back to the tent and seized Mackay's rück-sack. As I ran forward this time, what a sight met my gaze. There was the dear old *Nimrod*, not a quarter of a mile away, steaming straight towards us up the inlet, her bows just rounding the entrance. At the sight of the three of us running frantically to meet the ship, hearty ringing cheers burst forth from all on board. How those cheers stirred every fibre of one's being! It would be hard, indeed, for any one, not situated as we had been, to realise the sudden revulsion of our feelings. In a moment, as dramatic as it was heavenly, we seemed to have passed from death into life. My first feelings were of intense relief and joy; then of fervent gratitude to the kind Providence which had so mercifully led our friends to our deliverance.

A sudden shout from Mackay called me back to earth, "Mawson's fallen into a deep crevasse. Look out, it's just in front of you!" I then saw that Mackay was kneeling on the snow near the edge of a small oblong sapphire-blue hole in the névé. "Are you all right, Mawson?" he sang out, and from the depth, came up the welcome word, "Yes." Mackay then told me that Mawson was about twenty feet down the crevasse. We decided to try and pull him up with the sledge harness, and hurried back to the sledge, untoggled the harness, ran back with it to the crevasse, and let one end down to Mawson. We found, however, that our combined strength was insufficient to pull him up, and that there was a risk, too, of the snow lid at the surface falling in on Mawson, if weight was put upon it, unless it was strengthened with some planking. Accordingly, we gave up the attempt to haul Mawson up, and while I remained at the crevasse holding one end of the sledge harness Mackay hurried off for help to the *Nimrod,* which was now berthing alongside of the south wall of the inlet, about two hundred yards distant. Mackay shouted to those on board, "Mawson has fallen down a crevasse, and we got to the Magnetic Pole." The accident had taken place so suddenly that those on board had not realised in the least what had happened. A clear, firm, cheery voice, that was strange to me, was now heard issuing prompt orders for a rescue party. Almost in less time than it takes to write it, officers and sailors were swarming over the bows of the *Nimrod,* and dropping on to the ice barrier beneath. I called down to Mawson that help was at hand. He said that he was quite comfortable at present; that there was sea water at the bottom of the crevasse, but that he had been able to sustain himself a couple of feet above it on the small ledge that had arrested his fall. Meanwhile, the rescue party, headed by the first officer of the *Nimrod,* J. K. Davis, had arrived on the scene. The crevasse was bridged with a suitable piece of sawn timber, and Davis, with that spirit of thoroughness which characterises all his work, promptly had himself lowered down the crevasse. On reaching the bottom he transferred the rope by which he had been lowered to Mawson, and with a long pull and a strong pull and a pull altogether, the company of the *Nimrod* soon had Mawson safe on top, none the worse for the accident with the exception that his back was slightly bruised. As soon as the rope was cast free from Mawson, it was let down again for Davis, and presently he, too, was safely on top.

And now we had a moment of leisure to see who constituted the rescue party. There were the dear old faces so well known on our voyage together the previous year, and interspersed with them were a few new faces. Here were our old comrades Armytage and Brocklehurst, Dr. Michell, Harbord (the officer who—as we learned later—had sighted our depôt flag), our good stewards Ansell and Ellis, the genial boatswain Cheetham, Paton, and a number of others. What a joyous grasping of hands and hearty all-round welcoming followed. Foremost among them all to welcome us was Captain Evans, who had commanded the s.s. *Koonya,* which towed the *Nimrod* from Lyttelton to beyond the Antarctic Circle, and it goes without saying that the fact that the *Nimrod* was now in command of a master of such experience, so well and favourably known in the shipping world of New Zealand and Australia, gave us the greatest satisfaction. He hastened to assure me of the safety and good health of my wife and family. While willing hands packed up our sledge, tent and other belongings, Captain Evans walked with us to the rope ladder hanging over the bows of the *Nimrod.*

Quickly as all this had taken place, Mackay had already found time to secure a pipe and some tobacco from one of our crew, and was now puffing away to his heart's content. We were soon all on the deck of the *Nimrod* once more, and were immediately stood up in a row to be photographed. As soon as the cameras had worked their wicked will upon us, for we were a sorry sight, our friends hurried us off for afternoon tea. After our one hundred and twenty-two days of hard toil over the sea ice of the coast and the great snow desert of the hinterland, the little ship seemed to us as luxurious as an ocean liner. To find oneself seated once more in a comfortable chair, and to be served with new-made bread, fresh butter, cake and tea, was Elysium.

We heard of the narrow escape of Armytage, Priestley and Brocklehurst, when they were being carried out to sea, with only two days' provisions, on a small ice-floe surrounded by Killer whales; and how, just after the momentary grounding of the floe, they were all just able to leap ashore at a spot where they were picked up later by the *Nimrod.* We also heard of the extraordinary adventures and escapes of Mackintosh and McGillan in their forced march overland, without tent or sleeping-bag, from Mount Bird to Cape Royds; of the departure of the supporting-party to meet the Southern Party; and, in short, of all the doings at Cape Royds and on the *Nimrod* since we had last heard any news. Pleasantly the buzz of our friends' voices blended itself with the gentle fizzing of steam from the *Nimrod's* boiler, and surely since the days of John Gilpin " were never folk so glad " as were we three.

Here it may not perhaps be out of place to quote from Captain Evans' private log in reference to the relief of our Northern Party by the *Nimrod.* After hearing from the Western Party under Armitage that we were long overdue at Butter Point, and after consulting with Murray at Cape Royds, he decided to

The "Nimrod" held up in the Ice

CAPTAIN EVANS AND THE "NIMROD" AFTER A BLIZZARD

THE DECK OF THE "NIMROD" AFTER A BLIZZARD

commence to search for us, as suggested in Lieutenant Shackleton's instructions on February 1. He left accordingly at that date, and after looking for us in vain at the Butter Point depôt, and at Granite Harbour he sailed northwards for the Drygalski Ice Barrier Tongue, and when about three miles off our depôt island had sighted our little flag and cairn, but was not certain that it was a depôt. Nearer approach was precluded at the time by the pack-ice.

Captain Evans' private log reads as follows:

"*February* 3, 1.30 A.M.—Cleared belt of pack—proceeded westward along Drygalski Barrier edge. Moderate to strong south-west wind, force four to eight with snow drift; 7.30 A.M. to 9.30 A.M. off Barrier. (At this time the *Nimrod* must have passed within about three miles of the spot where we were at our last camp before reaching the inlet, and had it not been for a little falling snow our flag and tent would probably have been sighted on this occasion); 10 A.M. to 2.15 P.M.: coasted along the beach at distances of from one-fifth of a mile to three-quarters of a mile in water from ten to fifty fathoms; 1 P.M. wind dropped to calm; 2.15 P.M. bearing true north-20°-east, distant twenty-four miles Mount Melbourne. Came to top of bight (Gerlache Inlet) full of pack; sounded in sixty-four fathoms; took bearings and stood eastward to search Cape Washington; 3.30 P.M. entered the pack-ice.

Midnight: rounded Cape Washington at a distance of one and a half cables in eleven to twenty fathoms, both sides of the cape quite inaccessible—awful-looking ice cliffs northern side—crevassed ice slopes south side. Fresh south-south-west wind, force 5–5. . . . No sign of party or record anywhere.

February 4, 1909, 10 A.M.—Pack-ice stretching east and west to northward—turned back to try coast again to the southward. Fresh southerly wind, force 6—clear and fine, barometer 28.86, thermometer 17°. . . . Proceeding again along Drygalski Barrier; 3 P.M. sighted two flags on Barrier edge and a little back of it—small inlet developing. (The third officer, Mr. A. Harbord, first sighted these flags, and came to the captain and said, 'I think, sir, I see a flag,' and then Armytage, bringing his powerful deer-stalking telescope to bear on the object, said to the captain, 'It's a dead sitter, sir.' T.W.E.D.) 3.40 P.M. arrived at upper end of inlet—picked up Professor David, Mawson and Mackay, just arrived back from Magnetic Pole; 5 P.M. killed first Ross seal; 2 P.M. great depression, 4 P.M. great elation. . . . At 3.30 P.M., upon sighting the top of the Northern Party's tent, we fire a distress double detonator. Upon hearing this in the tent we learn that they all jumped up, and upset each other, and everything, including the tin of seal blubber and blood which they were drinking, and which Professor David pronounces good when you get used to it, and rushed out, Mawson first, who almost immediately went down a crevasse, from which Mr. Davis and a party from the ship soon pulled him up. A great meeting—a tremendous relief."

After afternoon tea came the joy of reading the home letters, and finding that the news was good. Later we all three had a novel experience, the first real wash for over four months. After much diligent work with hot water, soap and towel some of the outer casing of dirt was removed, and bits of our real selves began to show through the covering of seal oil and soot. Dinner followed at 6 P.M. and it is scarcely necessary to add that with our raging appetites, and all the new types of dainty food around us we over-ate ourselves. This did not prevent us from partaking liberally of hot cocoa and gingerbread biscuits before turning in at 10 P.M. None but those whose bed for months has been on snow and ice can realise the luxury of a real bunk, blankets and pillow in a snug little cabin. A few minutes' happy reverie preceded sound sleep. At last our toilsome march was over, the work that had been given us to do was done, and done just in the nick of time; the safety of those nearest and dearest to us was assured, and we could now lay down our weary limbs to rest.

Under Providence one felt one owed one's life to the patient and thorough search, sound judgment and fine seamanship of Captain Evans, and the devotion to duty of his officers and crew and no pen can describe how that night one's heart overflowed with thankfulness for all the blessings of that day. One's last thought in the twilight that comes between wakefulness and sleep is expressed in the words of our favourite record on the gramophone, the hymn so grandly sung by Evan Williams:

> "So long Thy power hath blest me, sure it still will lead me on."

If one may be permitted to take a brief retrospect of our journey the following considerations present themselves: The total distance travelled from Cape Royds to the Magnetic Pole and back to our depôt on the Drygalski Glacier was about 1260 miles. Of this, 740 miles was relay work, and we dragged a weight of, at first, a little over half a ton, and finally somewhat under half a ton for the whole of this distance. For the remaining 520 miles from the Drygalski Depôt to the Magnetic Pole and back we dragged a weight at first, of 670 lb., but this finally became reduced to about 450 lb., owing to consumption of food and oil, by the time that we returned to our depôt.

We were absent on our sledge journey for one hundred and twenty-two days, of which five days were spent in our tent during heavy blizzards, and five days partly in experimenting in cooking with blubber and partly in preparing supplies of seal meat for the journey from the sea ice over the high plateau, and three days in addition were taken up in reconnoitring, taking magnetic observations, &c. We therefore covered this distance of 1260 miles in 109 travelling days, an average of about eleven and a half miles a day.

We had laid two depôts before our final start, but as these were distant only ten miles and fifteen miles respectively, from our winter quarters they did not materially help us. We had no supporting-party, and with the exception of help from the motor-car in laying out these short depôts we pulled the sledges for the whole distance without assistance except, on rare occasions, from the wind.

The travelling over the sea ice was at first pretty good, but from Cape Bernacchi to the Nordenskjold Ice Barrier we were much hampered by screwed pack-ice with accompanying high and hard snow ridges. Towards the latter part of October and during November and part of December the thawing surface of saline snow, clogging and otherwise impeding our runners, made the work of sledging extremely laborious. Moreover, on the sea ice—especially towards the last part of our journey over it—we had ever present the risk of a blizzard breaking the ice up suddenly all around us, and drifting us out to sea. There can be no doubt, in view of the wide lanes of open water in the sea ice on the south side of the Drygalski Glacier, when we reached it on November 30, that we got to *glacies firma* only in the nick of time.

Then there was the formidable obstacle of the Drygalski Glacier, with its wide and deep chasms, its steep ridges and crevasses, the passage of this glacier proving so difficult that although only a little over twenty miles in width it took us a fortnight to get across. On the far side of the Drygalski was the open sea forcing us to travel shorewards over the glacier surface. Then had come the difficult task of pioneering a way up to the high plateau—the attempt to force a passage up the Mount Nansen Glacier—our narrow escapes from having our sledge engulfed in crevasses—the heavy blizzard with deep new fallen snow and then our retreat from that region of high-pressure ridges and crevasse entanglements—our abandonment of the proposed route up the snout of the Bellingshausen Glacier, and finally our successful ascent up the small tributary glacier, the " backstairs passage," to the south of Mount Larsen.

On the high plateau was the difficulty of respiration, biting winds with low temperatures, difficult sledging—sometimes against blizzards—over broad undulations and high sastrugi, the cracking of our lips, fingers and feet, exhaustion from insufficient rations, disappointment at finding that the Magnetic Pole had shifted further inland than the position previously assigned to it. Then, after we had just succeeded by dint of great efforts in reaching the Pole of verticity, came the necessity for forced marches, with our sledge, of from sixteen to twenty miles a day in order to reach the coast with any reasonable prospect of our being picked up by the *Nimrod.*

Then came our choice of the difficult route down the snout of the Bellingshausen Glacier, and our consequent difficulties in surmounting the ice-pressure ridges; then the difficulty of sledging over the "tile-ice" surface, the opposing ice barrancas formed by the thaw water while we were on the high plateau; the final heavy snow blizzard; our loss of direction when sledging in bad light and falling snow, and finally our arrest by the deep barranca of what afterwards was known as Relief Inlet.

But ours were not the only, nor the greatest, difficulties connected with our journey. There were many disappointments, dangers and hardships for the captain, officers and crew of the *Nimrod* in their search for us along that two hundred miles of desolate and, for a great part, inaccessible coast-line. How often black spots ashore, proving on nearer view to be seals or penguins, had been mistaken for depôt flags; how often the glint of sunlight off brightly reflecting facets of ice had been thought to be "helios," only the disappointed ones can tell; how often, too, the ship was all but aground, at other times all but beset in the ice-pack in the efforts to get a clearer view of the shore-line in order to discover our depôt! This is a tale that the brave men who risked their lives to save ours will scorn to tell, but is nevertheless true.

As the result of our journey to the Magnetic Pole and back, Mawson was able to join up in his continuous triangulation survey, Mount Erebus with Mount Melbourne, and to show with approximate accuracy the outline of the coast-line, and the position and height of several new mountains. He and I obtained geological collections, sketches and notes—especially on glacial geology—along the coast-line, and he also took a series

Party setting out from Ship

The Crow's Nest of the "Nimrod," as seen from the Deck

of photographs; while Mackay determined our altitudes on the plateau by means of the hypsometer. Mawson also made magnetic determinations, and I was able to gather some meteorological information. A summary of this work is given in the Scientific Appendix, and details will be supplied later in the Scientific Memoirs of the expedition.

Unfortunately the time available during our journey was too short for detailed magnetic, geological or meteorological observations. Nevertheless, we trust that the information obtained has justified the journey. At all events we have pioneered a route to the Magnetic Pole, and we hope that the path thus found will prove of use to future observers.

It is easy, of course, to be wise after the event, but there is no doubt that had we known that there was going to be an abundance of seals all along the coast, and had we had an efficient team of dogs we could have accomplished our journey in probably half the time that it actually occupied. Future expeditions to the South Magnetic Pole would probably do well to land a strong and well-equipped party, either at Relief Inlet, or better, as near to "Backstairs Passage" as the ship can be taken, and as early in December as the state of the sea ice makes navigation possible. A party of three, with a supporting-party also of three, with good dog teams and plenty of fresh seal meat, could travel together for about seventy miles inland; then the supporting-party might diverge and ascend Mount Nansen from its inland extremity. The other party, meanwhile, might proceed to the Magnetic Pole at not less than fifteen miles a day. This should admit of their spending from a week to a fortnight at the Pole, and they should then be able to return to the coast early in

February. Meanwhile, there would be plenty of scope for a third party to explore the foot-hills of Mount Larsen and Mount Nansen, search and map their wonderful moraines, and examine the deeply indented rocky coast-line from Nansen to the as yet untrod volcano Mount Melbourne.

Chapter Fourteen

ALL ABOARD: THE RETURN TO NEW ZEALAND

THE *Nimrod,* with the members of the Northern Party aboard, got back to the winter quarters on February 11 and landed Mawson. The hut party at this period consisted of Murray, Priestley, Mawson, Day and Roberts. No news had been heard of the Southern Party, and the Depôt Party, commanded by Joyce, was still out. The ship lay under Glacier Tongue most of the time, making occasional visits to Hut Point in case some of the men should have returned. On February 20 it was found that the Depôt Party had reached Hut Point, and had not seen the Southern Party. The temperature was becoming lower, and the blizzards were more frequent.

The instruction left by me had provided that if we had not returned by February 25, a party was to be landed at Hut Point, with a team of dogs, and on March 1 a search-party was to go south. On February 21 Murray and Captain Evans began to make preparations for the landing of the relief party. The ship proceeded to Cape Royds where Mawson was picked up. In accordance with my expressed wish he was offered the command of the relief party, and accepted it. If it became necessary to go south in search of the Southern Party the men landed at Hut Point would have to spend another winter in the south, as the ship could not wait until their return. It was therefore a serious matter for those who stayed, but there was no lack of volunteers. These arrangements being completed, most of the members of the expedition then on board went ashore at Cape Royds to get the last of their property packed ready for departure. The ship was lying under Glacier Tongue when I arrived at Hut Point with Wild on February 28, and after I had been landed with the relief party in order to bring in Adams and Marshall, it proceeded to Cape Royds in order to take on board the remaining members of the shore-party and some specimens and stores.

The *Nimrod* anchored a short distance from the shore, and two boats were launched. The only spot convenient for embarkation near the ship's anchorage was at a low ice cliff in Backdoor Bay. Everything had to be lowered by ropes over the cliff into the boats. Some hours were spent in taking on board the last of the collections, the private property, and various stores.

A stiff breeze was blowing, making work with the boats difficult, but by 6 A.M. on March 2 there remained to be taken on board only the men and dogs. The operation of lowering the dogs one by one into the boats was necessarily slow, and while it was in progress the wind freshened to blizzard force, and the sea began to run dangerously. The waves had deeply undercut the ice-cliff, leaving a projecting shelf. One boat, in charge of Davis, succeeded in reaching the ship, but a second boat, commanded by Harbord, was less fortunate. It was heavily loaded with twelve men and a number of dogs, and before it had proceeded many yards from the shore an oar broke. The *Nimrod* was forced to slip her moorings and steam out of the bay, as the storm had become so severe that she was in danger of dragging her anchors and going on the rocks. An attempt to float a buoy to the boat was not successful, and for some time Harbord and the men with him were

The Ship off Pram Point, just before leaving for the North

The Motor Car being taken aboard the "Nimrod" for the Return Journey

Ready to start Home

in danger. They could not get out of the bay owing to the force of the sea, and the projecting shelf of ice threatened disaster if they approached the shore. The flying spray had encased the men in ice, and their hands were numb and half-frozen. At the end of an hour they managed to make fast to a line stretched from an anchor a few yards from the cliff, the men who had remained on shore pulling this line taut. The position was still dangerous, but all the men and dogs were hauled up the slippery ice-face into safety before the boat sank. Hot drinks were soon ready in the hut, and the men dried their clothes as best they could before the fire. Nearly all the bedding had been sent on board, and the temperature was low, but they were thankful to have escaped with their lives.

The weather was bitter on the following morning (March 3), and the *Nimrod,* which had been sheltering under Glacier Tongue, came back to Cape Royds. A heavy sea was still running, but a new landing-place was selected in the shelter of the cape, and all the men and dogs were got aboard. The ship went back to the Glacier Tongue anchorage to wait for the relief party.

About ten o'clock that night Mackintosh was walking the deck engaged in conversation with some other members of the expedition. Suddenly he became excited and said, "I feel that Shackleton has arrived at Hut Point." He was very anxious that the ship should go up to the Point, but nobody gave much attention to him. Then Dunlop advised him to go up to the crow's-nest if he was sure about it, and look for a signal. Mackintosh went aloft, and immediately saw our flare at Hut Point. The ship at once left for Hut Point, reaching it at midnight, and by 2 A.M. on March 4 the entire expedition was safe on board.

There was now no time to be lost if we were to attempt to complete our work. The season was far advanced, and the condition of the ice was a matter for anxiety, but I was most anxious to undertake exploration with the ship to the eastward, towards Adelie Land, with the idea of mapping the coast-line in that direction. As soon as all the members of the expedition were on board the *Nimrod*, therefore, I gave orders to steam north, and in a very short time we were under way. It was evident that the sea in our neighbourhood would be frozen over before many hours had passed, and although I had foreseen the possibility of having to spend a second winter in the Antarctic when making my arrangements, we were all very much disinclined to face the long wait if it could be avoided. I wished first to round Cape Armitage and pick up the geological specimens and gear that had been left at Pram Point, but there was heavy ice coming out from the south, and this meant imminent risk of the ship being caught and perhaps "nipped." I decided to go into shelter under Glacier Tongue in the little inlet on the north side for a few hours, in the hope that the southern wind, that was bringing out the ice, would cease and that we would then be able to return and secure the specimens and gear. This was about two o'clock on the morning of March 4, and we members of the Southern Party turned in for a much needed rest.

At eight o'clock on the morning of the 4th we again went down the sound. Young ice was forming over the sea, which was now calm, the wind having entirely dropped, and it was evident that we must be very quick if we were to escape that year. We brought the *Nimrod* right alongside the pressure ice at Pram Point, and I pointed out the little depôt on the hillside. Mackintosh at once went off with a party of men to bring the gear and specimens down, while another party went out to the seal rookery to see if they could find a peculiar seal that we had noticed on our way to the hut on the previous night. The seal was either a new species or the female of the Ross Seal. It was a small animal, about four feet six inches long, with a broad white band from its throat right down to its tail on the under-side. If we had been equipped with knives on the previous night we would have despatched it, but we had no knives and were, moreover, very tired, and we therefore left it. The search for the seal proved fruitless, and as the sea was freezing over behind us I ordered all the men on board directly the stuff from the depôt had been got on to the deck, and the *Nimrod* once more steamed north. The breeze soon began to freshen, and it was blowing hard from the south when we passed the winter quarters at Cape Royds. We all turned out to give three cheers and to take a last look at the place where we had spent so many happy days. The hut was not exactly a palatial residence and during our period of residence in it we had suffered many discomforts, not to say hardships, but, on the other hand, it had been our home for a year that would always live in our memories. We had been a very happy little party within its walls, and often when we were far away from even its measure of civilisation, it had been the Mecca of all our hopes and dreams. We watched the little hut fade away in the distance with feelings almost of sadness, and there were few men aboard who did not cherish a hope that some day they would once more live strenuous days under the shadow of mighty Erebus.

I left at the winter quarters on Cape Royds a supply of stores sufficient to last fifteen men for one year. The vicissitudes of life in the Antarctic are such that such a supply might prove of the greatest value to some future expedition. The hut was locked up and the key hung up outside where it would be easily found, and we readjusted the lashing of the hut so that it might be able to withstand the attacks of the blizzards during the years to come. Inside the hut I left a letter stating what had been accomplished by the expedition, and giving some other information that might be useful to a future party of explorers. The stores left in the hut included oil, flour, jams, dried vegetables, biscuits, pemmican, plasmon, matches and various tinned meats, as well as tea, cocoa, and necessary articles of equipment. If any party has to make use of our hut in the future, it will find there everything that it requires.

The wind was still freshening as we went south under steam and sail on March 4, and it was fortunate for us that this was so, for the ice that had formed on the sea water in the sound was thickening rapidly, assisted by the old pack, of which a large amount lay across our course. I was anxious to pick up a depôt of geological specimens on Depôt Island, left there by the Northern Party, and with this end in view the *Nimrod* was taken on a more westerly course than would otherwise have been the case. The wind, however, was freshening to a gale, and we were passing through streams of ice, which seemed to thicken as we neared the shore. I decided that it would be too risky to send a party off for the specimens, as there was no proper lee to this small island, and the consequences of even a short delay might be serious. I therefore gave instructions that the course should be altered to due north. The following wind helped us, and on the morning of March 6 we were off Cape Adare. I wanted to push between the Balleny Islands and the mainland, and make an attempt to follow the coast-

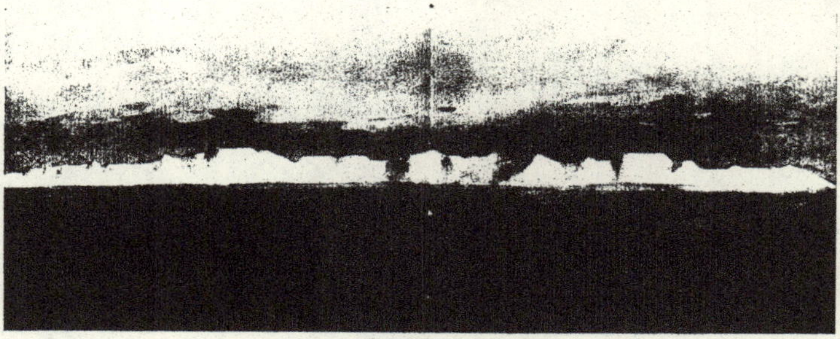

New Coast-Line West of Cape North
(*Sketched by G. Marston.*)

line from Cape North westward, so as to link it up with Adelie Land. No ship had ever succeeded in penetrating to the westward of Cape North, heavy pack having been encountered on the occasion of each attempt. The *Discovery* had passed through the Balleny Islands and sailed over part of the so-called Wilkes Land of the maps, but the question of the existence of this land in any other position had been left open.

We steamed along the pack-ice, which was beginning to thicken, and although we did not manage to do all that I had hoped, we had the satisfaction of pushing our little vessel along that coast to longitude 166° 14′ East, latitude 69° 47′ South, a point further west than had been reached by any previous expedition. On the morning of March 8 we saw, beyond Cape North, a new coast-line extending first to the southwards and then to the west for a distance of over forty-five miles. We took angles and bearings, and Marston sketched the main outlines. We were too far away to take any photographs that would have been of value, but the sketches show very clearly the type of the land. Professor David was of opinion that it was the northern edge of the polar plateau. The coast seemed to consist of cliffs, with a few bays in the distance. We would all have been glad of an opportunity to explore the coast thoroughly, but that was out of the question; the ice was getting thicker all the time, and it was becoming imperative that we should escape to clear water without further delay. There was no chance of getting further west at that point, and as the new ice was forming between the old pack of the previous year and the land, we were in serious danger of being frozen in for the winter at a place where we could not have done any geological work of importance. We therefore moved north along the edge of the pack, making as much westing as possible, in the direction of the Balleny Islands. I still hoped that it might be possible to skirt them and find Wilkes Land. It was awkward work, and at times the ship could hardly move at all.

Finally, about midnight on March 9, I saw that we must go north, and the course was set in that direction. We were almost too late, for the ice was closing in and before long we were held up, the ship being unable to move at all. The situation looked black, but we discovered a lane through which progress could be made, and in the afternoon of the 10th we were in fairly open water, passing through occasional lines of pack. Our troubles were over, for we had a good voyage up to New Zealand, and on March 22 dropped anchor at the mouth of Lord's River, on the south side of Stewart Island. I did not go to a port because I wished to get the news of the expedition's work through to London before we faced the energetic newspaper men.

That was a wonderful day to all of us. For over a year we had seen nothing but rocks, ice, snow and sea. There had been no colour and no softness in the scenery of the Antarctic; no green growth had gladdened our eyes, no musical notes of birds had come to our ears. We had had our work, but we had been cut off from most of the lesser things that go to make life worth while. No person who has not spent a period of his life in those "stark and sullen solitudes that sentinel the Pole" will understand fully what trees and flowers, sun-flecked turf and running streams mean to the soul of a man. We landed on the stretch of beach that separated the sea from the luxuriant growth of the forest, and scampered about like children in the sheer joy of being alive. I did not wish to despatch my cablegrams from Half Moon Bay until an hour previously arranged, and in the meantime we revelled in the warm

Floating Ice off Cape Adare

Last View of Cape Adare

The first Landing in New Zealand on the return of the Expedition.
A Bay in Stewart Island

sand on the beach, bathed in the sea and climbed amongst the trees. We lit a fire and made tea on the beach, and while we were having our meal the wekas, the remarkable flightless birds found only in New Zealand, came out from the bush for their share of the good things. These quaint birds, with their long bills, brown plumage and quick, inquisitive eyes have no fear of men, and their friendliness seemed to us like a welcome from that sunny land that had always treated us with such open-hearted kindliness. The clear, musical notes of other birds came to us from the trees, and we felt that we needed only good news from home to make our happiness and contentment absolutely complete. One of the scientific men found a cave showing signs of native occupation in some period of the past, and was fortunate enough to discover a stone adze made of the rare pounamu, or greenstone.

Early next morning we hove up the anchor, and at 10 A.M. we entered Half Moon Bay. I went ashore to despatch my cablegram, and it was strange to see new faces on the wharf after fifteen months during which we had met no one outside the circle of our own little party. There were girls on the wharf, too, and every one was glad to see us in the hearty New Zealand way. I despatched my cablegrams from the little office, and then went on board again and ordered the course to be set for Lyttelton, the port from which we had sailed on the first day of the previous year. We arrived there on March 25 late in the afternoon.

The people of New Zealand would have welcomed us, I think, whatever had been the result of our efforts, for their keen interest in Antarctic exploration has never faltered since the early days of the *Discovery* expedition, and their attitude towards us was always that of warm personal friendship, but the news of the measure of success we had achieved had been published in London, and flashed back to the southern countries, and we were met out in the harbour and on the wharves by cheering crowds. Enthusiastic friends boarded the *Nimrod* almost as soon as she entered the heads, and when our little vessel came alongside the quay the crowd on deck became so great that movement was almost impossible. Then I was handed great bundles of letters and cablegrams. The loved ones at home were well, the world was pleased with our work, and it seemed as though nothing but happiness could ever enter life again.

Appendix One

BIOLOGY

NOTES BY JAMES MURRAY, BIOLOGIST OF THE EXPEDITION

ON the calm evening of our departure for the south, while the New Zealand coast was still in view, the *Nimrod* was accompanied by a number of southern Black-back or Dominican gulls, with their beautiful white-bordered black wings. An occasional cormorant flew past. Next morning these shore birds had left us, and their place had been taken by the albatrosses and petrels which were to keep us company thenceforward till we reached the Antarctic Circle.

The pretty little speckled cape pigeon just crossed the Circle, and left us next day; the black-browed albatross was seen for a day longer; the sooty albatross went furthest into the Antarctic, and was last seen on January 18.

Just at the Antarctic Circle we were met by those peculiarly Antarctic birds, the snowy and Antarctic petrels. The wide zone of floating ice, a hundred miles or more across, would have been gloomy indeed without those two birds, which frequented this zone in large flocks. These beautiful petrels, the Antarctic, with strongly contrasting brown and white, the Snowy, pure white except for the black bill and feet, relieved by their bright plumage and sprightly flight the loneliness of this region. A few seals and penguins on the lower bergs were the only other living things among the ice.

The giant petrel and the little Wilson's petrel were the only birds that ranged right from the New Zealand coast to the shores of Victoria Land.

In the open sea to the south of the belt of ice even the Antarctic and Snowy petrels for the most part left us.

The desolation and lifelessness of the Antarctic were fully realised as we approached the great Ice Barrier. There was no living thing in sight as we steamed eastward, tracing the line of this immense glacier. Towards midnight there opened suddenly to our sight a scene of abounding life. The cliff of the Barrier terminated, and a wide bay opened up, extending far to the south, and partly filled by fast ice of one season's growth. Away to the eastward the cliff recommenced. This bay, which we afterwards referred to by the appropriate name of the Bay of Whales, was teeming with all the familiar kinds of Antarctic life. Hundreds of whales, killers, finners, and humpbacks, were rising and blowing all around. On the ice groups of Weddell seals were basking in the midnight sunshine. Emperor penguins were standing about or tobogganing in unconcerned parties. Skua gulls were flying heavily, or sitting drowsily on the ice. Only the Adelie penguin (busy nesting elsewhere), and the rarer kinds of seal were absent.

It could hardly be supposed that this was a chance gathering of all these animals. Passing the same spot on the return journey westward, there was still the same abundance of life. There was probably land near by.

Cape Royds.—To the biologist, no more uninviting desert is imaginable than Cape Royds seemed when we made our first landing, and for long afterwards. Here is absolute desolation, a black and white wilderness, rugged ridges of lava alternating with snowdrifts for a few miles, ending to the north and south in crevassed glaciers, and eastward in the snow-field stretching up to the rocky crags of the cone of Mount Erebus.

On the very edge of the sea, the little colony of Adelie penguins and the scattered skua gulls relieved the monotony. Beyond was no living creature, no blade of grass, or tiniest patch of welcome green.

Bleak and bare though it was, this stretch of two or three miles of broken country, where rocky peaks and ridges, moraines and snow drifts, diversified the surface, was the field of operations for the biologist. The white waste of glacier and snow-field was hopeless, the nearer country seemed little more promising.

The sea was there, known to be teeming with varied life, but it was inaccessible till the ice should bridge it over.

In the immediate vicinity of the camp were many little sharp peaks of kenyte, and short valleys filled with a cindery gravel derived from the decomposition of the same rock. Moraines covered the rock in places, and many of the valleys contained frozen ponds or lakes of various sizes up to half a mile in length.

The first walks over the hills did little to encourage the biologist. The rugged kenyte, with its hard projecting felspar crystals tearing the boots, supported no living thing. Little could we suspect that far beneath the thick ice of the lakes was plentiful life, dormant, it is true, but only waiting to be thawed out to spring into activity.

Gradually, as we came to know it, it began to appear that the barrenness was not so absolute as we at first supposed. On an early walk, Mr. Shackleton brought home a scrap of moss and lichen, but it was long afterwards that the melting of the snow in the next summer revealed that fact that on some of the moraines the growth of mosses and lichens was, comparatively speaking, luxuriant. A little dried-up pool, some two yards across, close by the penguin rookery, was quite covered by a film formed of bright green filamentous algæ. Around the edges of some of the smaller lakes, thick wrinkled sheets of a plant of a dingy green or brown colour were seen, resembling some of the large foliaceous lichens in form. Wild brought in some of the same plant which he had found embedded in the transparent ice of a lake afterwards known as Clear Lake. In this situation it was of a beautiful orange red colour. Later on, this plant was found to be abundant in all the lakes, and in many of them it formed continuous sheets over the entire bed. Many of the gravelly valleys showed a faint green tinge, showing where water had run in the summer-time.

Microscopic Life

It was while examining a bit of the orange-coloured weed under the microscope, that the first fresh-water animals were found. Two or three little red worm-like creatures were seen, creeping about like caterpillars, now and again coming to a stop and putting out their heads to feed, when they showed themselves to be rotifers, or wheel-bearing animalcules.

Having thus ascertained that there were microscopic animals on this plant, a large quantity of it was gathered and washed in water to remove the adhering organisms. The sediment obtained was concentrated by straining, and a drop put under the microscope. Myriads of living things appeared, animals and plants. The plant-life consisted of various spheres and threads of green and blue-green algæ, some of the latter resembling strings of beads.

The animals were more abundant, and in greater variety. The creeping rotifers were most plentiful, and there were several kinds of them besides the blood-red one first noticed. Some were of curious shapes.

Many were feeding, the two rotating wheels sweeping streams of minute particles towards the mouth.

Small bear-like creatures were scratching among the *débris* or fiercely "pawing the air" with great curved, dangerous-looking claws. These were the water-bears.

Miniature snakes (thread-worms) were twisting in and out, and lashing their tails. Some of the simplest of animals (protozoa), each consisting of a single cell, were there, the active infusoria swimming rapidly by means of their cilia, the slower-moving rhizopods putting out their little soft fingers to feed or creep about. Animals higher in the scale were not wanting, though these were never seen alive. Skins of some mites related to the cheese mite, and of some small shrimps (crustacea) were occasionally found.

Among all these animals the rotifers and water-bears were most important in point of numbers, and they lead such strange lives that they will be more fully described in later paragraphs. At any time during the winter an unlimited supply of these animals could be got for study by simply melting a piece of ice containing some weed. In summer, the ponds and smaller lakes were completely melted for weeks, and then they were still more easily got by washing some of the weed. A few animals not found in winter then hatched out from eggs and swam about in the water. A large and beautiful rotifer named *hydatina* appear in Coast Lake only.

In Coast Lake also a curious thing happened in summer. The stones at the margin became covered by bright red patches, as though they had been sprinkled with blood. These patches were found to be formed of rotifers, of the same kind which were commonest among the weed in winter. The red stains appear to rise owing to their rapid multiplication, and to their fixing themselves side by side as close as they can pack. The photograph shows a small part of one of these stains transferred by a brush to the microscope slide.

In Coast Lake the largest of these patches of rotifers would not be more than an inch in diameter, but Priestley tells us that in a lake on the west side of the Sound they formed patches " as large as a man's head " and of appreciable thickness. Though this rotifer usually attaches itself by the foot when feeding, many of them let go their hold and go swimming in the water, so that when water from any of these lakes is taken for driking it can be seen that there is a fair sprinkling of red grains in it which must be swallowed with the water.

In summer, too, in very shallow ponds and trickling streams an alga of a brownish-green colour grows in large translucent sheets.

Life in the Ice

As soon as animals were obtained from the weed enclosed in the ice in the manner described above, it was obvious that mere freezing did not kill them. They were first got in the shallow lakes, where the weed could be seen through the transparent ice at the margins. There were plenty in all the shallow lakes. A shaft was sunk through fifteen feet of ice to the bottom of Blue Lake. There was a film of yellow weed covering the gravel of which the bottom was composed, and on this weed several kinds of rotifers were found alive. This fact seemed more remarkable later, when we found that Blue Lake did not melt during the two summers that we spent at Cape

*

Open Water at Green Lake in Summer Time

Claws of a Water-Bear, magnified about 500 Diameters

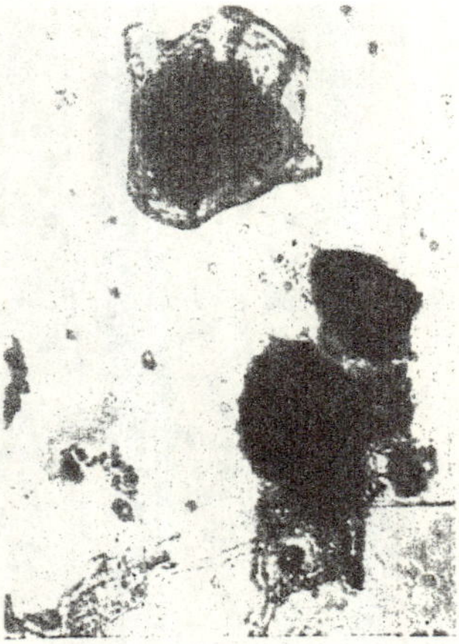

A new Species of Rotifer found at Ca Royds. Its distinctive feature is the p session of wing-like Protuberances at t Sides

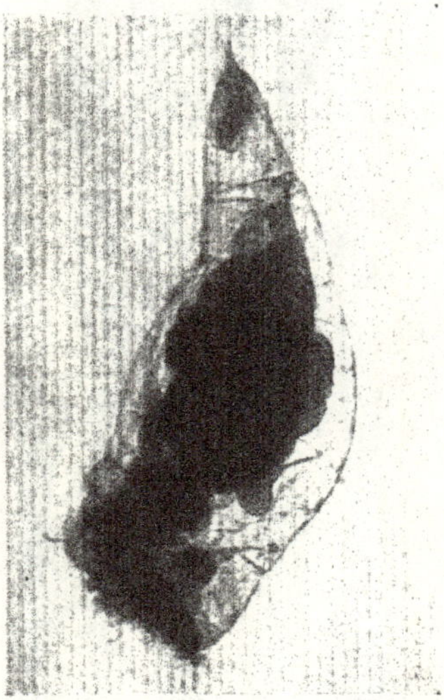

*A large, free-swimming Rotifer, called Hydatina. It is very plentiful in the Coast Lakes in the Summer

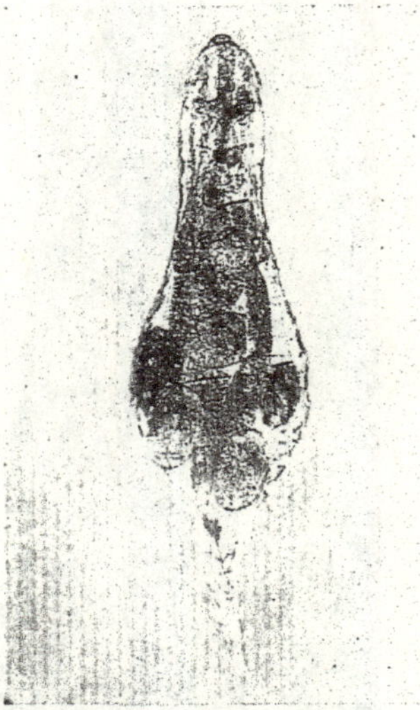

Viviparous Rotifer from the Salt Lak The oval Bodies seen in the Adult Anim are the Young

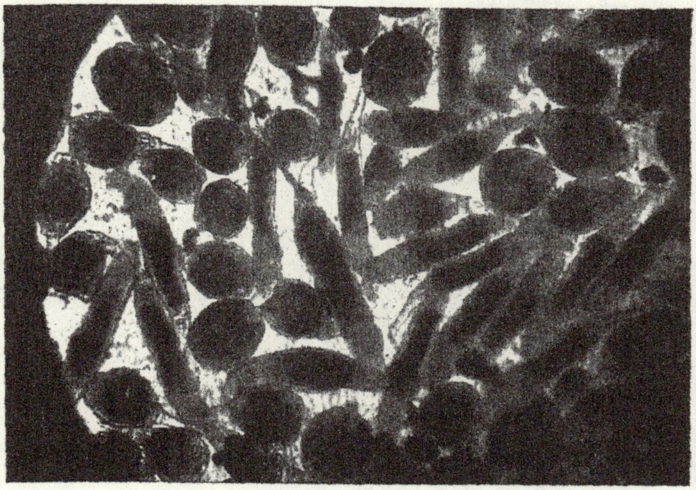

THE GREGARIOUS ROTIFER, WHICH FORMS BLOOD-RED PATCHES IN THE LAKES AT CAPE ROYDS

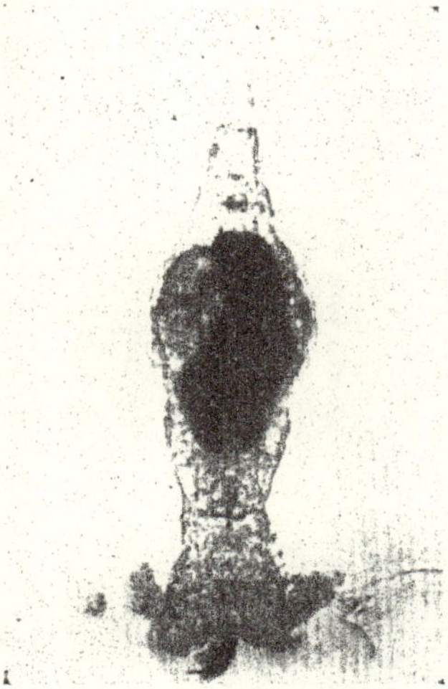

*A SINGLE SPECIMEN OF GREGARIOUS ROTIFER. THE DARK PORTION IS THE STOMACH, AND THE OVAL FORM OF AN UNBORN YOUNG ROTIFER CAN BE SEEN

THE COMMONEST WATER-BEAR IN THE CAPE ROYDS DISTRICT

Royds. This means that the animals must be capable of remaining frozen for years, possibly for many years, without being killed. Though enclosed in the ice, there was no means of knowing how low temperatures they could endure, for the ice in the lakes might never be so cold as the air.

A few simple experiments were carried out with the object of finding what degree of cold they could survive. Afterwards the experiments were extended in other directions; they were heated, they were immersed in various saline mixtures; in short, they were submitted to various tests such as they might be exposed to in nature. It must be admitted that we did not ascertain what limits of temperature they could endure. We only know that they *live* at a certain low temperature; the Antarctic was not cold enough to show us any temperature at which they *die*.

From facts previously ascertained we may predicate approximately what is the limit at the other end of the scale. Animal protoplasm is known to coagulate at a point well below the boiling-point of water. As the Antarctic was not cold enough, it was intended to use the resources of civilisation in order to get greater cold, by the use of liquid air. Unfortunately, the animals had to be subjected for some weeks to an almost tropical temperature, and were found to be all dead when they reached Sydney.*

Tenacity of Life

To test the degree of cold which they could stand blocks of ice were cut from the lakes and exposed to the air in the coldest weather of the whole winter. By boring into the centre of the blocks we found that they were as cold

* Since this was written, examination of the rotifers in London (in September 1909) has shown that they are still living.

as the air. A temperature of minus 40° Fahr. did not kill the animals.

Then they were alternately frozen and thawed weekly for a long period, and took no harm. They were dried and frozen, and thawed and moistened, and still they lived. At last they were dried, and the bottle containing them was immersed in boiling water, which was allowed to cool gradually and still a great many survived. Again they were put into sea-water, and into the brine from the bottom of Green Lake, which is so salt that it only freezes at about zero (Fahr.). They were left in these salt waters for a month, yet as soon as they were transferred to fresh water they began to crawl about as though nothing had happened.

Such is the vitality of these little animals that they can endure being taken from ice at a minus temperature, thawed, dried and subjected to a temperature not very far short of boiling-point, all within a few hours (a range of more than 200° Fahr.). It is not the eggs merely that survive all these changes, but the grown animals. These are animals comparatively high in the scale. The rotifers are worms, and the water-bears (which stood the same tests) are cousins to the insects and spiders. Some very lowly plants are not killed by being put in boiling water and doubtless many very simple animals can live through cold greater than we found in the Antarctic. Men can endure exposure for a time to very much lower temperatures, and to dry heat far above the temperature of boiling-water, but the case of the rotifer is very different. Its little body actually takes those different temperatures, man's body does not.

It is a curious fact that these animals, which can endure such extremes of heat and cold, and other unfavourable conditions, readily die when left in cold water at a mod-

THE EMPERORS' CONCLAVE

erate temperature. The water may get a few degrees warmer than they are accustomed to, and may be insufficiently aerated, but there is nothing to alarm them and induce them to make use of their remarkable means of protection (whatever they may be). Exposed to low temperatures or to salt water, they contract into little round balls, and in this state they are (somehow) safe. In the cold water, which is just slightly unfavourable, they see no cause for alarm, and so, as Mr. Shackleton aptly expressed it when told about it, "they go out without their Burberries," that being the great sin against prudence on the part of a polar explorer.

The rotifer is not as a rule a long-lived animal. I have heard of a patriach of five months, which then came to an untimely end. Generally their span is to be measured by days or weeks, or at any rate, is limited by a single season. The majority of the creeping rotifers can protect themselves against drying up, by coating themselves with a kind of varnish, and so they prolong what would be an ephemeral existence through a period of years. As all activity is suspended it may be questioned whether the animal gains anything by this marvellous protective capacity, but at least the chances of the race surviving are greatly increased by it. The Antarctic rotifers in like manner exist in a state of suspended animation when frozen in ice for a long period. If the lakes in which they live are only melted at long intervals and for short periods, it may be that some very ancient rotifers are alive beneath the ice, possibly scores of years old.

Viviparous Rotifers

Most rotifers lay eggs, but a good many kinds bring forth the young alive and very completely developed. At

Cape Royds, twelve kinds of creeping rotifiers were found. Eight of these were well-known kinds which elsewhere lay eggs, and at Cape Royds they are reproduced in the same way.

There were, however, two other kinds at Cape Royds, unknown anywhere else, and these are greatly more abundant than any of the others. It has been told how one of them forms blood-like patches in the lakes. Now, these two species bring forth living young, yet they belong to groups which usually lay eggs. One of them belongs to a genius (*adineta*), no other known member of which is viviparous; the other (*philodina gregaria*) belongs to a section of the genus in which all the previously known species are oviparous.

It is obvious that this mode of reproduction, being unusual in the groups to which they belong, must have been developed as an adaptation to the peculiar conditions, and their extraordinary abundance shows that it is perfectly successful. Yet this is contrary to our preconception of what will be favourable under severe conditions. It is usually supposed that eggs are better protected against evil conditions than the adult animals, and that production of winter eggs ensures the continuation of the race when the animals perish, and it appears to be generally so among the lower invertebrata. Yet here, in one of the severest climates in the world, the process is reversed. Not only are the animals viviparous, but parents and young are alike indifferent to prolonged interruption of their activity by freezing, and when thawing occurs, development proceeds from where it left off. Since the viviparous rotifers are the most successful in the struggle for existence at Cape Royds, it is rather curious to note the total absence of animals of the genus *Rotifer,* which are all normally viviparous.

*

Murray holding young Penguins

JOYCE AND THE DOGS IN THE PENGUIN ROOKERY

Two Emperor Penguins

ADELIE PENGUINS

BIOLOGY

Life in Salt Lakes

Green Lake is very saline. We do not yet know how much saline matter the water contains when the lake is entirely melted and mixed up by wind. The fluid obtained from under the ice in winter is a very dense and strong-smelling brine. There is abundance of life in this lake, but the number of kinds is much less than in the other lakes. Only two out of the twelve species of rotifers known at Cape Royds live in it. One of these (*callidina constricta*) is not very plentiful; the other (*adincta grandis*) is extremely so. This animal, while developing the power of enduring cold, has at the same time become accustomed to living in salt water. Though they were not killed by the Green Lake brine, which is so much salter than the sea-water, none of these rotifers have been found in the sea.

Water-bears in Ice

Water-bears were found to live while frozen in ice just as well as the rotifers did. It is an interesting fact that the only abundant species at Cape Royds is an Arctic species (*macrobiotus arcticus*) which was only previously known in Spitzbergen and Franz Josef Land, and which has not yet been detected in the various collections made on the other side of the Antarctic by Bruce's and Nordenskjold's expeditions.

Distribution of Rotifers, &c.

Most of the rotifers and other animals were found generally distributed in all the lakes visited. These covered a very limited area, the most distant being no more than thirty miles apart. The nature of the microscopic fauna of other parts of the Antarctic is scarcely known. Some of the Cape Royds lakes are richer than others, and the saline lakes are poorer, but the general distribution of most of the animals suggests easy dispersal from one to another. How are they conveyed? Only two methods suggest themselves to one acquainted with the local conditions. The skua gulls, which are so fond of bathing, may transfer a few adhering to their feet when they go from one lake to another. No other bird is at all likely to assist in this process. The other method, which seems likely to be the general one, is by wind. The weed at the margins of the lakes gets exposed and dried by the evaporation of the water, or more commonly by the ablation of the ice from the surface. It is then very light and easily blown about by the wind, carrying its freight of dormant rotifers, &c.

The experiments detailed above show that in course of this dispersal they are scarcely exposed to danger at all. The hardest frost or the hottest sun cannot harm them, and should they even fall into the sea, they will not be killed if the plant which forms a raft for them eventually reaches the shore, when they may be again dried and driven by the wind till they find a suitable place to resume living. A difficulty suggests itself in consid-ing this theory of dispersal by wind. The prevalent wind-storms are all from the southward. Northerly winds of force enough to move the dried plants are almost unknown. Though northerly winds prevail for a somewhat greater number of hours during the year than the southerlies, they are very light airs. Dispersal would then be easy in one direction only, and might be very slow, and dependent on the agency of the gulls in other directions.

Life Among Moss

The moss-dwellers are now well recognised among microscopic animals. There are rotifers and water-bears among the mosses at Cape Royds, but the mosses themselves are not very abundant, and the creeping rotifers have found a better home among the weed in the lakes.

Among the mosses the animals lead even a harder life than they do in the lakes. In the lakes, when they do melt, the rotifers enjoy a period of some weeks when they can move and feed and multiply. Among mosses they feel at once the lowest temperatures of the air. They are frozen during the greater part of the year. They are frozen even in summer for the greater part of every day. Only for a few hours daily, for a short time in the height of summer, the moss is thawed by the sun's rays. One wonders when the beasts get any time to grow. Yet they are there in abundance. They are all of different kinds from the lake dwellers. None of the rotifers were recognised, but some of the water-bears were.

Biological Problems

By what means do the rotifers survive freezing? It is not, as with higher animals, that they can keep warm in spite of the cold. They are too small for that. Their very blood, as the watery fluid filling the body cavity may be called, freezes very soon after the surrounding water.

Whence is the microscopic fauna derived? Are the rotifers and water-bears survivors from the remote time when a milder climate prevailed in Antarctica, when the country was covered with a vegetation of the higher plants and the coal-beds were in course of formation? Or are they colonists from the temperate regions, which have migrated across the stormy Antarctic Ocean under present-day conditions?

Some of the facts favour both theories. The small number of species, and the fact that the majority of them are widely distributed over the world, point to recent immigration.

PENGUINS

Though so much has been written about them, the penguins always excite fresh interest in every one who sees them for the first time. There is endless interest in watching them, the dignified Emperor, dignified notwithstanding his clumsy waddle, going along with his wife (or wives) by his side, the very picture of a successful, self-satisfied, happy, unsuspicious countryman, gravely bowing like a Chinaman before a yelping dog—the little undignified matter-of-fact Adelie, minding his own business in a way worthy of emulation. They are perfectly adapted to a narrow round of life, and when compelled to face matters outside of their experience they often behave with apparent stupidity, but sometimes show a good deal of intelligence.

Their resemblance to human beings is always noticed. This is partly due to the habit of walking erect, but there are truly a great many human traits about them. They are the civilised nations of these regions, and their civilisation, if much simpler than ours, is in some respects higher and more worthy of the name. But there is a good deal of human nature in them, too. As in the human race their gathering in colonies does not show any true social instinct. They are merely gregarious; each penguin is in the rookery for his own ends, there is no thought of the general good. You might exterminate an Adelie rookery with the exception of one bird, and he would be in no way concerned so long as you left him alone.

Some little suggestion of altruism will appear in dealing with the nesting habits of the Adelie. Thieving is known, among the Adelies at least. One very pleasing trait is shown, which they have in common with man. Eating is not with them the prime business in life, as it is with the common fowl, and most animals. Both Emperors and Adelies, when the serious business of nesting is off their minds, show a legitimate curiosity. Having fed and got into good condition they leave the sea and go off in parties, apparently to see the country, and travel for days and weeks.

The Emperor

We saw the Emperor only as a summer visitor. Having finished nesting, fed up and become glossy and beautiful, they came up out of the sea in large or small parties, apparently to have a good time before moulting. While the Adelies were nesting they began to come in numbers to inspect the camp. Passing among the Adelies, the two kinds usually paid no attention to one another, but sometimes an Adelie would think an Emperor came too close to her nest, and a curious unequal quarrel would ensue, the little impudence pecking and scolding, and the Emperor scolding back, with some loss of dignity. Though more than able to hold her own with the tongue, the Adelie knew the value of discretion whenever the Emperor raised his flipper.

They were curious about any unusual object and would come a long way to see a motor-car or a man. When out on these excursions the leader of a party keeps them together by a long shrill squawk. Distant parties salute in this way and continue calling till they get pretty close. A party could be made to approach by imitating this call. The first party to arrive inspected the boat, then crossed the lake to the camp. Soon they discovered the dogs, and thereafter all other interests were swallowed up in the interest excited by them. After the first discovery crowds came every day for a long time, and from the manner in which they went straight to the kennels one was tempted to believe that the fame of them had been noised abroad.

CEREMONIES OF MEETING

Emperors are very ceremonious in meeting other Emperors or men or dogs. They come up to a party of strangers in a straggling procession, some big important aldermanic fellow leading. At a respectful distance from the man or dog they halt, the old male waddles close up and bows gravely till his beak is almost touching his breast. Keeping his head bowed he makes a long speech, in a muttering manner, short sounds following in groups of four or five. Having finished the speech the head is still kept bowed a few seconds for politeness sake, then it is raised and he describes with his bill as large a circle as the joints of his neck will allow, looking in your face at last to see if you have understood. If you have not comprehended, as is usually the case, he tries again. He is very patient with your stupidity, and feels sure that he will get it into your dull brain if he keeps at it long enough. By this time his followers are getting impatient. They are sure he is making a mess of it. Another male will waddle forward with dignity, elbow the first aside as if to say, "I'll show you how it ought to be done," and go through the whole business again. Their most solemn ceremonies were used towards the dogs, and three old fellows have been seen calmly bowing and speaking simultaneously to a dog, which for its part was yelping and straining at its chain in the effort to get at them.

Left to themselves the Emperor penguins seem perfectly peaceable, and no sign of quarrelling was ever noticed. When a party of them was driven into a narrow space they resented the jostling, and flippers were freely used, making resounding whacks, which apparently are not felt through the dense feathery fur. The flipper strikes with equal facility forward or backward.

They seem to regard men as penguins like themselves.

They are quite unsuspicious and slow to take alarm, so long as you stay still or move very slowly. If you walk too fast among them, or if you touch them, they get frightened and run away, only fighting when closely pressed. As one slowly retreats, fighting, he has a ludicrous resemblance to a small boy being bullied by a big one, his flipper towards the foe elevated in defence, and making quick blows at the bully. It is well to keep clear of that flipper when he strikes, for it is very powerful, and might break an arm.

Emperors were killed by the dogs, but it is likely that they hunted in couples or in parties to do this. A long fight was witnessed between an Emperor and the dog Ambrose, the largest of our dogs native to the Antarctic. The penguin was quick enough in movement to keep always facing the dog, and the flipper and long sharp bill were efficient weapons, as Ambrose seemed to appreciate. Only the bill was used, and it appeared to be due to short sight that the blow always fell short. Many of the apparently stupid acts of both kinds of penguins are doubtless to be traced to their very defective sight in air.

The Emperor can hardly be said to migrate since he remains to breed during the winter darkness, and spends the summer among the ice or on shore in the same region. Yet he travels a good deal, and the meaning of some of his journeyings remains a mystery. The visits of touring-parties to the camp have been described. At the same season (early summer), when the motor-car was making frequent journeys southward to Glacier Tongue with stores for depôt laying, we crossed on the way a great many penguin tracks. Many of these were beaten roads, where large parties had passed, some walking, some tobogganing. They all trended roughly to the south-east, and the wing marks and footmarks showed that they were all outward bound from the open sea towards the shores of Ross Island. Some of the roads were twelve miles or more from the open sea. There were no return tracks.

We expected to find that they had gone in to seek sheltered moulting-places, but on a motor trip to the Turk's Head we skirted a long stretch of the coast, and found no Emperors.

On journeys they often travel many miles walking erect, when they get along at a very slow shuffle, making only a few inches at each step. In walking thus they keep their balance by the assistance of the tail, which forms a tripod with the legs. When on a suitable snow surface they progress rapidly by tobogganing, a very graceful motion, when they make sledges of their breasts and propel themselves by the powerful legs, balancing and perhaps improving their speed by means of the wings.

Eight of them visited the motor-car one day, near Tent Island, sledging swiftly towards us. Two of them were very determined fighters and refused to be driven away. One obstinate phlegmatic old fellow, who wasn't going to be hurried by anybody, did learn to hustle as the car bore down upon him.

The Adelie

The Adelie is always comical. He pops out of the water with startling suddenness, like a jack-in-the-box, alights on his feet, gives his tail a shake, and toddles off about his business. He always knows where he wants to go, and what he wants to do, and isn't easily turned aside from his purpose.

In the water the Adelie penguins move rapidly and circle in the same way as a porpoise or dolphin, for which they are easily mistaken at a little distance. On level ice

Emperors Visit the Adelie Rookery; Ceremonial Bowing

Emperors Bowing to one another

Coy

Adelie keeping her young One Warm

An Adelie inspecting a Dog

Group showing a Moulting Penguin

Building the Nest

* An Adelie calling for a Mate after Commencing the Nest

or snow they can run pretty fast, getting along about as fast as a man at a smart walk. They find even a small crack a serious obstruction, and pause and measure with the eye one of a few inches before very cautiously hopping it. They flop down and toboggan over any opening more than a few inches wide. They can climb hills of a very steep angle, but on uneven ground they use their flippers as balancers. They toboggan with great speed on snow or ice, or even on the bare rocks when scared, but in that case their flippers are soon bleeding. Very rarely they swim in the water like ducks. They lie much lower in the water than the duck. The neck is below the surface and the head is just showing.

The Adelie is very brave in the breeding-season. His is true courage, not the courage of ignorance, for after he has learned to know man, and fear him, he remains to defend the nest against any odds. When walking among the nests one is assailed on all sides by powerful bills. Most of the birds sit still on the nests, but the more pugnacious ones run at you from a distance and often take you unawares. We wore for protection long felt boots reaching well above the knee. Some of the clever ones knew that they were wasting their efforts on the felt boots, and would come up behind, hop up and seize the skin above the boot, and hang on tight, beating with their wings. One of these little furies, hanging to your flesh and flapping his strong flippers so fast that you can hardly see them move, is no joke. A man once stumbled and fell into a colony of Adelies, and before he could recover himself and scramble out, they were upon him, and he bore the marks of their fury for some time.

Some birds became greatly interested in the camp, and wanted to nest there. One bird (we believe it was always the same one) couldn't be kept away, and came daily, sometimes bringing some friends. As he passed among the dogs, which were barking and trying to get at him, he stood and defied them all, and when we turned out to try to drive him away, he offered to take us all on too, and was finally saved against his will, and carried away by Brocklehurst, a wildly struggling, unconquerable being.

The old birds enjoy play, while the young ones have no leisure for play, being engrossed in satisfying the enormous appetites they have when growing. Four or five Adelies were playing on the ice-floe. One acted as leader, advanced to the edge of the floe, waited for the others to line up, raised his flipper, when they all dived in. In a few seconds they all popped out again, and repeated the performance, always apparently directed by the one. And so they went on for hours. While the *Nimrod* was frozen in the pack, some dozens of them were disporting themselves in a sea-pool alongside. They swam together in the duck fashion, then at a squawk from one they all dived and came up at the other side of the pool.

Early in October they began to arrive at the rookery, singly or in pairs. The first to come were males, and they at once began to scrape up the frozen ground to make hollows for their nests, and to collect stones for the walls with which they surround them. The digging is hard work and is done by the feet, the bird lying prone and kicking out backward. As soon as any apology for a nest is ready the males begin displaying, as shown in the acompanying photograph. He points his bill vertically upwards, flaps his wings slowly, inflates his chest, and makes a series of low booming sounds, which increase in loudness, then die away again, the throat vibrating strongly. Then he slowly subsides into the usual attitude.

We supposed this to be a part of his courtship, or as some phrased it "advertising for a wife," but there is good reason to suppose that the pairing is done before the birds leave the sea. Generally the male's displaying passes entirely disregarded. He continues it all through the nesting-season, till the chicks are nearly fledged and the moulting-time is near. An epidemic of displaying often took the whole rookery at once, when the hens were mostly away disporting themselves in the sea.

When the rookery is pretty well filled, and the nest-building is in full swing, the birds have a busy and anxious time. To get enough of suitable small stones is a matter of difficulty, and may involve long journeys for each single stone. The temptation is too strong for some of them, and they become habitual thieves. The majority remain stupidly honest. Amusing complications result. The bearing of the thief clearly shows that he knows he is dong wrong. He has a conscience, at least a human conscience, *i.e.*, the fear of being found out. Very different is the furtive look of the thief, long after he is out of danger of pursuit, from the expression of the honest penguin coming home with a hard-earned stone.

An honest one was bringing stones from a long distance. Each stone was removed by a thief as soon as the owner's back was turned. The honest one looked greatly troubled as he found that his heap didn't grow, but he seemed incapable of suspecting the cause.

A thief, sitting on its own nest, was stealing from an adjacent nest, whose honest owner was also at home, but looking unsuspectingly in another direction. Casually he turned his head and caught the thief in the act. The thief dropped the stone and pretended to be busy picking up an infinitesimal crumb from the neutral ground.

The stone gathering is a very strong part of the nesting instinct. It was kept up while sitting on the eggs, and if at a late stage they lost their eggs or young, they reverted to the heaping of stones, which they did in a half-hearted way. Unmated birds occupied the fringe of the rookery, and amused themselves piling and stealing till the chicks began to hatch out.

After the two eggs were laid the males appeared to do most of the work. At any hour the males predominated, a very few pairs were at the nests, and relieving guard was rarely noticed. The females were never seen in the majority. Those which had been recently down to feed could be recognised by the fresh crustacea round the nests. Judging by this sign, it would seem that some birds never leave the nest to feed during the whole period of incubation. Many birds lost their mates through the occasional breaking loose of a dog. These birds couldn't leave the nests.

Rearing the Chicks

The rookery is most interesting after the chicks arrive. Many curious things happen as they grow. The young chicks are silvery or slaty-grey, with darker heads, which are for the first day or so heavy and hang down helplessly. As soon as they are hatched the mothers take equal share in tending them, whatever they may have been doing before that. For some weeks the nest cannot be left untended or the chicks would perish of cold or fall victims to the skuas. The parents keep regular watches, going down in turn to feed, and relieving guard is an interesting ceremony. The bird just arrived from the sea hurries to the nest. It is anxious to see the chick, and to feed it; the other is unwilling to resign, but at last reluctantly gets off the nest, evidently very stiff, stretches itself, and hangs about for a while before going down to the sea.

Mother Bird leaves the Nest

Strangers Displaying Interest in the Lonely Chicks

* Young Adelie and Parents

An Adelie Refusing to be Frightened

*

Murray's Advances Resented

Adelie trying to Mother a couple of well-grown Strangers

When the young ones can hold up their heads the feeding begins. At first the parent tries to induce its offspring to feed by tickling its bill and throat. The old bird opens its mouth and the chick puts its head right in and picks the food out of the throat. The bird can be seen bringing it up into the throat by an effort. If the young is unwilling to feed some food is thrown right up on to the ground and a little of it picked up again and placed on the chick's bill. After learning the way there is no need for such inducement, and the parents are taxed to satisfy the clamouring for more.

For some weeks after the young are hatched life in the rookery goes smoothly along. One parent is always on the nest and the young birds do not wander. Then the trouble begins. The young begin to move about and if anything disturbs the colony they run about in panic. As they don't know nest or parent they cannot return home. They meet the case by adopting parents, and run under any bird they come to. The old birds resent this and a chick is often pecked away from nest after nest till exhausted. The skuas get some at this time, but it is surprising how few. Most of the chicks take some old one unawares and get in the nest. She may have a chick already, or chicks, but as she doesn't know which is her own she cannot drive the intruder away. A sorely puzzled bird may be seen trying to cover four gigantic chicks. Some of the less precocious youngsters stay at home long enough to get to know the nest, and can find their way home after wandering a few yards. Such homes keep together a little longer.

The time comes when both parents must be absent together to get enough food for the growing chicks. Then the social order of the rookery breaks down and chaos begins. The social condition which is evolved out of the chaos is one of the most remarkable in nature, yet it serves its purpose and saves the race. A kind of communism is established, but the old birds have no part in it. They cherish the fiction that they have nests and children, and when they come up from the sea after feeding it is their intention to find the nest and feed their own young only. The young ones for their part establish a community of parents, and yet it isn't exactly that either, though it works out as if it were. It is each bird for itself. The chick assumes the first old one that comes within its reach to be its parent. Perhaps it really thinks so, as they are all alike.

An old bird, coming up full of shrimps, is met by clamorous youngsters before it has time to begin the search for its hypothetical home. They order it to stand and deliver. It objects and scolds, and runs off. It may be by the irony of fate that it is its own young which accost it, but it can't know that. The chickens are both imperative and wheedling. Then begins one of those parent hunts which were so familiar at the end of the season. The end is never in doubt from the first. Every now and again the old one stops and expostulates. This shows weakness. There is no indecision on the part of the young one. It never seems anxious as to the result, but in the most matter-of-fact and persistent manner hunts the old one down. The hunts are often long and exhausting. One chase was witnessed at Pony Lake beside the camp. Nine times they circled the lake, and the hunt was not over when the watcher had to leave. On that occasion they must have travelled miles. At the end the old one stops, and still spluttering and protesting, delivers up. One would think that in these circumstances the weaker chicks would go to the wall, but it does not appear to be so. There are no ill-nourished young ones to be seen. Perhaps the hunts take so long that all get a chance.

A few days after the eggs began to hatch there was a severe blizzard, which lasted several days. Snow was banked up round most of the birds. A snowdrift crossed the densest part of the rookery partly burying many birds. In the deepest part nests and birds were covered out of sight, and the only indication of the whereabouts of a bird was a little funnel in the snow, at the bottom of which an anxious eye could be seen. Many less deeply buried birds had freed one wing or both, which became stiff with cold, as they could not be got back again. The snow, melting by the heat of their bodies, and refreezing, made walls of ice round the birds. Many got alarmed and left the nests, when the snow fell in and buried them. In the warm sunny weather that followed the melting snow filled many nests with pools of water. Some birds showed ingenuity in dealing with these floods. They moved their nests, stone by stone (always keeping a hollow for the eggs or chicks) as much as their own width till they reached dry ground. While the snowdrift remained some birds whose nests were buried scraped hollows in the snow and collected a few stones. On a moderate estimate about half the young perished in this blizzard.

The old Adelies do not mind the cold. Their thick blubber and dense fur sufficiently protect them. In a blizzard they will lie still and let the snow cover them. Going to the rookery once after a blizzard I could see no penguins; they had entirely disappeared. Suddenly at some movement or noise I was surrounded by them; they had sprung up out of the snow.

Domestic Entanglements

While the Adelie appears to be entirely moral in his domestic arrangements, his stupidity (or his short-sightedness which causes him to seem stupid) gives rise to many domestic complications. No doubt the presence of our camp upset the social economy, and probably when undisturbed nothing of the kind would occur. He has little sense of locality and one little heap of stones is very like another, yet pairs seem to have no means of recognising one another but by the rendezvous of the nest. Husbands and wives, parents and children, do not know one another, but if found at the nest are accepted as *bonâ fide*.

All the birds go to their nests without hesitation when they come from the sea by the familiar route, but if taken from their nests to some other part of the rookery some find their way back without difficulty, others are quite lost. They are most puzzled when moved only a little away from home, and they will fight to keep another bird's nest while their own is only a couple of feet away. A bird will defend an egg or chick in the nest, but if it is removed just outside it will peck at it and destroy it.

Considering these facts it will be evident that if the rookery be disturbed confusion follows. A mere walk among the nests caused innumerable entanglements. One bird would leave the nest in fright, flop down a yard away beside a nest already occupied, or on a nest left exposed by another scared bird. Then one-sided fights would begin, one bird attacking another under the impression that it had usurped its nest, the rightful owner troubling little about the vicious pecking he was receiving, sitting calmly in conscious rectitude. A fight of this kind has been watched for an hour at a time, three neighbouring nests having been disturbed. One bird had got into another's nest, a second was trying to establish a claim to the occupied nest of a third, and meanwhile the chicks of number one were neglected in the cold. A bird which had no family came and covered the chicks, but looked conscious of wrong doing and kept ready to bolt on a second's notice. All these birds but the last wanted their own nests and were within a yard of them without knowing it.

In all such cases, even when a bird got established on the wrong nest, there was always an adjustment afterwards. When they calmed down they became uneasy, probably observing the landmarks more critically, and would even leave a nest with chicks for their own empty nest. A chick removed from the nest and put alongside was not recognised, and the old bird never seemed to connect the facts of the empty nest and the chick beside it. If a chick were taken from the nest under the old bird's very eyes and held in front of it, it was always the chick that was viciously attacked, not the aggressor.

Some experiments were tried on them in order to trace the workings of the penguin mind. If a man stood between a bird and its nest so as to prevent it from getting onto it, the bird would make many attempts to reach home, rushing furiously at the man. After a time it would appear to meditate, and then walk off rather disconsolately, make a tour of the colony to which it belonged, and approach the nest from another side. It appeared greatly astonished that the intruder was still there. This curious trait was often seen. It is like the ostrich burying its head in the sand and imagining it is safe, or like a man refusing to believe his own eyes. It appears to think that if it takes a turn round, or comes to its nest from the other side, the horrible vision will disappear.

A bird was taken from a nest which had a chick in it and put down at a little distance. Meantime the chick was put in a neighbour's nest. Presently the bird came running up. It started back on seeing the empty nest, not in alarm or fear, but exactly as if thinking: "I've come to the wrong house!" trotted off to a distant part of the rookery. Her reasoning seemed to be this: "There was a chick in my nest, therefore this empty nest can't be mine." She couldn't imagine the chick leaving the nest, and so never searched for it. It was only a yard from the nest all the time. After half an hour's searching in vain for any place like home she returned to the nest, and accepted the restored chick as a matter of course.

A lost chick was never sought for. There would be no use; it couldn't be recognised. On account of this peculiarity we were able to make many readjustments of the family arrangements. When the blizzard destroyed so many chicks we distributed the young from nests where there were two to nests where there were none. They were usually adopted eagerly and the plan was quite successful.

When both birds are at a nest that is disturbed, or when the mate comes up from feeding to relieve guard, there is an interchange of civilities in the form of a loud squawking in unison, accompanied by a curious movement. The birds' necks are crossed, and at each squawk they are changed from side to side, first right then left. The harsh complaining clamour which they make was for long mistaken for quarreling.

A bird returning from the sea came to the wrong nest and tried to enter into conversation with the occupant, who would have nothing to do with him. She knew her mate had just gone off for the day, and wouldn't be such a fool as to come back too early, so she sat still, indifferent to the squawking of the other. A look of distress came into his face as he failed to get any response, and he was slow to realise that he had made a mistake.

A small colony was found with about two dozen large chicks, unattended by any old birds. They were driven across the lake to a larger colony. Half-way over a few old birds were squatted, enjoying a rest. When the chicks saw them they ran up to them joyfully, saying: "Here's pa and ma, hooray!" To their surprise they got the reverse of a cordial welcome, being driven away with vicious peckings. They were driven on to the larger colony and were swallowed up in it.

The Adelies are not demonstrative of their affections. It is difficult to discover if they have any beyond the instinctive affection for the young. The pairing appears to be a purely business matter, and the mates don't even show any power to recognise one another. A penguin was injured by the dogs, but it seemed possible that it might recover, so we did not at once put it out of pain. In a couple of days it died. Shortly after we noticed a live penguin standing by it. We removed the dead bird to a distance, and after a while found the other standing beside it as before. It was the general opinion that it was the dead bird's mate which had found it out. Such an action is entirely opposed to what we expect after a long study of their habits. There are always plenty of dead birds about a rookery, and the living go about entirely indifferent to them. It is puzzling in any point of view, but it is less difficult to believe that the bird found its dead mate than that it took an interest in a dead stranger.

Altruism

When the young birds are well grown if there is an alarm they flock together, and any old birds present in the colony form a wall of defence between the young and the enemy. This habit has given rise to the belief that they are somewhat communistic in their social order, and that the defence of the colony is a concerted action. It is not so. Each bird is defending its own young one only, and will often fight with another of the defending birds, or peck at any young one which comes in its way.

There are real instances of altruism or kindness to strangers. Our passage through the rookery frightened away the parent of a very young chick. A bird passing at the distance of a few yards noticed it and came over to it. He cocked his head on one side and looked at it, as if saying: "Hullo! this little beggar's deserted; must do something for him." He tickled its bill, as the parents do when coaxing the very young chicks to feed, but it was too much frightened to feed. After coaxing it in this way for some time he turned away and put some food upon the ground, and, lifting a little in his bill, he put some on each side of the chick's bill. Just then the rightful parent returned and the helper ran off. This was not an isolated case, but was observed on several occasions.

One incident seemed to reveal true social instinct. From a small colony of about two dozen nests all the eggs but one were taken in order to find out if the birds would lay again. As it turned out they did not. The birds sat on their empty nests for some time, then they disappeared. When the time came for the solitary egg to hatch, about a dozen of the nests were reoccupied and the birds took their share in defending the one chick.

Departure of the Young

When they have shed most of their down the young birds congregate at the edge of the sea. They cease from hunting the old ones for food, and appear to be waiting for something. When the right time comes, which they seem to know perfectly, they dive into the sea, sometimes in small parties, sometimes singly, disappear for a time, and may be seen popping up far out to sea. They dive and come up very awkwardly, but swim well.

Birds Rising from Snow

A View of the Rookery after a Blizzard

Emperors on the March

Emperors at rest

It is marvellous how fully instinct makes these birds independent. The parents do not take them to the water and teach them to swim. They haven't even the example of the old birds, which stay behind to moult. At an early age they become independent of their own parents, and earn their living by hunting any old bird they find. Though they have spent their lives on land, and only know that food is something found in an old bird's throat, when the time comes they leave the land and plunge boldly into the sea, untaught, to get their living by straining crustacea out of the water in the same way as the whale does.

Some of our party reported that they saw penguins teaching the young to swim, but if this ever happens it is not general. Time and again the young have been watched leaving as described, entirely of their own accord. At that season nearly all the old birds are in the moult and never venture into the water.

Like the Emperor the Adelie is fond of travelling when family cares are off his mind. The great blizzard which wiped out half the rookery left hundreds of old birds free. They began to explore the adjacent country in bands. The round of the lakes was a favourite trip and broad beaten roads marked this route. Tracks also led to the summits of some of the hills, though the short-sighted Adelie could hardly go there for the view.

There was no general trek southwards, such as the Emperors made, yet the Southern Party found tracks of two at a distance of some eighty miles from the sea.

Nebuchadnezzar and Nicodemus

These names dignified two penguin chicks. While chaos reigned in the rookery I found them exhausted and covered with mire, having been hunted and pecked through the rookery. They were taken to the house, put in a large cage in the porch, and fed by hand with sardines and fish-cakes. The feeding was disagreeable. They didn't like the food and shook it out of their bills in disgust. So it was necessary to force it down their throats till it was beyond their reach.

In a few days they became quite tame and recognised those who fed them. Familiar only with our peculiar method of feeding them, one of them indicated when he was hungry by taking my finger into his bill. We shortened their names to Nebby and Nicky, and they answered to them, but they answered equally readily to the common name of Bill. The sounds of the rookery reached them and sometimes greatly excited them, and they made desperate efforts to get through the netting of their cage. At these times we would take them out for a walk. They made no attempt to go to the rookery, and were rather frightened.

Nebuchadnezzar was a very friendly little fellow, and would follow me about outside, and come running when called. The feeding was unnatural, and for this reason doubtless in a few weeks they died.

The Ringed Penguin

A single ringed penguin appeared at Cape Royds at the end of the breeding-season, just as the Adelies were beginning to moult. No ringed penguin had been seen in this part of the Antarctic before. It was evidently a stray one which had come ashore to moult. It is about the same size as the Adelie, but is more agile. It was at the season when the young Adelies go off to sea. At a little distance the ringed penguin, among a crowd of old

RETURN OF THE PENGUINS

Adelies, looked somewhat like a young Adelie with the white throat. I picked him up by the legs to investigate. To my surprise he curled round and bit me on the hand. An Adelie could not do so. A closer examination showed what he was.

The Skua Gull

Some hundreds of skuas nested in the neighbourhood of Cape Royds. At Green Lake and Coast Lake they were aggregated in what were known as skuaries. Coast Lake was the most populous skuary. Elsewhere they were widely scattered over the lower hills. There is no doubt that they are not social or even gregarious birds, and that they congregate at such places as Coast Lake out of their fondness for fresh-water bathing. They do not love their kind, nor have they any inducement to. Skua will prey on skua.

They are callous, greedy, vulturous and disgusting. They may be seen sitting in a circle round a sick or wounded penguin, and they have been suspected of meditating an attack on the eyes of a sleeping man. When the young penguins are hatching they squat down right among the nests waiting for a chance to snatch one.

When they have eggs or young they swoop down on an intruder in a menacing manner, which is rather alarming when one knows how sharp and strong are their bills and claws. They rarely strike, however, thinking better of it at the last moment, and swerving suddenly upwards. When they do strike it is probably through miscalculation. Some birds were known which did habitually strike at every swoop. These blows did no damage, as if they come in front one involuntarily ducks, and from any other direction they strike the cap only.

They were very bitter against dogs, and gave no attention to men when they were present. Several of the dogs had learned to eat skua eggs, and the birds no doubt knew of this. They have a harsh cry, and the shrill chorus that went up when we had failed to find a nest sounded very much like laughter. They are very quarrelsome. When the rookery was strewn with fresh penguin carcases, victims of some dogs which had broken loose, the skuas would fight viciously over one, when there were plenty for all.

The skua has no true courage like the penguin. On one occasion we snared a skua and dragged it by a string into the midst of a small colony of penguins in order to see how these ancient enemies would behave when brought together. The penguins nearest to the skua pecked at it without leaving their nests. It bit and scratched for a very short while, and then gave it up and made no attempt to defend itself, though it was not hurt. As it lay motionless some one called out that the poor thing was killed, and we began to feel remorse for our cruel experiment, but when thrown up into the air the skua did not come down again. The bird's one virtue is cleanliness. It enjoys its bath. Wherever there is a pool of fresh water the skuas may be seen splashing with great gusto, and they will bathe in a sea pool on occasion.

Whales

In summer whales were locally abundant, though nowhere else in such numbers as we saw in the Bay of Whales at the Great Barrier. As long as there was open water small schools of finners and larger ones of killers were seen daily in McMurdo Sound. Even when the Sound was densely filled with pack, they came to the little sea-pools. For a long time in winter no whales were reported.

Killer Whales Sounding

Seal suckling Young, and taking no Notice of the Motor Car

Joyce lifting a Baby Seal

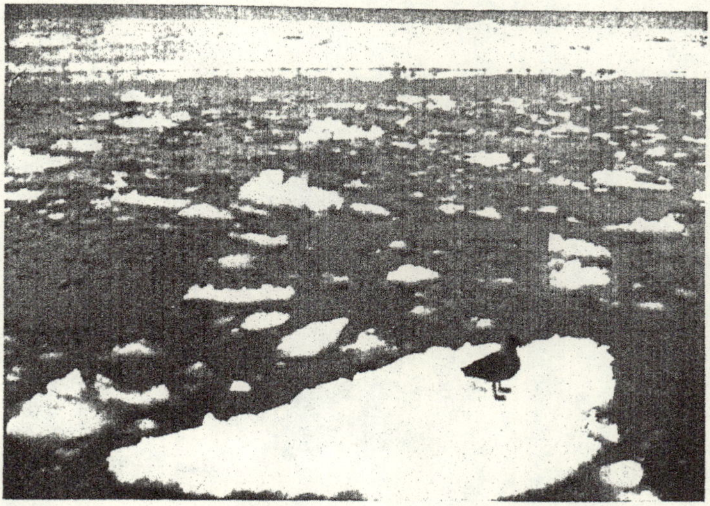

Skua Gulls

Seals Emerging from the Water at their Blow-holes

The finner with its little fin about half-way along the back, and its long pointed head, came very near, and often grazed the ship. One came vertically up close by the ship's side, the snout ten or twelve feet out of the water. As usual in such emergencies none of the cameras were ready.

The killers were often in family parties, or a few families together, some bulls of great size, with magnificent triangular fin, like a boat's sail, six or eight feet long, the cows with much smaller, often curved fin, the calves following close by their mothers' tails to avoid getting lost. Some very small calves were seen in January.

The humpback, with little rounded fin set far back, and the bottle-nose were rarely seen.

The killer sometimes rested his head on the edge of a floe and looked about with his wicked little eye for a seal or penguin. These he would try to knock off by rising under the floe, and on one occasion a party of three men, who spent an anxious twenty-four hours adrift on a floe, related that the killers were trying this experiment with them.

Seals

Of the four Antarctic seals only the Weddell was common at Cape Royds. The crabeater kept to the pack and rarely came on shore. The sea-leopard and the Ross seal were very rare.

The Weddell seal, or false sea-leopard, is a large and heavy animal. The skin and blubber of one large seal made a full sledge load. When it lies peacefully on the ice it is a shapeless lump. If disturbed, as for instance by putting your foot on it or throwing a dog at it, it shows ludicrous astonishment, curves both ends towards the intruder, open its mouth astonishingly wide, gasps and bleats with fear, its eyes starting from its head. It is slow to think of escaping, though its blow-hole may be close by, and prefers to roll over sideways.

Sometimes the Weddells pretend to be very fierce, and open the mouth to emit a kind of bellow, but they have no belief in their own fierceness and are usually looping away before the roar has time to frighten the enemy. In autumn as many as one hundred Weddells have been counted together on the ice of one little bay. Even in winter they came up through the tide-cracks occasionally, though sometimes several weeks would pass without any being seen. They used the holes which we kept open for dredging as blow-holes, and sometimes they reached there in a very exhausted condition, as evidenced by their distressed breathing. At these seasons there might be no other breathing-space for a long distance.

The nearest rookery of Weddells was at Inaccessible Island, about eight miles from the camp. When the first young were born early in November we ran out frequently by motor-car to watch them. They were in no way alarmed by the car.

The newly born calves, in their rough grey fur, already tried the intimidating roar, but there was more fear than fierceness in it. They can snap their lower jaws against the upper very rapidly, but without much force.

Some of the mothers were very cowardly, and made for the blow-holes, leaving the young to their fate. Others made a determined stand against the intruders and looked so much like business that we didn't care to venture too near, and a few carried the war into the enemy's country and ran at us and chased us off. The rough coat is cast and the smooth spotted skin like the adult appears at a very early age. In the middle of November we could scarcely find any that hadn't more or less moulted.

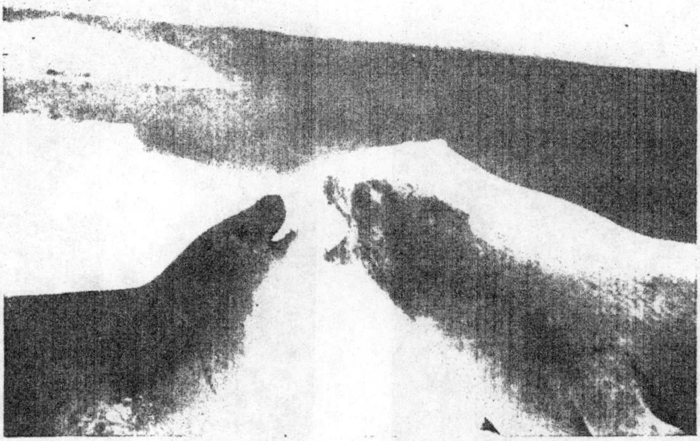

WEDDELL SEALS QUARRELLING

A WEDDELL SEAL ASLEEP

* Some of the Dogs

The crabeater can move much faster on ice or land than the Weddell. When not frightened it progresses in the same way as the Weddell, arching the back in caterpillar fashion very rapidly. When alarmed and excited it goes along for a short time in another manner, sweeping the tail end from side to side, much as a fish swims and actually gets forward a little in this way. A crabeater is able to hold its own against one dog, though it might be overcome by several. A fight between one and the dog Erebus lasted for an hour, if it could be called a fight where they never came to close quarters. The lighter dog circled about, snapping at the seal's neck and flippers. The crabeater always turned smartly enough to be ready for him and frequently made a feint of moving in one direction, then made a sudden turn and snap, drawing his head close down to his shoulders and shooting it out as he snapped, just as the sea-leopard is said to do. It was much more fatiguing for the heavy seal than for the dog, and it breathed heavily, making a continuous sound with its nostrils like snoring, but at the end of the hour it was still able to take care of itself. It became very angry as the dog's attack continued, whereas the Weddell after any amount of baiting only seems more and more astonished.

The Dogs

Our dogs, though of an originally Siberian strain, were reared in the mild climate of New Zealand, being the descendants of dogs left there some generations ago by a returning expedition. They were small, and showed evidence of crossing with ignoble races. Yet they showed no sign of degeneracy in their ability to endure the unaccustomed severe climate, and on the short journeys for which they were used they did splendid work.

They revelled in the cold, enjoyed nothing so much as a roll and a fight in a snowdrift, and wouldn't use the kennels provided for them, preferring to curl up in the snow, at most in the shelter of the kennels, or to lie on anything dark, such as a coal-bag. They showed some characteristics of the wild northern dog, and some had traces of their wolf ancestry. Gwen was purely wolfish in her wildness and impatience of restraint, and her son Terror was like her. It was attempted once to muzzle Gwen, after some penguin hunting exploit, but she nearly went mad in her efforts to get rid of the muzzle, turning and twisting so rapidly that the eye couldn't follow her, and she had to be freed from it.

The struggle for kingship was not so sanguinary as is common with such dogs. Old Scamp's authority was never seriously disputed, though Trip and Wolf occasionally fought him. Scamp was not the heaviest or strongest dog, but he downed them all by his vehemence. The females were very jealous, and were apt to eat one another's litters. They make very good mothers. A litter was born on the *Nimrod* while going south. After we landed one of these pups was killed by the fall of a house during a blizzard. The body was flung out on the hill-side, some distance away. The mother, Possum, discovered it and nursed it for a whole day. Though somewhat fierce and quarrelsome among themselves, the dogs were very friendly to man. They would take the severest beating when they had been misbehaving, and be friends the moment it was over.

The young pups, born in the Antarctic, were very self-reliant little things. When very young they used to issue from their shelter, run out in the snow and bark in defiance of everything. They got their drinking-water for a long time in the form of snow, and when summer came and water was given to them, they did not know what to make of it at first. They soon learned to appreciate it, and to consider it as the most valuable of all things, for of food there was always super-abundance. In late summer, when the snow had nearly all disappeared, and the lake by the house was frozen, then came a sort of water famine. We had no time to break the ice and give them water more than once a day. They got pretty thirsty in the sun, and some of them showed a good deal of intelligence in asking for and dealing with water.

One day, on going out of the house, I heard the dog Roland barking furiously. Roland was tied at a good distance from the house, and for some time I paid no attention to the barking. When at last I looked in the right direction, Roland picked up her water tin and waved it frantically over her head.

An old dog, Wolf, was so convinced of the value of water that, when a bowlful was given to him, he did not drink it all at once, but tried to keep a reserve for future use. He tried to bury it, as dogs habitually do with food. He carried the bowl carefully in his mouth, placed it in a hollow, and covered it up with gravel. Of course he lost the water, but the attempt deserved success.

In their hunting they showed much intelligence. They took much greater liberties with the helpless Weddell seal than with the more active and aggressive crabeater, and as for the sea-leopard they seemed to know enough to leave him alone altogether. They hunted in couples or parties, and so got the better of the penguins. The penguins appear to be provided with efficient defensive weapons, but the dogs learned where to seize them safely. While one dog kept the attention of a penguin in front, another slipped round behind it and snapped at its leg.

Some bolder dogs attacked the penguin in front, waiting for chance to bite at the neck. One snap finished the fight, the dogs usually leaving the disabled bird to chase a fresh one.

After learning that penguin hunting was a punishable offence the dogs became very cunning. They slipped away on their hunting expeditions, without attracting attention, and the first intimation we had of it was the distant barking as they surrounded some poor bird. Though they could have but little experience of the effect of shooting they stood in wholesome terror of a gun. Daisy and Gwen especially knew that a gun could hurt at a distance, and that flight was useless, so they slunk home when a shot was fired, keeping cover as far as possible, and hid below the house.

Daisy was the most inveterate hunter, and regularly took her children away to teach them to become self-supporting. At last her propensity led her, there is no reason to doubt, to a painful death. She took her whole family once out hunting on the pack-ice. The pack was blown out and the dogs were given up for lost. Some days later they all came back, having evidently had a trying experience, their faces matted with blood and sea salt. Emboldened perhaps by this escape, Daisy again went hunting on the pack, taking Roland with her, and again the pack went out to sea. Roland returned, but Daisy never did.

When taken for a walk through the rookery, the dogs bore themselves with a most virtuous air, looking with indifference at the penguins as if they had no idea what they were good for. When detected penguin-worrying, old Scamp made for his kennel, and sat there pretending he had never been away, looking very innocent, overlooking the fact that he was dripping with blood.

BIOLOGY

Marine Biology

Towards mid-winter the ice in a little bay, bounding Cape Royds on the south, was strong enough to permit of dredging at depths of from six to twenty fathoms. The bottom here was a fine black mud, with larger and small pebbles of kenyte in it. In the mud were embedded large shell-fish, the fragile purple Pecten, the siphon-bearing Anatina, and others. On the pebbles were growing bush-like sponges, and large sea-anemones adhered to them. One or two reddish-brown sea-weeds were plentiful and once or twice we brought up great turnip-like fixed tunicates. Ugly and greedy big-headed fishes (*Nototheia*), and equally greedy carnivorous whelks of a large size (*Neobuccinum*) crowded to any bait put down.

Among the sponges and seaweeds were numerous other forms of life, tube-dwelling worms with beautiful flower-like heads of tentacles; delicate shells, almost invisible to the naked eye, many larger and smaller crustacea, though none of any considerable size; hairy worms like the sea-mouse with a double row of phosphorescent lamps, flashing in succession from head to tail. When captured these worms have a spiteful habit of breaking themselves up into small pieces.

In this region there is a prevalence of yellow and orange-red colours. The commonest sponges are yellow; so is the sea-anemone. Some of the corals are very bright orange.

Most of the crustacea and many worms, star-fish, &c., are orange-red or yellow. All these are bottom forms of life. With the available methods of collecting very little was got in the open water of the sea, away from the bottom, only a few small crustacea, some diatirus, and occasionally a few sea-butterflies (*Pteropods*) of large size and red colour.

The phosphorescence remarked in some of the bottom worms was also found in the copepods of the open sea. The phosphorescence is displayed by cold-blooded animals, living in a temperature always some degrees below the freezing-point of fresh water, and it is shown equally throughout the winter.

Dredging at greater depths than twenty fathoms was rarely possible, owing to the nearness of open water in McMurdo Sound, always within a mile of the camp. From this cause we did no deep dredging at all, only on one or two occasions at nearly one hundred fathoms. From the mouth of the bay down to the depth of one hundred fathoms the bottom sloped steeply. Whether from this cause or owing to the strong current in the sound, there was no mud in this zone of the bottom. In the shallower parts there were large and small kenyte pebbles, but at fifty fathoms and upwards no pebbles were got. The bottom appears to be carpeted with a dense growth of living things, as if the dredge merely bit and was immediately drawn up it was usually full of stuff.

In this deeper region the animal life differed greatly from that in the muddy bay, though many kinds were found throughout both places. Here we first got the long-legged sea-spiders (*Pycnogonida*), glassy sponges, the white shells of Lima, the delicate lace-corals, &c. The sponges were especially abundant and in some variety, though rarely of much beauty. One glassy sponge resembled an egg, with bundles of long glassy spicules projecting at regular intervals from the smooth surface.

In this region there was less orange and yellow colouring, the tendency being to white. Most of the glass sponges and many of the horny sponges were white or pale cream-coloured, and the Lima shells and lace corals were white. There were still some yellow sponges and most of the sea-spiders were reddish. At this depth we got the same fish as in the bay. A fish-trap, baited, was put down at twenty-five fathoms. It caught some dozens of pretty large big-heads the first time it was drawn, and almost nothing afterwards, unless left for a considerable time. This seems to indicate that they are pretty plentiful, but that they grub very closely among the sponges and don't travel fast or far, so that the first haul exhausts the region immediately around the trap. When the trap was brought up great red worms hung like ribbons, one yard, or even two yards, below the trap. These could contract until they would lie on the palm of the hand.

*

Appendix Two

GEOLOGICAL OBSERVATIONS IN ANTARCTICA BY THE BRITISH ANTARCTIC EXPEDITION 1907–1909

By Professor T. W. EDGEWORTH DAVID, B.A., F.R.S.
and
RAYMOND E. PRIESTLEY, Geologist to the Expedition

Introductory.

THE conclusions provisionally adopted in these notes are based on the geological collections and observations obtained by the Southern Party, the Western Party, and the Northern Party of our expedition as well as by the whole party, when in winter quarters at Cape Royds.

The only determinable fossil as yet found in the great Beacon sandstone formation of Antarctica, the piece of coniferous wood, figured in these notes, was obtained from the collection made by the Southern Party.

As a result of the explorations, chiefly by Nordenskjold, Larsen, Gunnar Andersson, Bruce, Charcot and Arçtowski, we now know the following about the portion of Antarctica south of America.

In parts of Graham Land there must be a foundation platform of gneiss and gneissic granite, as boulders of these rocks, several metres in diameter, are found deposited on the plateau of Seymour Island, to the east of Graham Land, as recorded by Gunnar Andersson.

In 1903 the French Antarctic expedition, under the command of Dr. Charcot, landed on the South Shetland

Islands, and after exploring Palmer Archipelago and Gerlache Strait wintered at Wandel Island. An interesting and detailed account of the geological specimens collected has been given by Dr. E. Gourdon.* Amongst the rocks described are hornblende granites, quartz diorites, uralitic gabbros, trachyandesites with hornblende and mica, dacites and andesites with associated tuffs, labradorite rocks, diabase basalts, micro-granites with pyroxene and soda-hornblende. He also describes crystalline schists, quartzites and quartz veins.

Dr. Gourdon concludes that these rocks are part of the eruptive series of the chain of the Andes. Nordenskjold is of the same opinion in regard to the eruptives of Graham Land.

In the South Orkneys fossil graptolites, associated with radiolarian jaspers, were discovered by Bruce's expedition. These prove the existence there of older palæozoic rocks, considered to be of Ordovician age. As far as we can learn there is as yet no evidence of the presence, in that region, of rocks older than Ordovician, unless some of the crystalline rocks of south-western Graham Land, such as those of Borchgrevink Nunatak, &c., antedate that period.

The abundant fossil plants discovered by Nordenskjold's expedition at Hope Bay, at the north-eastern end of Graham Land, show that in that region, now continuously covered with ice and snow, there existed in Jurassic times a rich and diversified flora embracing ferns, cycads and conifers. Amongst the plants found at Hope Bay the genera *Sagenopteris, Thinnfeldia, Cladophlebis, Pterophyllum,* and *Otozamites* have been recorded as well

---

* Expedition Antartique Française, 1903–05, commandée par le Dr. Jean Charcot. Sciences Naturelles; Documents Scientifiques Géographie Physique—Glaciologie, Petrographie par E. Gourdon, Docteur-ès-Sciences de l'Université de Paris.

from the Trias-Jura rocks of Eastern Australia and India, some of the forms being found also in South Africa and in the Argentine Republic. The distribution is shown on the following table:

	S. Africa	India	Argentine	Australia
Sagenopteris * . . .	—	x	—	x
Thinnfeldia . . .	x	x	x	x
Cladophlebis . . .	x	x	—	x
Pterophyllum . . .	—	x	x	x
Otozamites . . .	—	x	—	x

So far no trace has been found in this flora of any representatives of the *Glossopteris* Flora of Gondwana Land, such as the *Phyllotheca* discovered by Gunnar Andersson in the Falkland Islands. Evidently in Jurassic time a mild and a moist climate prevailed in Antarctica.

The abundance of cretaceous *Ammonites* collected by the Nordenskjold expedition at Snow Hill Island, to the east of Graham Land, points to a continuance of mild conditions into cretaceous time. The fossil Araucaria, Beech, &c., unearthed by the Nordenskjold expedition at Seymour Island, adjoining Snow Hill Island on the northeast, prove that these mild conditions were further prolonged into some part of Tertiary time.

In marine strata, also of Tertiary age, and considered by Wilckens † to belong to Upper Oligocene or Lower Miocene, the Nordenskjold expedition found numerous bird bones since referred to five new genera, of penguins ‡

* This list has been kindly supplied by Mr. W. S. Dun, Paleontologist Geol. Sur. N.S. Wales, and of Sydney University.

† Die Meeresablagerungen der Kreide—und Tertiärablagerungen in Patagonien. Neues Jahr. f. Min. Beilage-Band 21. 1905.

‡ These are stated by Gunnar Andersson to be *Anthropornis Nordenskjoldi, Pachyteryx, Espheniscus Gunnari, Delphinornis Larsenii* and *Ichthyopteryx gracilis,* v. Bulletin of the Geological Institution of the University of Upsala. Vol. vii., 1904–5, No. 13–14, p. 45.

besides two vertebræ of a big mammal, referred to the genus *Zeuglodon.* The marine fossils associated with these remains enabled Wilckens to come to the above decision as to the geological age of the formation.

At Cockburn Island, to the north of Seymour Island, Gunnar Andersson describes a Pecten conglomerate 160 metres above sea-level. This marine formation he considers to be probably of Pliocene age, and the equivalent of the Parana beds of the north of the Argentine Republic or of the Cape Fairweather beds of Southern Patagonia.

Nordenskjold's expedition proved that during the maximum glaciation, in late Geological time, the inland ice rose 300 metres higher than it does at present, in the neighbourhood of Borchgrevink Nunatak, at the south-east end of Graham Land. This was proved by the maximum height of erratic boulders found on the slopes of the nunatak, above the present level of the surface of the inland ice sheet. Gunnar Andersson mentions the occurrence of raised beaches at Cockburn Island and also at Sidney Herbert Sound.

These pieces of evidence prove an emergence of the land, since the maximum glaciation, to the extent at all events of a few metres, possibly as much as forty metres.

In the portion of the Antarctic visited by the German expedition, 1902, under the leadership of Professor E. von Drygalski, the following information has been obtained:

In latitude 66° 48′ South, longitude 89° 30′ East, there rises at the edge of the inland ice a ridge-shaped remnant of a volcanic cone, the Gaussberg. This attains a height of 366 metres above the sea, and is formed of leucite-basalt tuff and leucite-basalt rich in olivine, lumps up to the size of one's fist being found in the lava. The top and slopes of the Gaussberg, as recorded by Dr. Philippi, are strewn with erratics.* These are formed of whitish garnet-bearing gneiss, a darker biotite-gneiss, mica-schist, fragments of red quartzite, &c. The crystalline rocks are considered to be probably of Archæan age.

In the Victoria Land region of the Antarctic the researches of Ross, Borchgrevink, and above all of Captain R. F. Scott and the geologist of the *Discovery* expedition, H. T. Ferrar, prove that there is developed in that region an ancient complex of gneisses and gneissic granites, with mica-schists, calc-schists and quartzites, and that these rocks are capped for a great distance by a formation almost horizontally bedded, called by Ferrar the "Beacon sandstone." A little argillaceous limestone was observed by him associated with this sandstone. Ferrar found plant remains in the sandstone, but in such an altered condition that they could not be determined. Ferrar has given a detailed and very valuable description of the geology of Victoria Land and Ross Island explored by him on this expedition. The petrology of the rocks collected has been worked out by G. T. Prior.†

Amongst volcanic rocks are comprised hornblende-basalts, olivine-basalts, dolerites, basalt tuffs, kenytes, phonolitic trachytes and phonolites. Amongst the foundation rocks of South Victoria Land, Prior records crystalline limestones with chondrodite, gneiss, granites, diorites, camptonites, kersantites and banakite. Amongst the sedimentary rocks he refers to sandstones, somewhat carbonaceous, as well as black shaly to slaty rocks.

The volcanic rocks, as pointed out by Prior, are closely allied in chemical composition and mineral constitution

* Veröffentlichungen des Instituts für Meereskunde und des Geographischen Instituts an der Universität, Berlin. Heft. 5 Octr. 1903. Deutsche Südpolar-Expedition auf dem Schiff "*Gauss.*"

† National Antarctic Expedition, 1901–1904, Natural History, Vol. i., Geology. British Museum, 1902.

to the volcanic series described by Dr. P. Marshall, from the neighbourhood of Dunedin, New Zealand. It may be noted that these volcanic rocks are developed partly along the coast-line of Victoria Land, partly in islands arranged in lines subparellel to this coast-line.

It is worthy of comment that the volcanic zones of Victoria Land were not definitely traced by Mr. Shackleton in the ranges reached by him in his furthest south journey this year. No trace whatever of volcanic rocks was noticed by him, either *in situ* or in the moraines of the vast coast range which bounds the Great Ice Barrier on its south-west side, near the latitude of 84° to 86° South.

Physical Geography

These observations relate wholly to the region between the meridians of 170° East and 150° West. The shoreline in this region of the Antarctic continent is deeply indented by the Ross Sea. This commences just south of the parallel of 70°, and extends to the parallel of 78° South. Ross Sea is bounded at its east side by dense belts of pack-ice and low snowbergs, which prevent any view of the coast-line being obtained excepting near the extreme south-east corner of Ross Sea; there, Captain Scott, in the *Discovery,* found a new range of mountains rising from a land afterwards known as King Edward VII Land. Southwards Ross Sea is bounded by the cliff of the Great Ice Barrier, which has an extent of about 470 miles in an east and west direction. This cliff averages about 150 ft. in height. In places it sinks to nearly sea-level in low gullies. The surface beyond this Great Barrier, except for certain broad shallow undulations and small snow ridges (sastrugi), is practically level. Mr. Shackleton, on his southern journey, proved that it extends southward for at least 350 miles. Westwards the

Great Barrier cliff terminates in high-pressure ridges against Cape Crozier, the easternmost point of Ross Island.

Ross Island with its towering volcanic cones rises like some vast castle at the end of this huge white wall. It is formed of four large volcanic cones, Mounts Terror, Terra Nova, Erebus and Bird. The three first volcanoes appear to be situated on an east and west line of fracture. Another fracture line probably passes in a southerly direction from Mount Bird through Mount Erebus. Thus, Erebus may be said to be at the junction of two important systems of earth fracture. Still further south several smaller craters are situated on what may be termed the Erebus Fracture Zone, including that of Crater Hill, near Hut Point, the winter quarters of the *Discovery* expedition. Still further south are the volcanic islands, White Island and Black Island, and somewhat to the south-south-west Mount Discovery with the long volcanic promontory trending from it to the east-south-east, known as Minna Bluff.

In the gaps between these islands and promontories the mass of the Great Barrier moves slowly, but surely, seawards towards the narrow south-westerly prolongation of Ross Sea known as McMurdo Sound. Pressure ridges of ice in this part of the Great Barrier, as well as actual measurements taken, prove that this part of the Barrier is moving seawards, both to the west and to the east of Ross Island.

McMurdo Sound is bounded on the south by the low terminal cliff of the Great Barrier only a few feet in height. This low ice cliff extends westwards across McMurdo Sound, for a distance of about thirty miles, to the magnificent coast range of Victoria Land. Majestic peaks of gneiss, granite, sandstone and limestone capped by eruptive rocks rise almost sheer from the coast

PRIESTLY BESIDE AN ERRATIC GRANITE BOULDER, LYING ON KENYTE AT CAPE ROYDS

GRANITE ERRATIC AT CAPE ROYDS

to altitudes of from 8000 up to 12,000 ft. Throughout its entire length from Cape North and the mountains recently discovered by our expedition further west, down to the parallel of 86° South, a distance of about 1100 miles, the ranges form a slightly elevated border to an inland plateau. The continuity of these plateau ranges is interrupted at intervals by wide valley-like depressions, occupied by vast glaciers. These glaciers slope steeply to the sea, or to the surface of the Great Barrier, and are heavily crevassed. Further inland they ascend by gentle slopes, interrupted occasionally by ice-falls, to the névé fields of the plateau.

As one traces the coast-line northwards, from opposite Ross Island in the direction in which the Northern Party travelled to the South Magnetic Pole, one encounters some very remarkable features which materially modify the form of the coast-line. The first of these is called on the chart of the Antarctic Ocean, prepared from observations under the direction of Captain R. F. Scott, the Nordenskjold Ice Barrier Tongue. It is about six miles in width, and projects twenty miles or more seawards from the coast-line. There is reason to suppose that this Barrier, as well as the one just to be described, is floating at its seaward extremity.

North of the Nordenskjold Barrier is the Drygalski Barrier or Ice Tongue. This is a huge glacier actively moving forwards into the sea. It is a true glacier at its landward end, with immense séracs, ridges and crevasses. The portion which projects seawards beyond the coast is about twenty miles in width, and thirty miles in length. Towards its seaward end, and also on its northern side, where it receives the bulk of the snow drifted by southerly blizzards, it partakes rather of a flat-topped barrier type than of the glacier type with its characteristic rugged surface.

Just inland to the north of the Drygalski Ice Barrier Tongue is a fine bay, Terra Nova Bay, inland from which, near its northern end, rises the majestic Mount Nansen. This is a flat-topped mountain, obviously capped by sedimentary rocks, and as a matter of fact, it has shed lumps of limestone and sandstone into the moraines beneath it.

Still further north the beautifully symmetrical volcanic cone of Mount Melbourne attains a height of 8337 ft. The volcanic rocks with which it is associated trend sharply to the south-west, terminating in the high rugged cliffs of Cape Washington.

To the north-east of Mount Melbourne is the deep indentation known as Wood Bay, and thence the coast bends abruptly to the east. It would seem indeed as though Mount Melbourne is probably situated on some east and west line of earth fractures, like Mounts Terror and Terra Nova. The coast-line then trends nearly north again, forming the west boundary of Lady Newnes Bay. Then it trends once more east to Cape Jones, an extinct volcano. Just off Cape Jones lies the large volcanic island, Coulman Island. From here the coast again trends chiefly northerly to Cape Adare. Volcanic rocks are extensively developed at this cape, but the ranges inland are formed of older rocks, such as granite, gneiss, schist, slate, &c., apparently still capped by the "Beacon sandstone" formation. The island known as Possession Island—also volcanic—lies to the south-south-east of Cape Adare. This long cape, where the *Southern Cross* expedition, under Borchgrevink, wintered, forms the north-east side of Robertson's Bay. From this bay the coast trends at first north-westwards for about 120 miles to Cape North. Near here, some hills, a little distance back from the coast, give one a strong impression of their being of volcanic origin, though it is possible that they are merely out-lying sugar-loaf hills, relics of a dissected plateau.

It was clear from the sight which we obtained of the part of the coast, beyond Cape North on March 8, 1909, that the hills were high, having an altitude of from 6000 to 7000 ft., as measured by sextant, and that they formed the abrupt termination seawards of a deeply denuded high plateau. This plateau is undoubtedly a northern prolongation of the one travelled over by the Northern Party of our expedition on their journey to the South Magnetic Pole. It is also certainly continuous with the plateau traversed by Captain R. F. Scott in his western journey, in 1903, and it is proved now that it is part of the same plateau to which Lieutenant Shackleton led the Southern Party, and over which they travelled to an altitude of 10,000 ft. when they reached their furthest point 88° 23′ South.

Throughout the whole of this magnificent coastal range the evidence of past ice action is extremely clear. Most of the valleys are wide, but a few, like the Ferrar Glacier Valley, are narrow. But whether wide or narrow, their rocky sides show most impressively the abrasive work of the great ice plough, indeed the rocky slopes bounding these glaciers are almost as even as the banks of a deep railway cutting. One is at once struck with the entire absence of those re-entering spurs and angles so characteristic of river-worn valleys.

A curious feature, already mentioned by Ferrar, is the development of an extensive coastal shelf, for at any rate about 150 miles northwards of the latitude of Cape Royds. This coastal shelf may be possibly ascribed to step faulting, but it is also possible that it may be due to an over-riding of the foothills of the coast range, and a ploughing of them out by the former great ice sheet of the Ice Barrier, at a time when its surface was fully

1000 ft. higher than it is at present, and when it spread northwards into the Ross Sea, probably at least 100 to 200 miles north of its present seaward termination.

As regards the reason for the plateau of Victoria Land terminating in such steep mountain slopes eastwards, it is of course possible, as Ferrar suggests, that this is due to a heavy fault or series of faults running parallel to the shore-line. Certainly the scenery, particularly in the neighbourhood of Mount Nansen, and between that and Mount Melbourne, suggests a comparatively recent change of base, down to which the base level forces have recently been working. In fact, these glacier-cut valleys appear to us to be distinctly young in their origin. The soundings in Ross Sea off this coast have some interest as bearing on this question, as also the presence of raised beaches in several places along the coast, and on Ross Island. McMurdo Sound, from Ross Island to the coast of Victoria Land, is only thirty miles wide, and yet the sea is nearly 500 fathoms in depth within a few miles of the coast. As evidence of crust movements raised beaches may be quoted. On Ross Island they were traced by us up to altitudes of 160 ft., and organisms were found in these beaches such as are found now living in the coastal waters, so that they probably indicate an uplift since the deposition of these organisms of a good deal more than 160 ft. At the Ferrar Glacier on the mainland, raised beaches extended up to at least 50 ft. above sea-level, and they reached apparently to an altitude of 20 to 30 ft. on the coast south-east of Mount Larsen, 200 miles further north. It is possible that the latter may be due to upthrust of the marine sediments by glacier ice. These changes in the level of the shore-line have taken place in quite recent geological time. It is, of course, possible that such

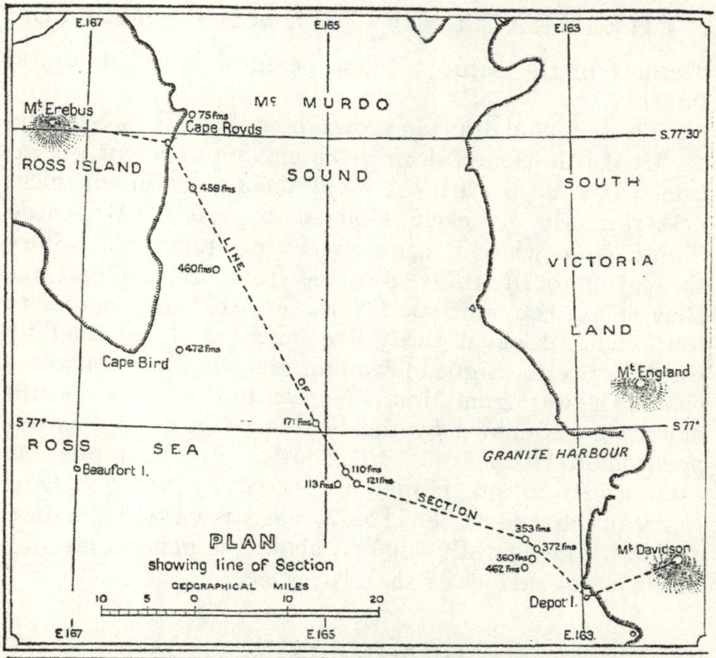
E.167
E.165
E.163
Mt Erebus
75 fms
Cape Royds
Mc MURDO
SOUND
S.77°30'
ROSS ISLAND
459 fms
LINE
SOUTH
VICTORIA
LAND
460 fms
472 fms
Cape Bird
OF
Mt England
S 77°
171 fms
S 77°
ROSS
SEA
GRANITE HARBOUR
Beaufort I.
110 fms
113 fms
121 fms
SECTION
353 fms
372 fms
360 fms
462 fms
Mt Davidson
PLAN
showing line of Section
GEOGRAPHICAL MILES
10
5
0
10
20
Depot I.
E.167
E.165
E.163.

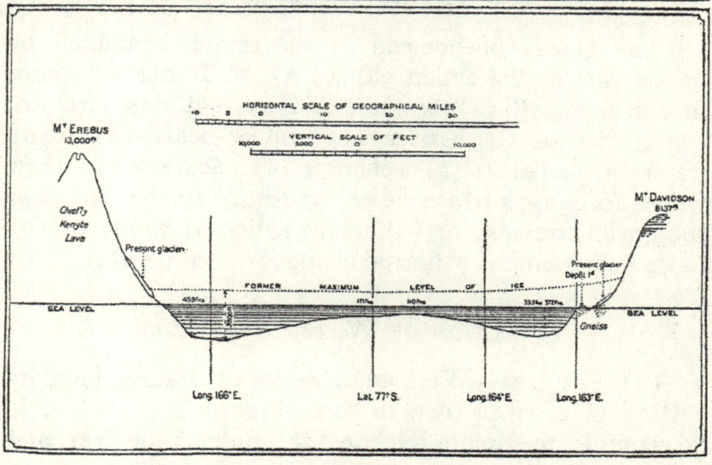
HORIZONTAL SCALE OF GEOGRAPHICAL MILES
10
5
0
10
20
30
VERTICAL SCALE OF FEET
10,000
5000
0
10,000
Mt Erebus
13,000 ft
Chiefly
Kenyte
Lava
Present glacier
Mt Davidson
8137 ft
Present glacier
Depôt Id
FORMER MAXIMUM LEVEL OF ICE
459 fms
171 fms
110 fms
353 fms
372 fms
SEA LEVEL
SEA LEVEL
Gneiss
Long. 166° E.
Lat. 77° S.
Long. 164° E.
Long. 163° E.

changes might occur without being due to geological faulting.

The sectional drawing shows the state of Ross Sea, and of McMurdo Sound during the maximum recent glaciation. It indicates that the ice attained a maximum thickness formerly of nearly 4000 ft. in parts of McMurdo Sound from which it has now entirely retreated. Since the voyage of Ross, in 1841, the front of the Great Ice Barrier has retreated southwards in McMurdo Sound to the extent of about thirty-five miles, as determined by Captain Scott. Signs of waning glaciation are conspicuous all the way from Mount Nansen to the furthest south mountains examined by the Southern Party of this expedition in latitude 85° 15′ South. For example, the summit of Mount Hope, discovered by the Southern Party in latitude 83° 33′ South, was strewn with erratics, at an altitude of fully 2000 ft. above the general level of the adjacent surface of the glacier ice.

GLACIOLOGY

The glacial phenomena of the region examined by us are due to the action either (a) of Water substance in the form of (1) sea ice, and ice-foot or shore ice; (2) glacier ice; (3) barrier snow and ice-fields; (4) inland ice and névé fields; (5) icebergs; (6) pack-ice; (7) thaw water forming surface lakes, and surface, englacial or subglacial streams; or (b) to the action of wind; or (c) to that of seasonal or diurnal changes of temperature.

(a) Action of Water Substance

(1) Sea Ice.—We made a series of observations, by cutting holes from time to time through the sea ice, to ascertain its maximum thickness throughout the area, and

also its methods of freezing. The maximum thickness measured by us amounted to about 7 ft., in the case of ice formed, in a sheltered position at Backdoor Bay near our winter quarters, between the middle of March and middle of September 1908. Ice had formed over the same area, a little earlier in March, to the depth of a few inches, but this was cracked up, and drifted away by the blizzards. This thickness of 7 ft. of ice was no doubt increased between the middle of September and early in December.

Ferrar states that the maximum thickness of sea ice which formed during the year 1903 at Hut Point, McMurdo Sound, was 8 ft. 5¾ in. We observed that in places the sea ice was fractured, through pressure of wind and tidal currents, and the broken slabs were forced over one another forming pressure ridges, from 10 to 20 ft. in height.

It was interesting to note the effect on the sea ice of a sudden fall of temperature. The contraction following on such a fall would put the sea ice, especially at its surface, into a high state of tension, and from time to time the surface would crack open with a loud report. These contraction cracks gaped to a width of 3 to 6 ft., and the sea water between the walls of the crack, of course, began to freeze over. Frequently after ice had formed to the thickness of a few inches a rise of temperature would expand the ice. This expansion would tend to expend itself on all weak spots, especially on the planks of thin ice formed between the walls of the contraction cracks. These would be buckled into small overfolds, until at last they became cracked through excess of thrusting, and overthrust faults resulted. In many of these cracks this process was frequently repeated.

Another feature worth noting in the sea ice is, that owing to the great difference between the temperature of the sea water below the ice and that of the air above it, as soon as a contraction crack opened, the sea water appeared to be steaming. A wall of thick vapour would rise along the whole length of each crack. To this the term of frost-smoke is sometimes applied. The water vapour, as it rose, was being constantly condensed and deposited on the walls of the narrow cracks, so that gradually the interspace became filled with ice, and not infrequently a ridge of soft ice would be built up along the line of the old crack, to a height of 6 in. or so above the general level of the surrounding ice. As the ice was often traversed by a perfect network of these cracks, the resulting ridges gave the ice-surface the appearance of Indian paddy-fields, with their dividing "bunds," or mud walls.

The sea ice was usually separated from the shore-ice, or ice-foot, by one or more well-marked tide-cracks. In McMurdo Sound, near our winter quarters, the tidal range of from 2 ft. to 3 ft. was quite sufficient to fracture the ice in contact with the land. The seals took advantage of these tide-cracks, and used them as blow-holes.

The chief geological work done by the sea ice, as far as we could ascertain, was the transport seawards of wind-blown rock detritus lodged on the shore-ice and ice-foot, in the manner about to be described.

Ice-foot or Shore-ice.—On first arriving on the shores of the Antarctic after the breaking up of sea ice, towards the end of the summer, one is puzzled to account for the low cliff, part snow, part ice, which almost everywhere fringes the coast and so makes landing from a boat difficult. This fringe is seen to be made up partly of ice at its base, resembling somewhat the stalagmites of limestone caves, partly of layers of compressed snow, in some cases alternating with bands of sand and gravel. The ice-foot generally rises to a height of 6 to 10 ft.

above sea-level. It is usually in the form of a flat narrow terrace from 20 to 100 ft. wide, sometimes in that of a sheer cliff, occasionally as much as 80 ft. to over 100 ft. in height, the summit of which ascends inland in a more or less steep snow slope. At its base, in summer, the ice-foot is almost invariably undercut by the sea, and from the overhanging roof thus produced there depend vast numbers of beautiful icicles. These icicles have, of course, been formed from the wash of the waves, and the sea water in the process of being frozen has extruded its brine, the salinity of which is such that the solution cannot freeze at a temperature above zero Fahr. These icicles are generally moist, and the moisture, in the form of concentrated brine, works downwards, under the influence of gravity, to the tips of the icicles, which thus become sticky. Hence when a blizzard springs up and drives snowflakes against them, the flakes stick on chiefly at the tips and gradually build out those foot-like structures which we have termed foot-stalactites, and which are illustrated.

During the winter of 1908 and the succeeding spring and summer, we were able to see clearly the mode of growth of the ice-foot. After the sea surface had been frozen over snow carried by the wind from the land, or from the surface of the Great Ice Barrier formed drifts of greater or less thickness over the sea ice close inshore. These, at their shoreward end where the cliffs are 80 or 100 ft. in height, may form drifts of equal thickness with the height of the cliff. These drifts, of course, thin out seawards. They are stratified and contain numerous dark bands formed of chips of rock, broken crystals of felspar, &c. When, during the summer, strong blizzards disrupt the sea ice, large rafts of ice are dislodged from near the shore, and these carry away on their surfaces portions of the old snow-drifts. As the work of destruction proceeds even the thick landward portions of the snow-drifts are cracked off in large slices, and float seawards, and thus in summer time is formed that almost universal low cliff known as the ice-foot. During heavy weather when the sea is open the waves wash over the lower portions of the ice-foot, with the result that it is being constantly bathed in salt water, which freezes in successive layers on its surface. Thus, when the temperature is low the old masses of snow-drift, of which the upper part of the ice-foot is formed, become cased over with ice much in the same way as snowbergs become encased as the result of their being splashed by sea spray.

(2) Glacier Ice and Névé.—The glacier ice of the portion of the Antarctic area examined by us either terminates inland in glaciers, some of which are hanging glaciers, and some piedmont glaciers, or ice-slabs, or it comes down to the sea where it is broken off from time to time to form true icebergs, close to the shore-line; or—and this is a feature emphasised already by Mr. H. T. Ferrar—the ice may advance for a considerable distance from the shore-line into the sea, in some cases from 20 to 30 miles, probably far more in the case of the Great Ice Barrier, and thus discharge icebergs from its sides as well as its snout. Such glaciers were described by Ferrar as piedmonts-afloat, and we propose to retain this term for them.

Glaciers.—A good example of this type was to be seen a little over two miles southerly from our winter quarters, just south of Cape Barne. The glacier there, called by us the Cape Barne Glacier, terminates seawards in a cliff about 100 ft. in height, and some three miles in length. It has its source in the névé fields of the western slope of Mount Erebus. These are fed, not only by new falling snow, but also by large quantities of drift snow swept over by the south-east blizzards on to

Summer effect on a Berg; Icicles forming

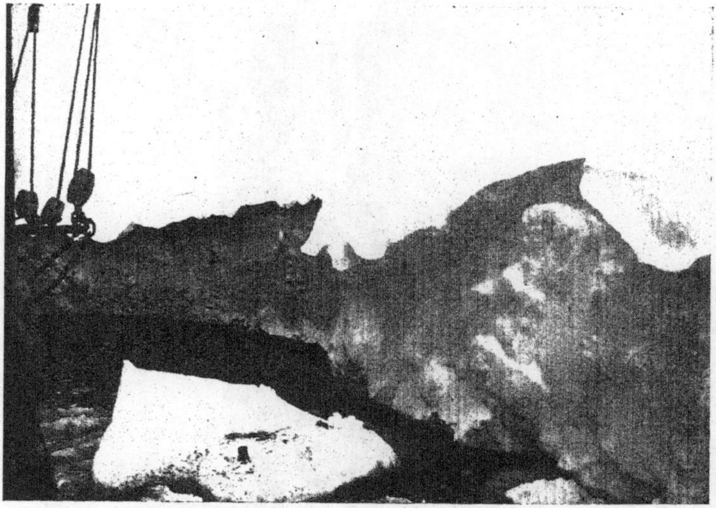

The Barrier Edge south of Hut Point, after the Sea Ice had broken away

* Two erratic Boulders of Granite on the Slopes of Erebus

this, the lee side, of Erebus. The glacier was considerably crevassed at its seaward extremity, and passed up gradually, at a distance of some four or five miles inland, into the névé field. This glacier was not moving actively, as we never observed any trace of buckling or crushing of the sea ice, where it abutted against the foot of the glacier cliff. Had there been any appreciable forward movement it could not have failed to ridge up or crush the opposing sheets of continuous sea ice. At the same time the crevassed state of this glacier ice proved that movement was still in progress.

A glacier of a Greenland type, also on a large scale, is the Mount Nansen Glacier, occupying the wide depression between Mounts Nansen and Larsen. This great glacier is from 12 to 20 miles in width, and 60 to 70 miles in length. It is very heavily crevassed, and its surface is extremely irregular towards its seaward end. Where the surface falls steeply, it has raised immense pressure ridges in the sea ice along the shore, and bristles with hummocks and séracs. So difficult was this surface for sledging that we were forced to abandon it, after attempting to take our sledge by way of this glacier on to the Magnetic Pole plateau.

On striking the upper end of this glacier, some 60 miles inland, we found that there was still a little ice present here and there underneath the wide névé field. This névé field spread out into a wide plain, and for a considerable distance before reaching the Magnetic Pole, the latter being over 220 miles inland at right angles to the coast-line—our horizon on all sides was bounded by these same vast névé fields. It is obvious that the Mount Nansen Glacier is moving steadily seawards, as shown by the great pressure ridges which it has raised in the sea ice opposing its advance. It must, therefore, still form an outlet for the névé-formed ice of the inland plateau.

The interesting question here suggests itself, Is there sufficient snowfall annually, on the area of the névé fields drained by this glacier, to compensate for the ice which is lost by ablation, or by being discharged as icebergs into the sea? Until more data are available no accurate quantitative answer can be given to this question. At the same time it may be remarked that there is a tolerably heavy snowfall along this part of the coast, and for a distance of at least 50 miles inland. Portions of the high plateau, at a greater distance inland from the shore than 50 miles at present probably receive only a very small snow supply. It may be doubted whether the surface of this névé field far inland is not on the whole being reduced in level through the snow being drifted off it by the wind, or removed by the slow process of ablation.

It is interesting to note that in front of the termination of the Mount Nansen Glacier there is an immense old moraine of the nature apparently of a medial moraine. We could trace this for fully 23 miles in advance of the present glacier snout. It follows that in comparatively recent geological time the Mount Nansen Glacier has retreated by at least the amount quoted above.

Piedmont Glaciers on Land.—A curious feature observed along the greater part of the coast-line of Victoria Land, from near Mount Discovery up to the Drygalski Ice Barrier Tongue, is the development, on the great coastal shelf, at an altitude of about 1000 ft. above sea-level, of a massive covering of blue glacier ice. This, in some cases, reaches the sea and breaks off to form bergs. In other cases the sheets do not reach the sea, and, therefore, are probably on the wane. For the latter Ferrar suggests the appropriate name of ice-slabs.

Several theories might be advanced to account for them. They may represent actual relics of the old Barrier ice sheet, which once filled McMurdo Sound and Ross

Sea for probably fully 100 miles north of Ross Island. Another view is that they may be local developments of ice resulting from a coalescing of a number of small névé fields developed in the cirques among the foot-hills of the plateau ranges. These foot-hills are frequently as much as 10 to 20 miles back from the edge of the coast-line.

PIEDMONT GLACIERS AFLOAT.—Three well-marked examples of this type of glacier came under our notice. The first was Glacier Tongue, between our winter quarters and the old winter quarters of the *Discovery* at Hut Point. Glacier Tongue, as shown on the Admiralty Charts and the Reports of the *Discovery* expedition, is an elongated mass of ice stretching from the shore-line into the sea for a distance of about five miles. It has a width of about half a mile near its seaward end, and about a mile where it rests against the land. Both on its north and south side the Tongue is deeply indented with a number of bays. Its height above sea-level varies from about 40 ft. up to nearly 100 ft. While the *Nimrod* was lying alongside this remarkable piedmont in February 1908, Captain England took soundings at about a mile east of its seaward end, and got a depth of 157 fathoms. As the maximum height of the glacier above sea-level does not here exceed about 40 ft., and the sea is 940 ft. deep, if the ice were aground it would have only one twenty-third of its volume above water, which of course is physically impossible. We must, therefore, conclude that this part of Glacier Tongue is afloat. At the same time it should be mentioned that alongside of this glacier there are traces of cracks, which some observers have considered to be tide-cracks. There may be true tide-cracks near the shoreward end of the glacier, but we were not satisfied that the cracks noticed near its seaward end were really of the nature of tide-cracks. While waiting for the arrival of the Southern

Party, the *Nimrod* lay in a snug natural dock, formed by one of the bays on the north side of this glacier tongue, at about a mile from its seaward end. We carefully watched for any evidence of rise and fall of the tide in relation to the shore-line of the glacier, but were unable to observe any. We concluded from this circumstance that the glacier must be rising and falling in unison with the tide. The sounding-tube of our Lucas sounding-machine brought up a quantity of *serpulæ* and sponge spicules from the sea bottom beneath the edge of the glacier where our ship was moored.

The second piedmont-afloat is the Nordenskjold Ice Barrier Tongue. This Ice Barrier Tongue is about 20 miles in length, and 5 to 6 miles in width. Its southern edge is formed of ice and polished névé. Fierce blizzards have swept any loose snow off this southern edge of the Barrier. The northern edge was formed largely of snow, being chiefly of the nature of snow-drift, from 40 to 50 ft. in thickness. The latter terminates in a vertical cliff with overhanging snow cornices. Obviously this cliff was the combined result of the blizzard winds driving snow northwards to the lee side of this piedmont-afloat, and to the breaking away in summer time of the sea ice supporting the northernmost portions of this snow-drift. Slices are thus removed from time to time from the northern edges of the drifts, and so the cliff of the portion left behind becomes higher in proportion as the thicker ends of the wedges of snow-drift become broken away. Certainly no true tide-crack was visible on the south side of this Barrier, and only a small crack was seen on its north side. Strange to say, this big mass of ice and consolidated snow, which rises at its centre a little over 100 ft. above sea-level, does not appear to communicate directly with a névé field at its inland end. Apparently then the Nordenskjold Ice Barrier is not now being directly fed from the inland névé fields. It appears to represent an old piedmont-afloat, which is in the act of dwindling away from want of supplies of ice from the interior.

Thirdly, the Drygalski Ice Barrier Tongue is also of the nature of a piedmont-afloat. It is probably floating for at least three-quarters of the distance of 30 miles to which it projects from the shore into the sea. The surface of this glacier, where it leaves the shore-line, is extremely rough and rugged, being traversed, as stated in the narrative, by an immense number of chasms, pressure ridges and crevasses. On the south side, where the ice was still unbroken when we reached the glacier on November 30, 1908, the old sea ice was forced up into strong pressure ridges. The whole appearance suggested to us that this glacier is moving actively from inland seawards. We could see with our field-glasses that at a distance of about 50 miles inland it descended by steep ice-falls from a high plateau beyond. At the point where we crossed it, the glacier rose to an altitude at its centre of about 200 ft. above sea-level. It was here 12 miles in width. Further eastwards and therefore seawards, the glacier ice was more and more levelled up with snow, until eventually it passed into a true barrier type with a comparatively smooth surface.

Captain Evans, after he brought the *Nimrod* into "Relief Inlet," where he picked up the Northern Party just returned from the Magnetic Pole, sounded alongside of the Drygalski Glacier and found a depth of 655 to 668 fathoms, at a distance of only about 18 miles from the rocky shore-line.

As the Barrier here rises to a height not exceeding 50 ft. above sea-level it must surely be afloat.

During the few weeks of thaw, in December and January, torrents of water must rush off from this glacier in the form of englacial or subglacial streams. These in some cases cut deep open valleys with more or less precipitous sides; in other cases they tunnel channels for themselves under the covering of hard snow and ice, and the roofs of these tunnels collapsing through want of support produce rugged ravines, very difficult to cross with sledges.

This Drygalski Ice Barrier, on its northern side, contained in places a considerable amount of moraine material. It was evident that at the time when the glaciation of this region was at its maximum it must have been continuous with the Mount Nansen glacier. These two glaciers, when united, doubtless formed a huge piedmont-afloat.

(3) BARRIER SNOW- AND ICE-FIELDS.—The structure of the Nordenskjold and Drygalski Barriers throws considerable light on one of the most difficult problems of the Antarctic—the origin of the Great Ice Barrier. To ascertain the amount of annual snowfall on this Great Barrier is of very great importance, but we found this a hard problem, chiefly on account of the difficulty of distinguishing between true newly fallen snow and old snow which has been drifted along by blizzards. We tried, during our observations in Antarctica, to eliminate the drift snow from the true snowfall, and our general conclusion now is, that at Cape Royds the annual snowfall is equal to about 9½ in. of rain. On the journey of the southern depôt party under Joyce, when laying a depôt for the relief of the returning Southern Party in January 1909, the fortunate discovery was made of Captain Scott's old Depôt A. The sharp eyes of Day discerned, at a distance of several miles, the top of the depôt bamboo pole with just a wisp of the old black flag still attached to it. Knowing the importance, from a scientific point of view of estimating the extent and direc-

tion of movement of this depôt, in the six years four months and a half that had elapsed since Captain Scott left it there, the party visited it and Mackintosh took a series of angles and measurements, which enabled him to determine that the depôt had moved bodily to the east-north-east at the rate of a little over 500 yards a year for the past six years and six months. The party also dug down through the hard snow to a depth of 8 ft. 2 in. when they came upon the original snow-surface on which the depôt was formed. They were thus able to show that, during the above period, October 1, 1902, to February 15, 1909, on the average about 13 in. per year of hard snow had accumulated. In order to determine the density of this snow they melted down a considerable quantity of it, and measured the volume of the thaw-water resulting. This showed that the annual accumulation of snow on this part of the Great Ice Barrier is equal to about 7½ in. of rain.

This depôt is in the latitude of Minna Bluff, about 78° 40′ South. Further north, as, for example, in the neighbourhood of Mount Nansen, the snowfall appeared to be considerably heavier, and it seemed to be heavier still nearer to the Antarctic Circle. As 7½ in. of rain is equal to 7½ ft. of snow it is obvious that the accumulation of snow, even as far south as between latitude 78° and 79° is not inconsiderable, but on account of its great density this compressed snow, near Minna Bluff, formed a layer annually 13½ in. thick, instead of 7½ ft. thick.

For the sake of simplicity it may be assumed that the rate of accumulation over the Great Barrier generally is about 1 ft. annually. Now it has been proved that the Great Barrier extends inland for fully 300 miles in places. From the observations at Minna Bluff, and the rate of movement of Captain Scott's Depôt A, as measured by Captain Scott and again by Joyce, it may be inferred that the Great Barrier there is travelling seawards at the rate of about one-third of a mile per year. From this it may be argued that a snowfall on any part of the Barrier 300 miles inland would take 900 years to reach the edge of the Great Ice Barrier, where bergs are discharged into the sea. At this rate, if 1 ft. of snow is added to the Barrier every year a layer of snow, formed 300 miles inland, 900 years ago, if it reaches the Great Barrier cliffs at the present day will be covered by a thickness of 900 ft. of snow. Obviously this theory gives a vast thickness of snow to form the seaward end of the Great Barrier. Theoretically then as the result of the calculations from the observations of the southern depôt party, it might be argued that a considerable thickness of the berg material derived from the Great Ice Barrier was formed of consolidated snow and névé rather than of true glacier ice. Practical proof of this was afforded us by another series of observations. At the end of the breaking up of the sea ice in the summer of 1907–8, three bergs drifted into McMurdo Sound, and grounded between our winter quarters and Cape Barne. During the following winter the sea was frozen over around these bergs, and we were able to go over to them and study them. Fortunately they were much tunnelled by sea-worn caves. This enabled us to see their internal structure. We found that all around the edge, particularly along the line of the wave-worn groove which surrounds all bergs, a good deal of ice was developed. This ice resulted evidently from the freezing of sea water as the waves washed and dashed against the foot of the berg cliff. In heavy weather a large quantity of spray would be flung high up against the cliff faces of the berg, and the spray freezing would encrust the exterior of the berg with ice. There was no evidence, however, of the existence of any solid ice inside the berg, this portion of it being formed purely of compressed snow.

From this fact we were led to speculate as to whether the whole of the berg might not be formed of hard snow, its submerged portion saturated, but only superficially, with sea water. That this was actually the case was proved later by Captain F. P. Evans. He saw in these bergs an excellent shelter for his ship from the blizzards, and moored the *Nimrod* to one of the larger bergs. While here he took soundings around the most typical of these tabular bergs, and found that whereas its cliff face rose to a height of 80 ft. above sea-level, the berg was aground in only thirteen fathoms of water; that is, the berg was submerged to a depth of 78 ft., so that practically half of it was out of the water and half immersed. This direct observation is obviously of great importance as bearing on the mode of origin and structure of the so-called icebergs of the Antarctic. There can, we think, now be little doubt that a great proportion, in some cases the whole, of the material of typical Antarctic bergs is formed of consolidated snow rather than ice. These observations may now be considered in their bearing on the origin of the Great Ice Barrier. Captain Scott has shown that the Great Ice Barrier for the greater part of its length, probably for 400 miles at least along its edge, is afloat. Wherever we got near to the cliff face of the Barrier, and we were at times very close to it, we were unable to see anything of the nature of true glacier ice, even in cases where the cliff rose to a height of over 150 ft. above sea-level. On the other hand, there was every appearance of the Barrier being formed of numbers of superimposed layers of snow. On the line of argument previously given it is not improbable that a thickness of 900 ft. of snow, or thereabouts, may accumulate on a large proportion of the Barrier near its terminal cliff, so that obviously, a great part of the thickness of the Great Ice Barrier is probably due to this compressed snow.

The question still remains, as to what becomes of the glacier ice which undoubtedly does feed the Barrier at many spots along its western and southern boundaries. For example, the great glacier, 50 miles wide, up which the Southern Party travelled from the spot where they were compelled to diverge from the Barrier, latitude 83° South, must be discharging vast quantities of ice into the Barrier. This same glacier had raised pressure ridges on the Barrier surface for 20 miles out from its junction with the Barrier. It is clear, too, from the fact established both by the *Discovery* expedition and our own, that the Great Ice Barrier is moving seawards. The propelling force can be no other than that of glacier ice. This glacier ice descending from the inland plateau must also move seawards, but as it gets nearer to the Great Barrier ice cliff it becomes weighted down with a vast thickness of superincumbent snow, and it is quite possible that under these conditions a great deal of it may be thawed off from below by the sea water.

The question here suggests itself, does the water circulating beneath the Great Ice Barrier ever have a temperature high enough to thaw fresh water ice? It does, of course, thaw the sea ice quite rapidly.

(4) INLAND ICE AND NÉVÉ FIELDS.—Reference has already been made to this type of ice under the head of "Glaciers," in the description of the Mount Nansen Glacier. The great glacier discovered by the Southern Party between 83° 33′ South and 85° South, over 100 miles in length and 50 miles in width descended about 6000 ft. in that distance from a vast inland snow plateau. This plateau is identical with that traversed by Captain Scott's party of the *Discovery* expedition, on their western journey in 1903. It is identical also with the plateau

The Barrier Edge Breaking away four miles south of Hut Point

travelled by the Northern Party of our expedition in their journey inland to the South Magnetic Pole, as well as with the new land discovered by our expedition to the west of Cape North. This vast plateau extending, it is practically certain, for over 1200 miles from north to south, and over 200 miles from east to west, 7000 ft. high at its northern end, and at least 10,000 ft. at its southern, is formed superficially of névé. Doubtless beneath the névé is glacier ice. The névé is possibly of no very great thickness, for the horizontally bedded or gently inclined plateau rocks of the Beacon sandstone formation rise to heights of 8000 to 10,000 ft. above sea-level along the eastern border of the plateau.

This structure of the plateau is illustrated on the diagram (p. 306).

(5) Icebergs have already been described under the heading "Barrier Snow- and Ice-fields."

(6) Pack ice has also been referred to under the heading "Sea Ice." It may be noted that in the Ross Sea the bulk of the pack ice, formed chiefly of fractured masses of sea ice, partly of small snowbergs and icebergs, impelled by the south-easterly winds drifts past Cape Adare to the part of the Antarctic Ocean which lies between Cape North and the Balleny Islands. This region appears to be permanently beset with very old pack ice and icebergs. As most of the blocks of sea ice have been twisted and piled on one another, this pack may be described as "screwed pack."

(7) Thaw-water forming Surface Lakes, and Surface, Englacial or Subglacial Streams.—Some of the streams formed by thaw-water have already been described under the head of glaciers, in the case of the Drygalski Glacier. In the latitude of this glacier in 75° South, the thaw set in about December 10 and lasted to about the third week in January.

Lakes and Lake Ice.—We found it difficult during the short period of our stay in the Antarctic to ascertain to what thickness this ice formed during a single season. The difficulty arose from the fact that in the summer of 1908–1909 the fresh water ice of these lakes did not entirely thaw.

Some of the lakes were slightly saline, and some of these, such as Green Lake and Coast Lake, thawed completely during the summer and during the winter the ice froze over them from top to bottom until, in the case of Green Lake, in August 1908, only a very little saline water, a few inches in depth, remained unfrozen, below a thickness of 5 ft. of ice. At the same time of year the water of " Coast Lake," also somewhat saline, was frozen solid, the lake being a little under 5 ft. deep.

In the case of the fresh-water lake, known as Clear Lake, it was noticed that during the summer the ice thawed chiefly on the south side of the lake, where it was in contact with the black rock, and where that rock was specially warmed by the sun's rays. The sun being highest when it is due north, has its greatest heating effect on southern slopes. The ice towards the middle and north side of the lake did not wholly thaw. The same remark is true of the ice of the Blue Lake near our winter quarters.

In the case of the Blue Lake we found, as the result of the shafts sunk, that in the south-west division of this lake the ice was at least 15 ft. in thickness, while in the north-western division Brocklehurst sunk a shaft to 21 ft., but in his case a little water was found beneath the ice, whereas in the first case the ice was solid down to the bed rock. We are of opinion that this Blue Lake ice had not been thawed for probably at least three seasons.

As many of these small lakes were encircled by steep

banks of hard rock, the ice, expanding as it formed, was forced to curve itself upwards in order to make room for itself, as it could not burst the sides of the rock basin. Thus its surface was frequently convex upwards. In this process of upward bulging of the ice towards the lake's centre each successive fresh layer of ice as it froze beneath the older and overlying layers, would buckle them and bend them. The latter would at last crack open, and so in the final stage of freezing of a small shallow lake, by the time that the whole of the water had frozen from top to bottom, the basin would be occupied by a biconvex cracked lens of ice, the cracks being widest at the top and tapering away to nothing below.

A curious feature which we observed in the lake ice was the presence of what we termed "snow tabloids." We found that in some cases these were merely empty bubble-like spaces in the ice filled with air. In other cases, however, where the bubbles were larger, 3 to 6 in. in width, they were occupied by snow. In some cases patches of thin rippled snow were inter-stratified in this lake ice.

Most of this fresh-water lake ice exhibited at its surface a very beautiful structure, which we termed coralloidal structure.

The mode of origin of this curious structure will be discussed in the Scientific Memoirs of this Expedition.

(b) Action of Wind in Relation to Antarctic Glacial Phenomena

An explanation has already been given of how vast quantities of finely divided rock material, chiefly in the form of sand, are constantly being blown on to the sea ice by the wind. For some distance seawards from the shore such wind-blown material must form an appreciable

These Sections prove that there cannot be any very great thickness of Ice and névé at the South Pole or South Magnetic Pole

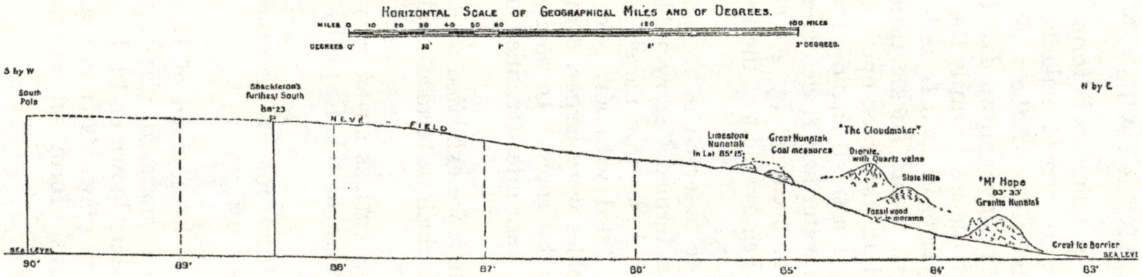

Section from Great Ice Barrier to Shackleton's Furthest South.

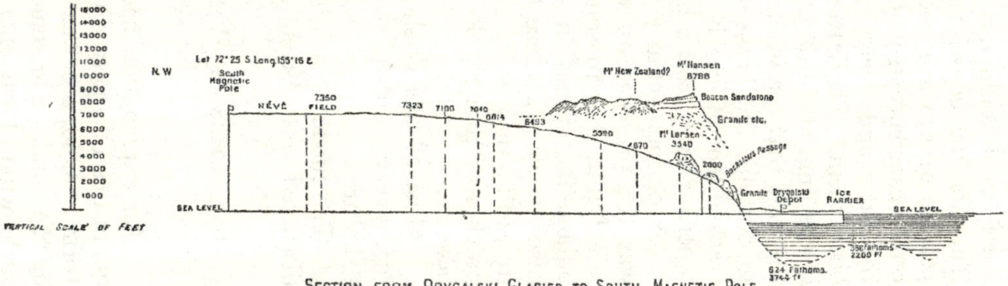

Section from Drygalski Glacier to South Magnetic Pole

* A Weathered Kenyte Boulder near the Winter Quarters

Felspar Crystals from summit of Mount Erebus (natural size)

amount of the sediments now forming on the sea floor. Ferrar has already commented on the great importance of the wind in these regions as a destructive agent.

By accelerating evaporation of snow and ice, and by its mechanical erosive force on the surfaces of snow-fields, the blizzards are important contributors to the present deglaciation of Antarctica. The amount of snow annually blown out to sea must be very great, inasmuch as during blizzards, often of several days duration, the air is frequently so thick with fine particles of snow that one cannot see more than a few yards in front of one. We observed that sledge tracks and footprints on the snows of the coastal areas, or of the inland plateau, were nearly always, after the expiry of a few weeks, left in relief. This suggests that, at present, in many parts of Antarctica the general surface of the snow and ice is being continually lowered by ablation and wind drift.

(c) Changes of Temperature

During spring and autumn when sunrise and sunset replace the perpetual sunlight of summer, and the perpetual darkness of winter, the range of temperature between noon and midnight is most marked. On March 10, 1908, when at an altitude of about 9000 ft., on Mount Erebus, and with the thermometer at about 10° Fahr., we observed that snow in contact with black lumps of kenyte lava exposed to the sun's rays thawed rapidly, so that we were able to get water to drink by laying a lump of snow in saucer-shaped hollows on the surface of this lava. At night these same rocks became very cold. There was no question here of the survival of any original volcanic heat in the lava, as there was no thaw whatever of the snow where it touched the lava in spots shaded from the sun's heat. This absorption of heat by black rocks partly explains the survival of lichens high up on the slopes of Erebus. It also has the effect of prolonging a superficial local thaw from summer far into spring, on the one hand, and autumn on the other. Such a great diurnal range of temperature, combined with the effects of summer thaw followed by the severe frosts of winter, exerts a powerful disrupting force upon the rocks, and accounts for the extensive rubble banks and sheets and patches of loose and broken felspar crystals, which are spread over such a large area of country near Mount Erebus, &c. At our winter quarters at Cape Royds we at first mistook these for beds of volcanic tuff.

VOLCANIC ROCKS

ROSS ISLAND.—As the chief varieties of volcanic rocks met with in Ross Island have already been described by Messrs. Ferrar and Prior, a brief description of these will suffice.

At Ross Island we particularly studied the relations to one another of the three principal types of rock there developed, viz., kenyte, trachyte and basalt. We are now in a position to say that, on the whole, the trachytes appear to have been the oldest rocks, the kenytes to be of intermediate age, and the basalts the newest. The evidence for this is as follows:

On the western slopes of Mount Erebus, above our winter quarters, specimens were not infrequently found of what at first sight appeared to be fragments of sandstone enclosed in kenyte lava. A closer inspection of these showed that they were in reality varieties of trachyte. Similar specimens were met with in the kenytes near Cape Barne. It would seem from this that the oldest lavas in this area were trachytes, and that later kenyte eruptions followed, which partly destroyed the trachytes, and thus the disrupted trachyte fragments subsequently became embedded in the kenyte lava. In the next place we found that at Cape Barne the kenyte had been very powerfully intruded by the basalt. Large fragments of kenyte were frequently found entangled in the basalt of the comparatively recent volcanic cone at Cape Barne, and on a line trending inland in a south-easterly direction. In the case also of the long spur which extends from Mount Erebus to the old winter quarters of the *Discovery* expedition at Hut Point it is clear that the latest volcanic products of that locality are scoriaceous basalts. These basalts are obviously newer than the trachytes of Observation Hill; they are even newer than the olivine basalts of Sulphur Hill in the same area. In our ascent of Erebus we found that not only were there old kenyte lavas developed on its flanks, but that at intervals on the way up the rock was still kenyte, becoming of newer and newer age until the modern active crater was reached. This crater was partly filled with molten lava from June to September 1908. It is evident from this that some of the kenyte is amongst the newest of all the volcanic products of Ross Island. The following table shows the probable chronological relations of these lavas in descending order:

Kenyte of modern crater.	Scoriaceous basalt.
	Olivine basalt.
Pre-basaltic kenyte.	
	Trachyte.

East Coast of Victoria Land.—The succession, therefore, on the whole has been from trachyte through kenyte to olivine-basalt. There can be no doubt that the whole of the trachyte eruptions, the pre-basaltic kenyte, and the olivine basalts antedate the epoch of greatest recent glaciation.

The occurrence of large blocks of kenyte at Gneiss

Point, a few miles to the east-south-east of Granite Harbour, suggests the possibility that some of the dark sheets of rock near the highest portions of the plateau may perhaps be formed in part of kenyte. At the same time it is possible that these kenyte boulders of Gneiss Point have been drifted northwards from Ross Island amongst the moraine material of the western branch of the former gigantic Great Ice Barrier, when it occupied the whole of McMurdo Sound. Dykes of dark rocks, resembling tinguaites, were not uncommon from Granite Harbour northwards to near Geikie Inlet, south of the Drygalski Glacier. We also observed dyke rocks full of small black bright prismatic crystals of hornblende. These appear to be of the nature of hornblende lamprophyres. Small pieces of scoriaceous volcanic rocks were found by us as far north as Cape Irizar. These fragments may have come from the mountains of the western plateau.

Foundation Rocks

The oldest rocks seen by us in the Antarctic belong to the series already described by Ferrar and Prior, and consist of banded gneiss, gneissic granite grano-diorite and diorite rich in sphene. In some spots, as at Cape Bernacchi, masses of very coarse white crystalline marble are interspersed in the gneiss.

These foundation rocks have their planes of foliation sharply folded in places, as is the case at Depôt Island, the axis of folding there being approximately parallel to the trend of the coast-line. Near the same spot huge enclosures can be seen in the gnessic-granite. These are partly greenish grey quartzites in masses ten to twenty feet in diameter, partly large lumps of blackish green coarsely crystalline hornblende rock, with much sphene and a white mineral, apparently either saussuritised felspar, or scapolite. In other places, as at Cape Bernacchi, black tourmaline schists with epidote were frequently interspersed through the gneiss, and the gneiss was also traversed by veins of white aplite, with small crystals of garnet. The coarsely crystalline belt of marble in the gneiss at Cape Bernacchi contained abundant graphite in the form of small flakes. It appears to us that the marble and the quartzite represent an old sedimentary formation, and the large enclosures of hornblende-and-sphene rock an old amphibolite or gabbro, both the former and the latter types of rock disrupted by the intrusive gneissic granites.

Ferrar was of opinion that in the neighbourhood of the hill marked (*d*) on his map showing the valley of the Ferrar Glacier, the grey granite of these hills is older than the dolerite which rests upon its even upper surface, but that the pink granite of (*d*) is intrusive and later than the dolerite.* This is an important observation. We did not see this spot, but in other areas, as near Granite Harbour, the dolerite appeared to be newer than the granites.

Older (?) Palæozoic Sedimentary Rocks

Apart from the large enclosures of quartzites, &c., in gneissic granite, already referred to, the next oldest sedimentary rocks appear to be greenish grey slates brought back by the Southern Party from the surface of the great glacier up which they were travelling between Mount Hope and "The Cloudmaker," in approximately latitude 84° South. These fragments, as Lieutenant Shackleton informs us, were blown on to the surface of the ice from what appeared to be mountains of slate further west. The approximate relative position of these slate

* Nat. Ant. Exped. Nat. Hist., Vol. i., Geology, p. 38. Brit. Museum.

hills and of the granite hill of Mount Hope, and the nunatak of coal-bearing Beacon sandstone further south is shown in the photograph.

At Cape Royds we found occasionally, but sparingly, erratics of radiolarian chert. The radiolaria appear to be of older palæozoic type, but we were unable to discover any rock like it *in situ*. Obviously at Cape Royds the erratics have travelled from some spot to the south and west. It is possible that these black and grey cherts belong to the limestone series discovered by the Southern Party in 85° 15′ South. This limestone varies in colour from pink to dark grey. The pink limestone is banded with some dull green earthy mineral; it contains numerous obscure casts resembling those of radiolaria. The dark bluish grey portion of this limestone does not show any trace of organisms. Apparently it has been too much metamorphosed to retain the outlines of any of its original fossils. It is traversed in all directions by veins of white calcite. The limestones appear to be almost horizontally bedded, and are several hundreds of feet in thickness. The Southern Party were unable to determine the relation of this massive limestone to the adjacent Beacon sandstone, as unfortunately there was a break in the continuity of the section which prevented the junction of these two formations being seen. This limestone, 7000 ft. above sea-level, is higher geographically than the Beacon sandstone formation, but as the latter dips away from it towards the north-east, the limestone may be stratigraphically below the Beacon sandstone. The Southern Party discovered large blocks of limestone breccia in the moraines near "The Cloudmaker." The fragments in the morainic breccias near this mountain are formed of limestones, not unlike those of the great nunatak, but whether these breccias belong to the basal beds of the Beacon sandstone formation, or to the base of the massive limestones, or to crush breccia zones in the massive limestones is not as yet apparent. It is even possible that they may be stratigraphically above the Beacon sandstone, but this is improbable. Fragments of old limestones were observed by us in the ancient moraines of the Mount Nansen Glacier. These were associated with pieces of sandstone and fragments of grey clay shale with obscure impressions of fossil roots. This limestone appears to have been derived from the Beacon sandstone formation. It is very much altered through recrystallisation, and we have been unable to recognise in it any fossils.

It may be mentioned that when journeying to the South Magnetic Pole at a spot about twenty miles southeast of Granite Harbour we found on a small island, Terrace Island, a large fragment of argillaceous limestone. This had evidently been derived from the Beacon sandstone formation in the adjacent hills. When broken open it was found to contain small oval bodies, pointed at one end, and about one-third of an inch in length resembling seeds of fossil plants, possibly coniferous seeds.

This specimen was subsequently left at "Depôt Island," and has not since been recovered.

The Beacon sandstone formation has now been proved to extend from at least as far as Mount Nansen in the north to latitude 85° South, where it was explored *in situ* by the Southern Party, a distance of fully seven hundred miles. As described by Ferrar its thickness in the Ferrar Glacier Valley amounts to fully two thousand feet, and even then the base of the formation was not seen.

The following is a generalised section of the Beacon sandstone formation in 85° South, in descending order, from information supplied by Mr. Shackleton and F. Wild of the Southern party:

500 ft. sandstone.

300 ft. sandstone with bands of shale, and about seven seams of coal, or seams formed of black shale alternating with laminæ of bright coal. In this belt the following seams were seen *in situ:*

1 ft. to 1½ ft. coal seam.
Strata.
7 ft. coal seam with bands of grey shale.
Strata.
5 ft. coal seam apparently formed of clean coal.
Strata.
3 ft. (about) coal seam.
Strata.
3 ft. (about) coal seam.
Strata.
3 ft. (about) coal seam.
Strata.
3 ft. (about) coal seam.

Seven-hundred-foot sandstones with numerous water-worn quartz pebbles in the lower beds. These pebbles are from one to two inches in diameter. Total 1500 ft.

In the medial moraine, below the great nunatak in 85° South, the Southern Party obtained, amongst several specimens of sandstone with much mother-of-coal, or mineral charcoal, one specimen of special interest. It was a fragment of fine grained hard sandstone, evidently derived from the Beacon sandstone formation higher up, showing a black band one-quarter of an inch thick running through it. Micro-slides of this examined at the University prove that it is a coniferous wood. The following description of it has been written by Mr. E. J. Goddard, B.Sc., Macleay Research Fellow of the Linnæan Society, New South Wales:

GEOLOGICAL OBSERVATIONS

Specimens from Medial Moraine, December 11, 1908 (No. 101)

"Longitudinal sections of the included dark masses give a homogeneous banded appearance of a distinctly organic nature. The banded appearance is due to the vascular nature of the organic elements composing the mass. The whole structure recalls to one's mind the appearance given by longitudinal sections of the xylem portion of the vascular area of a gymnosperm, such as *Pinus*. Only the xylem area is represented in the specimen, no traces of medullary, cortical, or phloem tissue being visible. Medullary rays are present as shown in the micro-photograph.

"The xylem itself is composed of a homogeneous mass of vessels, tracheidal in nature, no differentiation as regards the vascular elements being present. In places one may readily make out in longitudinal sections dark opaque bands of much greater size individually than the tracheides. These in all probability represent resin passages belonging to the xylem. It would seem, further, that these masses might be considered as being nothing more than an aggregation of material similar in nature to that of the walls, and due to changes under the process of petrifaction. This, however, is opposed by the fact that they occur even in these small sections fairly commonly, and at the same time are all of exactly the same size as regards width. At all events they represent some definite structure, and in all probability resin passages.

"The walls of the tracheids themselves seen under the high power of the microscope appear to be pitted, but the preservation is by no means good enough to warrant any remarks on this beyond that in the common wall of adjacent tracheides occur clear spaces of the same relative importance as the bordered pits of such a gymnosperm as *Pinus*. These clear spaces occur regularly along the length of the tracheides, and stand out strongly against the dark colour of the walls in their preserved condition.

"The nature of the xylem itself leads to the conclusion that it is a portion of a gymnospermous plant, resembling strongly in nature the same portion of a coniferous plant."

If the conclusion as to the coniferous character of the wood is correct, and there seems little reason to doubt its correctness, the lower limit of the age of the Beacon sandstone is perhaps lower carboniferous or Upper Devonian, unless conifers in the Antarctic had a deeper range in geological time than elsewhere.

The plate shows the general appearance under the microscope of this fossil wood. The medullary rays are fairly distinct.

This is the first determinable fossil plant that has been obtained from the Victoria Land portion of the Antarctic.

Although a date not older than carboniferous or Devonian is suggested by the presence of this coniferous wood, it is of course possible that the Beacon sandstone is of higher geological antiquity, and if the radiolarian rocks, already referred to, are conformable with the Beacon sandstone, it may even go far down into the older palæozoic.

The degree of induration of these unfolded plateau sandstones and the general absence of fossils from the limestones is suggestive of a high geological antiquity.

RAISED BEACHES

Raised beaches were observed at two distinct localities on the western slopes of Ross Island, also near the mouth

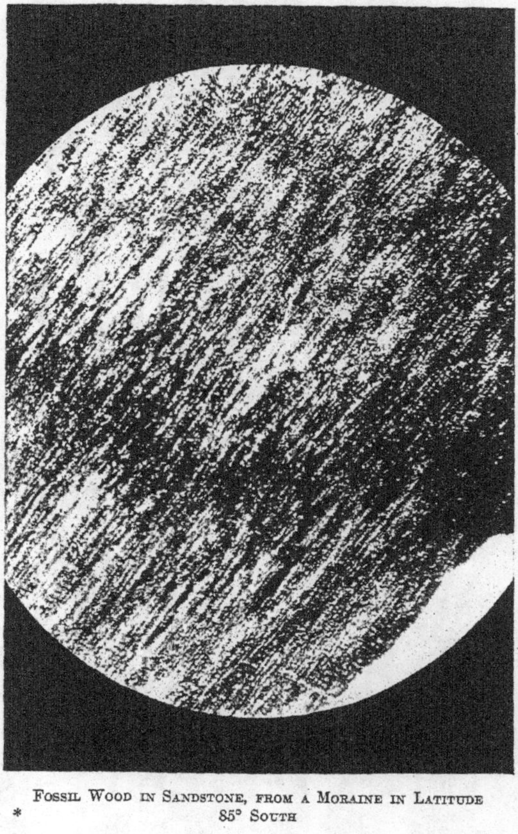

Fossil Wood in Sandstone, from a Moraine in Latitude 85° South

*

of the Ferrar Glacier, as well as to the south-east of Mount Larsen.

The first locality on Ross Island was discovered by Armytage; the second locality by one of us; the raised beaches near the Ferrar Glacier also by one of us; while the Mount Larsen raised beaches were observed by the Northern Party on their journey to the South Magnetic Pole. These four localities may be referred to as:

(1) Back Door Bay deposit.
(2) Cape Barne deposit.
(3) Ferrar Glacier deposit.
(4) Mount Larsen deposit.

Deposit (1) lies at a height of 160 ft. above sea-level. It was found at the bottom of a shallow flat-bottomed gulley sloping down eastwards into a small arm of McMurdo's Sound, called by us Back Door Bay. The area of the deposit was only a few square yards. It consisted of a brownish earthy material with abundant remains of crushed tubes of serpulæ. Diatoms were fairly abundant in it. We were at first in some doubt as to whether this serpula deposit was a genuine raised beach, or merely a mass of sandy material from the sea floor pushed up by the ploughing action of the old McMurdo Sound ice sheet. There has, of course, been a similar question raised in regard to the so-called raised beaches of Moel Tryfaen in North Wales. The subsequent discovery by one of us of an extensive raised beach at an altitude of about 180 ft. near Cape Barne (deposit 2) is confirmatory evidence as to the genuine raised beach character of the first-mentioned deposit. At this Cape Barne deposit the material is largely formed of spicules of siliceous sponges and molluscan shells, as well as the remains of serpulæ. The nature of the organisms there present is such as to render it probable that the deposit was formed in water of some depth. It may therefore indicate an elevation of this part of the coast near Cape Barne, not merely of 180 ft., but perhaps of double that amount.

Deposit (3) near the entrance of the Ferrar Glacier Valley. These deposits chiefly consist of brownish sands with very numerous shells of a large species of pectens. They were discovered by one of us, and traced to altitudes of from 50 to 60 ft. Further north at Cape Bernacchi there are strongly marked terraces suggestive of raised beaches, which were observed by the Northern Party to extend up to altitudes of 100 ft. At Terrace Island, about twenty miles to the north of Cape Bernacchi we observed well-marked terraces, ranging up to about 80 to 85 ft. above sea-level. These had every appearance of having been laid down by the action of the sea, though no sea-shells were found amongst the sands and coarse gravel constituting this deposit.

Deposit (4) south-east of Mount Larsen. A very curious type of what may be termed a raised beach or possibly an upthrust area, was observed by us on our journey from the sea ice near the Drygalski Glacier to the foot of the small branch glacier at what we called Backstairs Passage. This moraine deposit consisted of greenish grey muds underlying coarse moraines of granite blocks and quartz and felspar porphyry. Beneath the grey muds was ice; the ice was just very slightly saline in places. These muds contained remains of serpulæ in great quantity, an enormous number of shells of that well-known arctic type amongst the foraminifera, *Biloculina,* numerous representatives of horny polyzoa, siliceous sponges, and a perfect form of solitary coral allied to a perfect shell of lyothyrina, &c., dentalium, chiton, &c.

The height of this deposit above sea-level could not be

definitely ascertained by us. It was at least 20 ft., possibly as much as 30 ft., above the sea.

The evidence generally of these raised beaches is very interesting, as showing the probability that when the ice, in recent geological time, had its maximum extension in this region, there was probably a subsidence, possibly due to the load, on this part of the earth's crust, of such a vast extra thickness of ice, and that after the subsidence, which probably lagged behind the epoch of maximum glaciation, there has been a gradual re-emergence to the amount of 150 to 200 ft. The brachiopod and polyzoon in the raised marine mud from the Larsen area have kindly been identified for us by Mr. C. Hedley and Mr. E. F. Hallman respectively, of the Australian Museum.

Peat Deposits

Amongst formations of recent origin may be mentioned peat. A deposit of peat was discovered by one of us formed on the bottom of the lake called Coast Lake.

This peat is formed from the remains of a large fungoid plant, which grows in profusion in the water of these coastal lakes when the ice thaws in midsummer.

A deposit of mirabilite was discovered by one of us at the spot shown on the map near Cape Barne, after the summer thaw had set in. The white colour of the salt during winter rendered it indistinguishable from the surrounding snow, but after the thaw the difference of course was noticeable between this white salt and the edges of the dark volcanic rock.

Apparently this mirabilite forms one of the products of evaporation of an old lake. Numbers of these dried lake basins were noticed by us amongst the glacial moraines on these western slopes of Erebus near Cape Barne and our winter quarters.

In regard to the important theoretical question as to whether the west coast of Victoria Land is of an Atlantic or of a Pacific type, the following considerations present themselves.

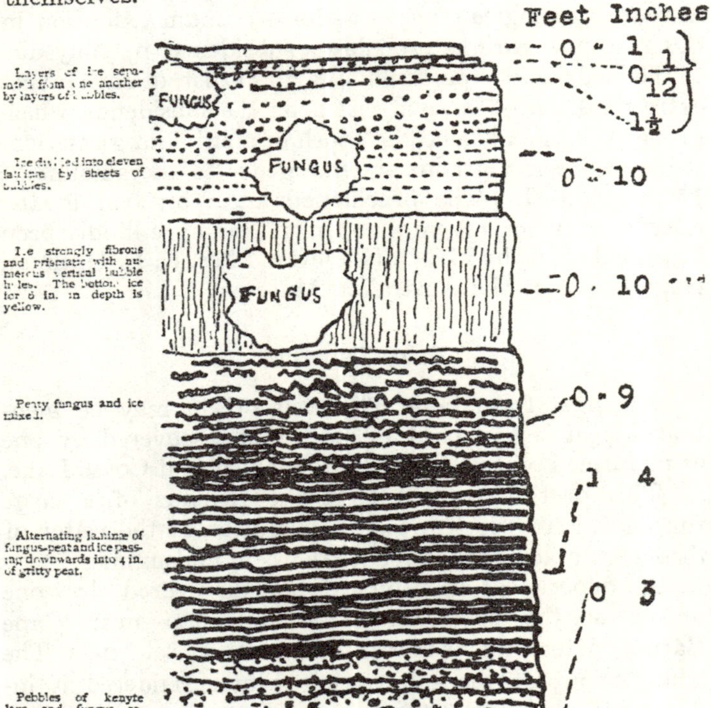

As pointed out by H. T. Ferrar, the massive Beacon sandstone formation terminates in steep, and in places precipitous, slopes along the whole line of coast from beyond Cape North southerly to Mount Discovery. From there to where Mr. Shackleton and his Southern Party, after ascending the granite and slate mountains of this coastal range in latitude 83° 33′, longitude 170° East, reached 85° South at an altitude of 6000 ft. sedimentary rocks were found to cap the ranges. They were disposed in gently dipping or nearly horizontal strata, and extended south-easterly to at least 86° South.

At the Great Nunatak in latitude 85° South, longitude 165° East, the sandstone beds containing the seven seams of coal dip in a north-easterly direction at an angle of about 6° to 8°. There was no evidence there, nor as far northwards as our explorations extended, that is, to Mount Nansen, of any folding in the Beacon sandstone formation.

At Depôt Island, to the north of Granite Harbour, the ancient gneiss platform is there folded parallel with the coast-line, but this structure appeared to be the exception rather than the rule.

The Beacon sandstone, for reasons already given, is perhaps as old as palæozoic, possibly older palæozoic. So, if the above conclusions are correct, there has been no appreciable folding in the part of Victoria Land examined by us since palæozoic time.

The coast-line is, therefore, in our opinion, of the Atlantic rather than of the Pacific type, and probably owes its trend and position to a powerful fault or zone of faults, with a down-throw to the east. The volcanoes Mount Melbourne, Mount Erebus, Mount Discovery, &c., are probably on this fracture, or zone of fractures. If this view as to the Atlantic type of this part of the coast is correct, what has become of the great Andean folds developed on the west side of Graham Land?

Possibly as Wilckens has suggested, west of Alexander I Land the cordillera is submerged through faulting.*

* Centralblatt für Min. Geol. und Pal., 1906, No. 6, p. 179.

In this case the Ross Sea subsidence area (if such it be) would have approximately the same relation to the Andean trend-lines in the Antarctic that the Gulf of Mexico and the Antillean and West Indian fracture zones bear to the trend-lines of the tropical Andes. It must, nevertheless, be admitted that the Great South Polar Shield of ancient and practically incompressible crystalline rock, intensely folded in the past, would be incapable of being further folded now, and if the Andean zone of disturbance traversed this shield it would be likely to traverse it as a zone of fractures with local lava effusions, rather than as a fold range of the Pacific type. This important matter will be discussed by us in detail in the Geological Memoir of this Expedition.

SUMMARY.—The following inferences are tentatively suggested in regard to the geology of Antarctica:

(1) The majority of the tabular bergs of this region are largely, in some cases wholly, snowbergs, not icebergs.

(2) True icebergs are also found.

(3) Glaciers in the Antarctic push out in some cases thirty miles from the coast, and must be afloat, as argued by Ferrar, for the greater part of this length.

(4) The Great Ice Barrier is formed of true glacier ice at its sides and inland extremity, but the centre and seaward portion is formed, in its upper part, chiefly of snow. We agree with Captain Scott's conclusions that the Great Barrier except at its edges and perhaps at some distance inland must be afloat. At its eastern side it has been moving seawards at the average rate of about five hundred yards a year for the past seven years.

(5) Throughout the whole of the region of Antarctica examined by us, for 16° of latitude, there is evidence of a recent great diminution in the glaciation. In McMurdo Sound this arm of the sea now free from land ice was

formerly filled by a branch of the Great Ice Barrier, whose surface rose fully 1000 ft. above sea-level, and the Barrier ice in this sound, in areas from which the ice has retreated, was formerly about 4000 ft. in thickness.

(6) The snowfall at Cape Royds from February 1908 to February 1909 was equal to about 9½ in. of rain.

(7) The névé-fields of Antarctica are probably of no great thickness.

(8) The southern and western sides of the sector of Antarctica south of Australia is a plateau from 7000 to 10,000 ft. high, which may possibly extend across the South Pole to Coat's Land and Graham Land.

(9) Ross Sea is probably a great subsidence area.

(10) The Beacon sandstone formation which extends for at least 1100 miles from north to south in Antarctica contains coniferous wood associated with coal seams. It is probably of palæozoic age.

(11) Limestones, pisolitic in places, in 85° 15′ South, and 7000 ft. above sea-level contain obscure casts of radiolaria.

Radiolaria, in a fair state of preservation, occur in black cherts amongst the erratics at Cape Royds. They appear to belong to the same formation as the limestone. These radiolaria appear to be of older palæozoic age.

(12) The succession of lavas at Erebus appears to have been first trachytes, then kenytes, then olivine basalts. Erebus is, however, still erupting kenyte.

(13) Peat deposits, formed of fungus, are now forming on the bottoms of some of the Antarctic glacial lakes near 77° and 78° South.

(14) Raised beaches of recent origin extend at Ross Island to a height of at least 160 ft. above sea-level.

THE HEART OF THE ANTARTIC

NOTES IN REGARD TO MOUNT EREBUS

By Professor T. W. EDGEWORTH DAVID
and RAYMOND PRIESTLEY

Volcanic Eruptions

We observed that the eruptions of Erebus, like those of Stromboli, were most frequent during a low barometer. The following is a description of the chief eruption of Erebus witnessed by us on June 14, 1908:

This morning, about 8.45 a.m., as the small blizzard of the preceding night was subsiding we noticed that Erebus was more than usually active; the steam cloud over its summit was broader and taller than usual; and there were frequent outbursts of steam.

At 11.30 a.m. it was noticed that an eruption of altogether unprecedented vigour, as far as our experience of Erebus went, was in progress. Immense masses of steam rushed upwards to at least 2000 ft. above the summit in about half a minute, and spread out to form a vast mushroom-shaped cloud. This rapidly became asymmetrical; while the main steam column was bent over to the left (northerly) by the return air current from the Pole, the higher ascending portion, at about 2500 to 3000 ft. was carried by the upper current in a southerly, or more probably south-easterly direction. At about 2.30 p.m. there was a specially grand outburst of steam. It rushed upwards nearly vertically, just a trifle bent to the north, and dashed its head with great violence through the mushroom-shaped cloud or canopy, emerged on the other side, and must have ascended there from 1000 to 2000 ft. higher; in all about 5000 ft. above the summit of the mountain.

At 3.15 p.m. a bright glow was seen on Erebus by

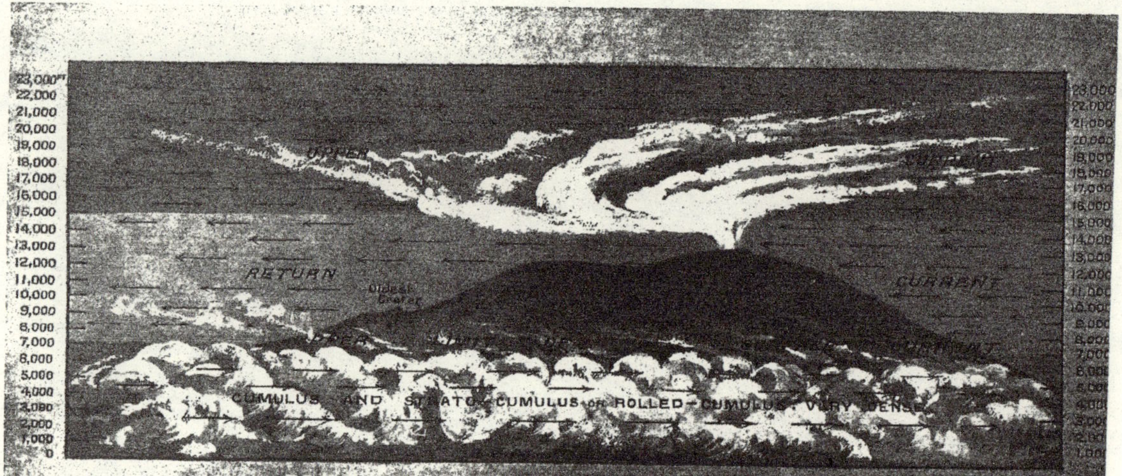

EREBUS ERUPTION, June 14TH 1908, 3-45 p.m.
Showing upper air currents.

Messrs. Priestley, Murray and Mackay. It illuminated the whole steam column to the base of the mushroom. We observed that the top of this column then spread out gradually, first into a club-headed lump, then into a mushroom-like form.

At 3.25 P.M. a remarkably bright glow suddenly lit up all the lower part of the steam column above the crater. This was seen by Shackleton, Priestley and Day.

By 3.45 P.M. the steam cloud had spread out much more and had besides risen higher, and the first formed portions of the cloud trailed away in long streamers to the south with one extensive branch going first north by east, then north-westerly.

At 3.50 P.M. there was another bright glow. As regards the uprushes of steam the interval may not have been constant, but it seemed to be about four to five minutes.

Towards 6 P.M. the bend over the steam column to the north-north-west or north-west was more strongly marked, and the asymmetry of the steam cloud was due to a preponderating amount of steam gathering to the north-west side of Erebus. By this time the older part of the steam cloud had formed a species of thin cirrus cloud, about 20,000 ft. to 30,000 ft. above sea-level.

At 6.40 P.M. Shackleton observed a very bright glow on the steam cloud.

By 8 P.M. the eruption appeared to be subsiding, and the steam cloud meanwhile stretched across the sky, now in a direction from east-south-east to west-north-west (the surface wind was at the time about north-north-east), and appeared to be passing almost over Sandy Beach and Horseshoe Bay. It was now delicately draped in the form of sinuous thin folds like a thin muslin skirt. Evidently the wind direction had changed at the summit level to about east-south-east to west-north-west.

That night there was a full moon. The sky a lovely deep blue near the zenith, paler towards the horizon. When the moon came over the great steam cloud of Erebus (22½° of arc) the scene was sublime. Its light was brightly reflected from the small glaciers at the south-west foot of the cone of Erebus; all the rest of the west and north-west slope was in deep shadow. In middle distance below the base of the cone soft white mist swathed the mountain. Near Backdoor Bay and Cape Royds brightly moonlit patches of snow showed up the black and dark brown patches of rock; our hut and stable in the middle of the foreground with the dazzling white surface of our little lake with its winding bays and coves.

On June 17, at 8 P.M., we noticed remarkable white cloud-like dense white cumulus to north-north-east over the ridge at the back of our hut and towards Horseshoe Bay.

At 11 P.M. (about) Mawson came running down from Anemometer ridge to say that an eruption had broken out from a new quarter. We rushed out and witnessed a distinct eruption amongst the huge masses of steam hanging in the air to the north-north-east. The new mass of steam, of great volume, rolled up rapidly, starting at perhaps 2000 ft. above sea-level, to probably at least 5000 ft. above the sea. A photograph was taken of it, and afterwards more distinct steam eruptions were witnessed.

The eruption appeared to have its origin on the southerly and south-south-west slope of Mount Bird, at perhaps about 2000 ft. above the sea-level. The steam cloud appeared to ascend to a height of about 5000 ft. above sea-level in an incredibly short space of time. These fumaroles or intermittent geysers must be among the most powerful known. They appear to be developed near the meridianal earth crack which extends from Mount Bird southerly through Mount Erebus.

GEOLOGICAL OBSERVATIONS

A description of the craters of Erebus has already been given in the narrative of this work, and the fact is emphasized that the interior of the old crater is largely filled with layers of large felspar crystals and pumice, alternating with beds of snow. Five of these felspar crystals are figured on the accompanying plate.

ADDITIONAL NOTES ON ERUPTIONS

By JAMES MURRAY

We were not so fortunate as to witness an eruption of any great moment. The activity of Mount Erebus showed in the form of increased volume of the steam-cloud or " smoke-cloud " ejected from the crater, and in a red glow or flare at the crater, often visible at night. There were also steam eruptions from fumaroles more or less distant, sometimes many miles distant from the active crater.

The Smoke-Cloud.—We referred to the cloud of steam which issued from the crater as the " smoke-cloud," to leave no chance of confounding it with the great laminated cloud which commonly hung over the mountain. The variations in the volume of the smoke-cloud were associated with high and low barometer. Occasionally, for a few days, no smoke was visible, and this was usually at times of high pressure. Even then we could not be sure that there was no smoke issuing, as there is a gap in the crater not visible from our camp. Great activity in the ejection of steam occurred with the glass as high as 29.5 (very high for this region), and as low as 28.3 (pretty nearly our minimum).

A large volume of cloud was often thrown up with great force, to a height of several thousand feet, sometimes as much as 5000 or 6000 ft., where it spread out

ERUPTION OF JUNE 14, 1908

ERUPTION OF NOVEMBER 27, 1908

and formed a "mushroom." Commonly in these eruptions two columns could be distinguished issuing from the crater, one white (steam), the other brown. When the two columns were soon caught by the wind, the white one went higher, and the dark one remained lower. Several times there were three columns in the crater, and on April 11 there were two in the crater, and one outside,

GREAT STEAM ERUPTION, SEPTEMBER 8, 1908

well to the right. This last might be of steam which had rolled out through the gap.

One of the largest eruptions happened on June 14, when the steam was repeatedly blown up to a great height, and getting into a still region of several thousand feet above the crater, spread out on both sides into a great mushroom cap. At one period, while we were looking at it, a more forcible eruption took place, and the ascending column penetrated the mushroom and formed a cone above it, as shown in the figure.

A beautiful eruption occurred on November 27, while a party was breaking camp at a height of about 1600 ft. on the mountain. Many diverging columns of steam shot up simultaneously, forming a design like Prince of Wales feathers. It was accompanied by varied and curious cloud forms. An attempt to depict it is made in the second illustration.

From May till September, when it was sufficiently dark to allow it to be seen, a red glow was common at the mouth of the crater. Usually it only lighted up at long intervals (ten minutes or more), and appeared to be simply an illumination from some heated matter hidden within the crater. Sometimes it flared up frequently, flashing out every few seconds, with brighter glows at longer intervals.

On several occasions (as on June 14 and 25), when the voluminous smoke-cloud showed unusual activity, there appeared to be red-hot or incandescent matter thrown out to a considerable height (1000 to 2000 ft. perhaps), and the glow lighted up the lower side of the mushroom cap. Nothing was ever seen to fall outside the edge of the crater.

Dr. Mackay pointed out some fumaroles on the old crater, which were visible from the camp at Cape Royds. They would not have been detected had they not been seen by the party which ascended the mountain. Occasionally little jets of steam could be seen issuing from these, and on May 21 quite a large cloud of steam came from one of them.

The greatest steam eruptions did not come from Mount Erebus, but from a point low down between it and Mount Bird. In this direction steam-clouds were seen rising in April, but in June Mawson twice reported considerable eruptions. On June 17 many jets of steam were seen, extending over a considerable area.

On September 8 a single jet of steam shot up to an immense height, apparently to about twice the height of Mount Erebus. A strong gale was blowing at the time, and the steam was carried away in a few seconds, so that there was no time to take an angle. In spite of the gale the steam shot up with such velocity that the column appeared quite vertical. The figure shows it too near Erebus, for the sake of comparison.

Appendix Three

SCIENTIFIC RESULTS OF THE WESTERN JOURNEY

Section I

GEOLOGICAL AND GEOGRAPHICAL

By RAYMOND E. PRIESTLEY, Geologist of the Expedition

THE stratigraphy of that portion of the western mountains which forms the cliffs on either side of the Ferrar Glacier has been well described by Mr. Ferrar in his account of the geological work carried out by the National Antarctic Expedition, 1901–3. In its broad outlines the sequence of rocks may be sketched as follows: (1) at the base lies a thick series of gneisses and schists; (2) a mass of granite with intrusive sills of dolerites lies unconformably on the schistose series; (3) this again is capped unconformably by the series of sedimentary rocks and interbedded volcanics to which Mr. Ferrar has given the name of the Beacon sandstone.

In the foothills on either side of the lower slopes of the glacier a series of schists and gneisses, with occasional beds of limestone, is exposed, and the most striking feature of this series is the great variation in texture and mineral composition of these metamorphic rocks. Further up the valley of the glacier these are capped by great masses of granite and porphyry, which are traversed by one sill of dolerite and capped by another.

The intrusive nature of these dolerites is abundantly

proved by the fact that in places thin veins arising from it can be seen cutting across the granite both above and below, and even in one or two cases enclosing large masses of that rock, which thus appear to have been caught up in the dolerite.

On the north side of the Bluff, which forms the western and upper extremity of the Kukri Hills and separates the east fork of the glacier from the Dry Valley, this structure is very conspicuous and the lower sill of dolerite appears to divide, one branch running upward and joining the upper sill, whilst the other and main branch runs downward at a slight angle for a short distance, is lost for fifty yards behind a large talus heap, and reappears at a lower level where it becomes first horizontal and then can be traced as far as the horizon as a broad black band with a very slight upward trend for the last mile or two. From the upper dolerite sill small veins extend downward into the granite below, and in the Bluff, at the upper end of the Kukri Hills a branch runs from the lower to the upper sill.

The main object of our journey up the Ferrar Glacier was to examine the Beacon sandstone at any accessible exposures with a view to the discovery of any traces of former organic life, and with this object I carefully examined at every halting-place the sandstone blocks in the moraines amongst which we frequently encamped, but without success. Had there been originally any organic remains the probabilities are great that they would have been in nine cases out of ten obliterated by the intense amount of weathering that the rocks had undergone. The more resistant of the blocks were coated, in many cases, with a hard crust of carbonate of lime derived from the original cementing material, and in places as much as an eighth of an inch in thickness, and those which were originally held together by a soluble cement were so friable that I was able to take blocks as big as my head and break them up in my fingers with as much ease as I could a similar-sized block of friable marl from the English trias; whilst in many cases large masses, when struck fairly hard with the haft of my geological hammer, crumbled to a powder of individual quartz grains. Each stone, owing to its greater capacity for absorbing heat than that of the ice around it, had melted a hollow round it of a size dependent on its own, and in cases where the hollows had not been tapped by some of the many interlacing cracks with which the face of the glacier was seamed, they were full of thaw-water covered only when in shadow by a thin sheet of ice, and the saline taste of this water, which we frequently were compelled to use for cooking purposes, was an eloquent tribute to the amount of matter filched from the boulders.

Having examined the medial and lateral moraines up as far as the Cathedral Rocks, and now approaching a region where, judging from the map, there seemed some likelihood of reaching accessible exposures of large masses of the Beacon sandstone, I asked Armytage to make for the passage between the Solitary Rocks and the north side of the glacier, and to make a *détour* on the way in order to examine the foot of the Obelisk Range if possible.

In pursuance of this plan we left the Cathedral Rocks on December 20 and made for the north side of the glacier, examining morainic matter on the way, and after successfully negotiating a series of ice-falls by keeping well to the north, we camped practically on the site of Scott's first camp after leaving the Dry Valley, namely, at the north-western end of the Bluff, which forms the upper end of the Kukri Hills. We had a good view of the Solitary Rocks from here and of the islands beyond them, and, in view of subsequent discoveries on closer inspection, I may here say that from our camp of this night I would myself have declared both that the Solitary Rocks were an island, that there were two islands off the south-eastern end of them, and that in all probability they were of Beacon sandstone, all three of which suppositions were made by the *Discovery* sledge-parties, and all three of which eventually turned out to be incorrect.

On the other hand, even from here the lowest three or four thousand feet of the Obelisk Range, marked on the map as Beacon sandstone, looked much more like the granite bluff under which we were camped, and I was more than ever anxious to examine it as closely as possible.

It was at this date that Armytage informed me that he was afraid his orders would not allow us to reach the Depôt Nunatak as he was due to meet the Northern Party at Butter Point on January 3, and as that removed my chance of getting fossils from the carbonaceous bands which occur in the sandstone of the nunatak, I determined to give fuller attention to that lobe of the glacier we were now entering and to examine the Obelisk Range, if possible, and the Solitary Rocks, together with the islands off them.

We therefore broke camp on the 21st and reached the north wall of the glacier late in the afternoon, but found ourselves cut off from the cliff exposure as the ice terminated in a sheer precipice between two and three hundred feet high and with a considerable and swift river running at the bottom. It was, however, possible from where we were to see the grain of the coarser varieties of the rock, and I had no hesitation whatever, even then, in naming it in my diary as a similar granite, with intrusions of a dark green porphyry, to that seen in the Kukri Hills.

On continuing round the lobe of the glacier, the next big fact which forced itself on my recognition was that the Solitary Rocks were not islands, but were connected with the north wall of the glacier by a ridge of the same granite over a thousand feet high.

We were obliged at the time that this first became clear to us to move further towards the centre of the ice, as the edges of the glacier had been swept clear of snow, and Armytage did not care to camp on the bare ice with a strong wind blowing straight towards the ice-cliff. After the camp had been pitched, Armytage volunteered to go with me, and we walked down to to the edge of the cliff opposite the north wall of the glacier and followed this cliff round until we were opposite the Solitary Rocks, thus clearly demonstrating that there is no passage for the glacier ice to the north of these rocks, but that they are the butt end of a peninsula banking up the ice above and forcing it round to the southern side, where it descends as a series of ice-falls. At the foot of the cliff, in the corner made by the isthmus and the main mass of the Obelisk Range, there lies a lake of considerable size which was fed by a stream from under the Solitary Rocks and by many torrents from the gullies in the mountains. This lake gives rise to the main drainage river, which follows the north wall of the glacier. I subsequently saw this river at its mouth, and found that in reaching the sea it splits into a number of sluggish distributaries which have deposited a considerable delta of fine detritus at the eastern extremity of the Dry Valley. The amount of river denudation, although accomplished in a very small portion of the year, must be of no mean geological result, since all the many streams I saw during this thaw were choked with fine sediment, and the lake was of a yellow colour owing to the amount of fine material held in suspension in its waters.

At two points I was able, during the next day, to gain access to the Solitary Rocks. These consist of alternate bands of black and yellow rock which are identi-

Side of Ferrar Glacier. Figure of Man in Lower Left Corner

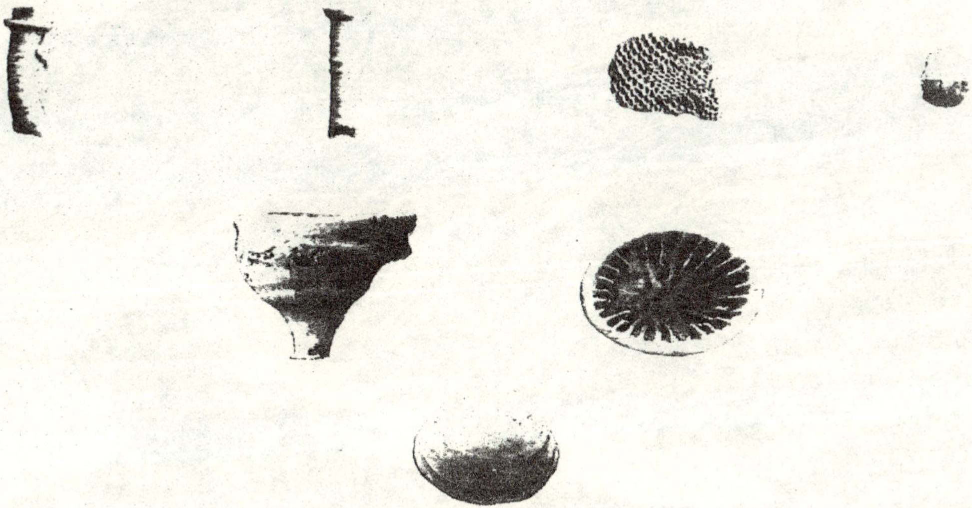

Solitary Coral — Solitary Coral seen from above

Lyothyrina Blockmann (ventral valve)

MARINE ORGANISMS FROM RAISED BEACH S. E. OF MOUNT

fied on Ferrar's map as dolerite and Beacon sandstone; but on December 22 I climbed down the ice-cliff, crossed the frozen stream separating me from the exposure, climbed up the talus scree and collected from the lower yellow band of the Northern Solitary and found the rock to be a granite similar to that of the Kukri Hills.

There were numerous specimens of the dark porphyry in the scree, but I saw none *in situ.* I also picked up several pieces of a dark fine-grained basaltic rock in the moraines, a fact which points to the occurrence of basalts among the volcanic rocks of the upper regions of the glacier. The large height at which these last ice-worn specimens were found indicates to some extent what shrinking there has been in the amount of ice overflowing through this great glacier. It was at this exposure that I observed many great blocks that had evidently fallen from the upper yellow band, and as these proved to be a granite differing from that of the lower band only in being of somewhat finer and more compact texture, I decided to ask Armytage to allow me to prolong my stay on this side of the glacier for a few hours longer in order to collect if possible from this upper band itself. In this attempt I was partially thwarted as I was unable to find any accessible place where the scree reached right to the upper band, but fortunately I was able at one other place to collect from some blocks of the rock which had fallen on the top of a scree sufficiently high to cover up the lower granite completely and also half the dolerite band above it. I have not the slightest hesitation, therefore, in mapping both the light bands of the Solitary Rocks as granite.

Specimens were also collected from the lower dolerite band, but the upper one proved inaccessible.

After leaving these rocks we decided to make a day's march round by the islands mapped at their south-eastern extremity, and from there to the bluff end of the Kukri

Hills before proceeding to the south side of the glacier to look for fossils in the Beacon sandstone of the Knob Head Mountain and the Terracotta Range. At the point where the island (for the two islands indicated on the *Discovery* map proved to be one) was struck, it proved inaccessible, but was sufficiently close for a fairly accurate description to be noted down and a rough plan and sketch were also taken. At the north-western end the sill of dolerite still remained as a sheet of considerable thickness capping the granite, and cutting across the granite were two pipes of dolerite several feet thick which persisted until they were lost beneath the ice. The middle of the island is much lower than either end, and this causes the island, when viewed from some distance, to have the appearance of two smaller ones. The south-eastern end of the island is a low granite hill with one shoulder capped by a thin portion of the dolerite sill, the majority of which has been removed by denudation. A specimen picked up proved the dolerite to be a much finer-grained variety than that of the Solitary Rocks. The plan will be published in the geological volume of the results of the expedition, together with other sections illustrating different details.

Subsequently we moved across to Knob Head Mountain, and the rest of the time before it was necessary for the party to return was occupied by an exhaustive examination of the rocks of that mountain and the Terracotta Mountains, which, though they yielded no fossils, produced many fine specimens in different varieties of the Beacon sandstones, and of junctions between the sandstone and the dolerite. Specimens of many other types of rocks were obtained from the moraines at the foot of the cliffs.

The cliffs at this spot are seamed with gullies, which must be almost always swept by the strong winds which blew practically continuously during the three or four days we stayed there. A geological result of this overflow of cold air from the inland plateau is that several beautiful instances of mechanical weathering through the agency of the sand blast are to be seen. The most marked instance of this type of weathering was to be seen in the occurrence of numerous cups of sandstone beautifully polished and concentrically striated on the outside and hollowed out more or less perfectly inside, ranging in size from that of a large cocoanut to that of an ordinary glass marble. We were for some time puzzled to account for the origin of these pot-holes, as we named them, but the mystery was cleared up when a block of weathered sandstone was found which proved to be full of round patches, anything up to half a foot in diameter, which were of a different colour and consistency from the rest of the rock. Several of these round nodules had weathered out and were lying at different distances from the parent rock, and when one or two of them were collected and broken across the middle with a hammer they were seen to consist of a very hard and indurated outer shell from an eighth to half in inch thick, enclosing a dark green ferruginous-looking sandstone of larger grain and with less cohesion between the grains. The mode of origin of these structures was now clear for they were to be seen in all stages of formation from the nodule which was left intact and only polished on the outside, through the various intermediate stages, to the perfect cup where the inside had been completely cleaned out and only the polished outer shell was left, containing many of the quartz grains which had been used by the wind as a file to get rid of the less resistant portion of the stone. Even a further stage was to be seen, for in time the outer shell itself began to wear away and the cup became sufficiently light to be moved by the wind and many had been lifted up and smashed to pieces.

One other testimony to the power of the plateau wind as a denuding agent was well shown at a height of 6000 ft. on the Terracotta Mountain, where the talus became sufficiently thin for the Beacon sandstone to be seen *in situ*. One of the beds of sandstone was a fine grained white rock and this stood out in ledges eighteen or twenty inches wide, and the underside of these ledges had been weathered into a series of thin, roughly hexagonal columns from one to nine inches long, and from a quarter to half an inch thick.

Where these columns hung close together their original structure seemed, as before mentioned, to be hexagonal, and it appears probable that the weathering has been assisted in producing this particular type by some secondary structure due to alteration and secondary crystallisation in the rock itself. When the columns hung further apart, owing to the gaps made by the entire removal of some, they had lost all definite shape and resembled nothing so much as stone icicles. Several specimens of both types of weathered sandstone were collected, but those of the latter type were extremely difficult to preserve during a sledge journey over rough ice, and one or two also were damaged in the transition from the Butter Point Depôt to the ship.

On the way down the glacier any specimens were collected which had peculiar markings on them, and these and the other specimens secured will be carefully examined when the detailed petrological work is done, but it is very improbable that any of them will reveal the presence of any organic remains sufficiently unaltered to be identifiable.

At the entrance of the east fork of the glacier and on the north side we passed and examined moraines which were essentially different from the ordinary moraines met with higher up, and there seems no doubt that the sea has played a considerable part in the heaping up of this material at the mouth of the glacier. It is proposed to deal more fully with this subject in that section of the chapter having reference to the recent elevation of land.

Section II

DESCRIPTION OF THE STRANDED MORAINES AND DRY VALLEY, WITH SPECIAL REFERENCE TO THE RECENT ELEVATION OF THE LAND BORDERING McMURDO SOUND

We first visited the moraines on December 4, when I collected a few specimens of rock and some moss and fungus for Murray, but was not able to take much extra weight as we were already pulling 215 lb. per man. Subsequent visits were made on December 13 and January 5, and it is from notes taken on these three visits that I have compiled the description which follows.

The moraines are several miles long and of considerable breadth, while many of their numerous small hills reach a height of between a hundred and a hundred and fifty feet. They consist of a heterogeneous collection of *débris* of numerous varieties of rocks, and the material ranges in size from blocks containing many cubic feet of rock to the finest dust. They are separated from the Piedmont glacier which here fringes the mountains by a stream-channel cut out almost to sea-level, and the water which has accomplished this erosion is evidently the result of the summer thaw, the stream being fed during that season both from the glacier and from the snowdrifts on the western side of the moraines. This stream is undercutting the ice, and from the exposures of morainic material on its western bank it appears probable that the mantle of *débris* continues right up to the flanks of the foothills to the west.

The *débris* being mostly of a dark colour the amount of thaw in summer is considerable, and the whole district is seamed with stream-channels which, during the few sunny days in the height of summer, are filled with running water. Every basin-like hollow between the ridges and peaks is filled with a lake corresponding in size to that of the hollow. In spite of the loose nature of the *débris* the lake basins are enabled to hold water because at all periods of the year the ground at a depth of a few feet is frozen hard. Even in summer, the whole mass of the *débris,* except an outer mantle, is firmly cemented, in consequence of the freezing at a slight depth of the percolating water supplied by the melting snowdrifts.

Proof of this is seen where the streams have cut fairly deep channels in the moraines. In the walls of these channels lenticles of opaque ice fairly free from gravel and varying from an inch or two to a couple of feet in thickness are to be seen in many places, while in other places, if a few feet of the outer mantle of the stream-cliff are removed, the gravel behind is found to be firmly cemented.

Most of the streams run northwards, and at the northern end of the moraines quite a thick alluvial deposit, having a strong resemblance to a series of miniature deltas, is to be seen along the ice-foot, awaiting subsequent removal to the sea. The amount of material removed from the moraines in this way must be very considerable. Another agency which must be fast reducing the size of the moraines is the direct heat of the summer sun on the cliffs at the northern end. Frequently while we were camped near the moraines small avalanches of gravel and mud fell on to the ice-foot, and many tons of material brought down in this manner must be carried away when the ice-foot breaks up in the late summer months. The wind also plays an important part in the transport of finer material; the snow for several miles to the north of the moraines being full of grit, which is so abundant that, as I have observed in my diary, it accelerates considerably the melting of the drift-snow and surface of the ice.

There is a much greater proportion of pebbles and larger boulders in the upper layers of the morainic material than in the lower, as seen in section in the northern cliff exposures. This phenomenon, I think, is partially explained by the fact that the finer material would gradually be carried down by the thaw-water and used to increase the compactness of the lower layers at the expense of the upper ones; but in addition I am inclined to believe that when the glacier which borders the moraines was actually providing an outlet for the ice accumulating on the mountains above it, it brought down its quota of morainic material from the local sources, which local material would reach the moraines rather in the condition of large fragments than as the finely divided *débris* which is essentially the result of the prolonged trituration for which a long journey is necessary. This latter explanation finds support in the abundance of local erratics on that portion of the moraines nearest the shore, erratics which are identical with the different great formations which crop out in the sides of the glacier valleys furrowing the western mountains. As for the material which makes up the main mass of the moraines, a great proportion of it must have come across the Sound, because while there is no evidence of any other great outburst of kenytic material besides that of Mount Erebus, large quantities of kenyte and kenytic fragmental rocks were picked up by my companions and myself during our short stay here.

With regard to the distribution of the local erratics, I will quote a note from my diary of January 6, 1909, when I observed that: " The southernmost portion of the moraines is almost entirely composed of angular basaltic and kenytic *débris* on the seaward side, the local boulders becoming more common as the landward side is approached, though at the northern end of the moraines these boulders become very numerous even on the extreme seaward side."

An important characteristic in addition to those already noted is the very isolated occurrence of some of the erratics.

Some small conical heaps consisting entirely of fragments of one kind of rock were undoubtedly, from the angular nature of the fragments, the final results of the frost weathering of very large original blocks. Not so in all cases, however. In the case of one basalt tuff particularly I noticed that it was found entirely covering two or three small hills at the south-eastern corner of the moraines, and I found it nowhere else. The pieces were all rounded or subangular and they were too scattered to have been the result of the weathering of a few large boulders.

I have noticed in my diary a similar occurrence of olivine basalt on one of the mounds fringing the Terraced Lake at Cape Barne, within two miles of winter quarters.

One fact points to recent elevation of these moraines. At the north-eastern end of the moraines a number of flat-topped hills and ridges were of the same height and all capped by several inches of a brownish deposit which proved, on examination, to be a fungus similar to those found in the lakes at winter quarters. The whole district seems, therefore, to have been at quite a recent date a lake bed. The lake has been elevated and drained, and its bed has been dissected by streams, whilst the higher land which formerly existed to the east and constituted the boundary of the lake, has been worn down and removed during the recent elevation of the moraines by a combination of the successive summer thaws and marine erosion.

I have given a fairly full description of these moraines because their characteristics are interesting, and they are a duplicate of those we afterwards examined in the north-eastern fork of the Ferrar Glacier, to which Captain Scott gave the name of Dry Valley. In describing the latter I wish to pay particular attention to the raised beach there.

Specimens of moss, fungus and alga were secured, as also numerous specimens of a peculiar rotifer living in the lakes, but a striking feature of the life of the moraines was the absence of any type of lichen.

The Dry Valley of New Harbour

It was originally intended that a sledge-party of three should make an exhaustive examination of the area which was originally occupied by the north-eastern arm of the Ferrar Glacier, but from which the ice has at present retreated, and which is now covered by a thick deposit of morainic *débris*. The plans were, however, decidedly modified through the Northern Party failing to return to Butter Point, and our operations being much restricted by our orders, we were unfortunately unable to spend more than two days in the Valley. On January 12 we left Butter Point for New Harbour, and reached our objective at 6 P.M. the same day.

The 13th and 14th were spent examining the moraines, and on the 15th we returned to our permanent camping-place at the Point, and resumed our wait for the Northern Party.

We had hardly been on shore half an hour when Armytage picked up some fragments of *Pecten colbecki*, the shell of a mollusc, which is still found living on the sea-floor at Back Door Bay. Encouraged by this find we devoted all our energies towards making similar discoveries at different points along the foreshore, and were successful beyond my wildest hopes. Not only were numerous fragments and many whole valves of the *Pecten* obtained at all elevations up to fifty feet above sea-level, but in several spots fragments of the shell of an *Anatina,* a mollusc which still flourishes abundantly in the sea on the other side, were very numerous, and in one place, by digging down several inches into the sand with the lid of a cocoa-tin, I secured two double-valve shells of another species of bivalve which I have seen commonly in our biologist's dredge.

From the abundance of the shells I have no doubt at all that this was a genuine raised beach, and the recent nature of the shells points to the probability of the rise of the land being still in progress. A few shells might have been cast up by the waves of a storm during the short time that the bay is free from ice or ice-foot, but the vast number of the organic remains precludes the possibility of their having all been cast up by this means, while a still stronger point is the nature and wholeness of the shells: both the *Pecten* and the *Anatina* possess shells of extreme fragility, and we cannot suppose that they could be hurled about by waves capable of throwing them hundreds of yards up the gravel banks without being smashed to atoms. Yet hundreds of valves of the *Pecten* especially were practically intact, ears and all, and the only way to account for their presence so far out of their native element is on the supposition of a gradual rising of the land, such as, from evidence recently brought to light, appears to be taking place along large stretches of the western coast of the Sound.

There were large mud-flats bordering the ice-foot in this valley, and these reached their greatest extent where they were augmented by the deltaic material of the many streams draining the valley, and on them, a few feet above the present sea-level, I secured numerous specimens of amphipods, and sea-spiders, and one small fish, all in a much desiccated condition. It is not necessary to postulate a recent rise in the level of the land to account for these specimens, since, in the short period when the sea is released from the control of its icy winter covering, a strong wind blowing directly into the bay would inevitably cause a rise in the level of the water sufficient to assure the submergence of those portions of the mud-flats immediately adjacent to the ice-foot. Upon the recession of the sea, numerous animals would be left stranded in any slight depression in the recently covered flats, and evaporation and ablation would remove the sea water during the late summer and autumn, leaving the desiccated remains of the animals and giving rise to an efflorescence of salt on the surface of the mud. Indeed, this sequence of events might very well have been caused by the very blizzard which raged from February 20 to the 22nd, 1908, when the *Nimrod* was driven north.

The remaining features of the Dry Valley moraines are very much a repetition of those of the stranded moraines, but on a much larger scale. I was unfortunately only able in the limited time at my disposal to cursorily examine a few square miles of the moraines, and in my longest excursion only penetrated three or four miles inland. This particular walk, however, resulted in one interesting observation, for I reached a height of between five hundred and six hundred feet, and found that the abundance of what might be called erratics foreign to the valley, namely, kenyte and basalt and fragmental rocks appertaining to these two types, had in no wise diminished in quantity, their proportion to those rocks which obviously might have been derived from the sides of the valley or from the higher reaches of the glacier remaining much the same as at sea-level.

Only two of the many stream-channels which furrow the district immediately round the camp are now active, the rest being at the best only occupied by a series of stagnant pools, while numerous depressions of smaller extent, with a heavy efflorescence of salt coating the gravel, mark the site of former pools.

The most northerly and most flourishing stream in the Dry Valley has cut a channel fifty feet deep through the stratified gravel, the sides of which slope steeply at angles between 45° and 75°. The water here was unfit to drink, owing to the amount of fine sediment held in suspension. As the stream became sluggish when breaking up into numerous branches and meandering across the alluvial stretches of land at its mouth, this fine sediment could quite easily be observed settling down in sufficient quantity to add appreciably to the delta even in the course of a few days.

During our stay at this camp I collected numerous specimens of the more interesting erratics and bags of the finer material of the deposits, and Brocklehurst secured specimens from New Harbour Heights, including two pieces of fairly pure calcite from one of the limestones interbedded with the Archæan schists.

Biological Field Notes

Specimens of moss, fungus and lichen were obtained for Murray. One of these specimens is worth individual notice, namely, a fungus-like substance growing as a thick layer on the black mud of the tide-flats just above the present level of the water and well inside the region which must be covered either at an ordinary high tide or during any unusual rise of the water.

The peculiar point about the plant which first struck me was the formation of a series of cones, as much as six inches high, and formed by the gas from the organic mud underneath collecting as large bubbles under the impervious skin of fungus.

We saw altogether three whole specimens of Crabeater seals and a like number of Weddells dead on the hills hereabouts, and the moraines are covered with scattered remains. One of the Crabeater carcases was lying at a steep angle on one of the banks bordering the northern stream I have mentioned earlier in my notes, and at a height above sea-level of considerably over two hundred feet, but this does not compare with the heights at which we found similar skeletons in the Ferrar Glacier Valley, no less than three being found there between the two thousand- and three thousand-foot levels.

Moraines in the East Fork of the Ferrar Glacier

The last ten miles of the east valley of the Ferrar Glacier below the first ice-falls is occupied by a peculiar variety of ice which, at its highest point, is not forty feet above sea-level, and is apparently composed largely of frozen slush formed by the inundation of the winter snow-drifts by the summer thaw-water. With this are intermingled patches of macrocrystalline ice exactly comparable with the ice so common on the lakes at Cape Royds, especially Blue Lake, and evidently caused by the freezing of pure thaw-water. I propose to give the detailed evidence regarding the nature of this stretch of ice when the scientific results of the expedition are published in full. At present I will only mention certain deposits, a portion of which is visible as a series of small hills protruding above the ice-surface, which hills, from their occurrence in a fairly straight line, appear to be the more prominent peaks of a partially submerged ridge. These deposits I examined carefully on December 31, 1908, on our return from the glacier, and was able to secure numerous specimens which rendered it certain that they are exactly comparable with the stranded moraines and also with those afterwards examined in the Dry Valley. Concerning these deposits I have down in my diary the following notes: "Large quantities of the *débris* are composed of the local granites and schists, but in the space of half an hour I have picked up specimens of two or three varieties of tuff (consolidated volcanic ash), basalt, kenyte and an olivine and augite kenyte which occurs sparsely at Cape Royds, and which I have so named because it contains large porphyritic crystals of augite and olivine besides the more common anorthoclase felspar, and also various porphyries which I have not seen or heard of in the valley itself."

Thus in the case of all three of these moraines, the stranded moraines, those of the Dry Valley, and those of the East Fork of the Ferrar Glacier, the conclusion is irresistible that a considerable portion of the material composing them has either been brought many miles up the coast of the Sound, or has been carried right across the Sound. The three agencies which alone could be responsible for this transport to any large extent are: (1) Shore-drift. (2) A considerably greater extent of the ice-sheet and all its affluents, such as glaciers, barriers, &c. Of such an extension there are abundant evidences, of which I may here mention the finding of granite and schistose erratics at a height of 1100 ft. on the slopes of Erebus by Professor David, Armytage and myself in the autumn of last year. (3) The third agency, and one which I have myself observed in operation, is the transporting power of icebergs and pieces broken off the ice-foot. At Cape Royds I have frequently seen, especially around Flagstaff Point, large boulders of kenyte being carried out on pieces of the ice-foot, and I have seen several icebergs which were full of a fairly fine sediment. While the ice is at its present extent, however, this last agency can only be a minor one.

Section III

EFFECT OF THE SUMMER SUN ON DIFFERENT VARIETIES OF ICE AND SNOW

The melting of large stretches of normally pure snow is carried out by the increase of some grains to the detriment of others, and this, in cases where the air temperature is low and the upper crust is cooled by a cold breeze, results in the formation of crusts of snow delicately equipoised over considerable areas, for the vapour from the lower snow crystals is cooled and condensed on those of the upper layer, and in time the crystals of the latter become firmly joined together. Underneath this layer the ablating process still continues, leaving as a final product a powder of much larger granules of snow than the particles of the original drift, whilst the upper crust is only immediately supported from beneath in places few and far between. Quite commonly, when the leader of the sledge-party stepped on the edge of one of these areas, the whole crust would shatter and fall to the ground with a soft sibilant sound.

It is this process which causes the lessening in the amount of true drift in the summer blizzards. In those winds during this season which were free from snow the only drift was a low one, seldom reaching above our thighs, and I should decidedly attribute this to the fact that the granules of ice produced by the summer ablation are sufficiently large and heavy to prevent an ordinary wind from carrying them very far or high.

When this process is going on in the drifts above the sea-ice it is somewhat modified, because the sea-ice melts at a temperature between 28° and 29° Fahr., causing actual thawing of the drift to take place from below and the ice-crust over the top of the drifts to become much thicker. The thickness in cases of deep drifts attains as much as a quarter of an inch. When this crust breaks beneath the foot, an unwary explorer is liable to be let down six or nine inches into a pool of salt water several inches deep, and this rendered sledging over the sea ice in the late summer particularly objectionable.

On December 21 the temperature of the air in the Ferrar Glacier Valley reached 40° Fahr., and in the few hours during which the air remained calm two inches of snow which had fallen the previous night had either been entirely removed as thaw-water or converted into a thin coating of rough ice, opaque through the inclusion of air, which formed a distinct help to us in our subsequent marches, as it very much reduced the slipperiness of the ablation ripples.

The principal characteristics of this big thaw on an Antarctic glacier are interesting, and some of them may be mentioned here. Every stone in the glacial moraines is surrounded by a large hollow, scooped out by the heat radiation from the boulder, and these hollows are all, more or less, full of thaw-water, which is also overflowing through every deep ablation ripple and crack. A number of the small rivulets from these sources join up at a little distance from their source to form a stream, and innumerable streams are flowing down over the convex face of the glacier to join the river which can be heard roaring at the foot of the cliff which ends the ice-face. In searching for a camp on the night of the 21st we passed across large areas of the glacier face which were slightly depressed below the general level and at the bottoms of these depressions we were compelled to wade through stretches of water, three or four inches deep, for hundreds of yards.

Stream of Running Water in the middle of the Ferrar Glacier in Midsummer

Other characteristics of the thaw are described in one of the extracts taken from my diary and drafted into the general report of the journey in a previous part of the book.

One other effect of the sun's heat on snow under peculiar conditions, which I should like to mention here, is the case I noticed of the snow surrounding many boulders in the lowest reaches of the glacier below the first ice-falls. Much of the larger morainic *débris* was here surrounded by a coating of pure ice under the snow of the drift, and between each boulder and the drift snow was to be seen a transitional granular stage like that already mentioned, but much more exaggerated, many of the grains nearest the boulder being a quarter of an inch or more in diameter.

Finally, in closing this short report of the scientific results of the western journey, I should like to pay a tribute to the generous help I received from my companions, who thought nothing of climbing screes many hundreds and even thousands of feet in height at the end of a long day's march in order to assist me in obtaining a representative geological collection, so that many of the most valuable specimens secured were collected by them. To Armytage especially, the leader of the party, my thanks are due for the unfailing consideration he showed for the scientific work of the party and for the way he accompanied me on every excursion I made after specimens of any description. For reading the manuscript of this article and suggesting many alterations in the text I am much indebted to Mr. T. F. Sibly, D.Sc., F.G.S.

Appendix Four

NOTES ON PHYSICS, CHEMISTRY, AND MINERALOGY

ICE AND SNOW

NOTES BY DOUGLAS MAWSON, B.Sc., B.E.

THE most interesting results were obtained in a study of ice in all its forms, the temperature conditions in such a climate producing phenomena hardly to be realized by the student of more temperate regions. At temperatures even far below freezing-point, snow is able to compact itself and become quite hard by a slow process of vaporisation and recrystallisation. The tendency is for the smaller snow particles to disappear and the larger to be added to at their expense, and when other conditions are suitable well-formed crystals are built up in this way. Thus well-developed hexagonal barrel-shaped crystals of ice were found developed on a seal's liver, which had been buried during the winter.

Variations in atmospheric temperatures were always accompanied by phenomena such as this; for example, at times of falling temperature, no matter through what limits, a development of spicular ice formed over all exposed objects. This accompanied a fall of even, say, from 50° Fahr. to 60° Fahr. below freezing-point. Serious interference with spectroscopic work was thus introduced on account of the camera lens becoming thickly coated with ice. A similar phenomenon was continually produced in a small laboratory built as an adjunct to the hut. This

Ice Crystals on Roof of the Hut Porch

Ice Formations

opened indirectly by the outer passage to the interior of the hut. Moist air from the living-quarters, kept at a mean temperature just above freezing-point, circulated through the cracks around the door and entered the laboratory, which for the most of the time maintained a mean temperature of 40° Fahr. to 50° Fahr. below freezing-point; there the water vapour crystallised out, coating the walls and passages with a thick formation of ice blades. The result was exquisitely beautiful, but most undesirable, finally making it necessary to abandon the room for everything but ice work.

In the same way ice formations appeared on the colder portions of the interior of the hut. In crevasses beautifully formed crystals up to four inches in length were found developed on the walls from circulating vapour. Along cracks in the sea ice during winter such formations were abundantly produced. In such situations the sea-water actually "smokes," an effect due to the freezing out of abundant water vapour present in the warmer air ascending from it. This condensation produces beautiful fern-like crystal formations, not only on the sides of the cracks but also on the frozen sea surface itself; these ice flowers are best developed when the surface waters freeze most rapidly.

During the formation of the surface ice some of the sea salts are squeezed upward through capillary cracks to the surface and there in the form of concentrated brine eventually freeze as cryo-hydrates and form nuclei for additions from atmospheric-water vapour. The net result is the production of little rosette-shaped aggregates of radiating crystal blades, which were met with up to two inches in height.

As a rule, however, the moisture is precipitated from the atmosphere in the form of snow. Rain is quite uning-point the snow fell as large six-rayed feathery flakes. It appeared that even at the same temperature, if the chilling of the vapour-laden air was more rapid, the snow fell as tiny felted spheres, one-tenth of an inch and more in diameter. This we referred to as sago, or tapioca snow, according to the coarseness of the grain. Snow falling in colder conditions, appears as tiny compact six-rayed crystals. At times when low temperatures prevailed, the air was filled with tiny crystals of ice about one-hundredth of an inch in diameter descending from a cloudless sky.

Recently fallen snow is quite soft, but soon compacts itself. This may take place in summer-time within a few hours should a cold period succeed a warmer one. A piecrust-like surface is produced in this way where a superficial hardening of the soft snow has preceded, though not sufficiently strong to support a heavily laden sledge. Very soon, however, the snow becomes more firmly bound together, and the usual smooth-sledging type of winter snow-ice surface is produced. What is known as barrier ice in the Antarctic is an immensely thick formation of this compacted snow horizontally stratified. This is what one sees in the face of the Ross Barrier and numerous other small barriers along the coast. Its characteristic horizontal surface, vertical fractures, and dazzling whiteness distinguish it from other varieties of ice even at a great distance. Nearer at hand, the stratification lines, appearing every few inches apart, are readily noticeable. Such barrier ice is really a snow-field afloat and may be expected in tongue-like forms jutting out into the sea at the debouchure of glaciers, or forming the cliffs along the sea-fronts of piedmont glacier-slopes. The granules increase in size and are more intimately interlocked in the lower portions of such formations. The tabular icebergs typical of the Antarctic which have originated by breaking off from the barriers, are, owing to the nature of their formation, less dense and far more buoyant than the transparent glacier-icebergs of the Arctic. The plateau ice, existing under conditions of great cold, sparse snowfall, and continuous strong winds, is predominantly hard. At a short distance below the surface it is still finely granular, but very compact. This is seen where the blizzards have abraded the superficial softer formation and exposed the deeper ice in channelled sastrugi; it then resembles polished Carrara marble. The more typical glacier-ice consists of interlocking crystals an inch or more in diameter.

In the vicinity of Cape Royds there are a number of small lakes occupying rocky basins. The smaller of these thaw out each year during summer, others are only partly thawed, or not at all. In these, refrigeration during the autumn proceeds under tranquil conditions. As a result the ice appears prismatic with the longer axes of the prisms parallel and vertical. Winter snow, falling on this surface, slowly consolidates, building at the same time similarly oriented additions to the surface ends of the lake ice prisms. Should several years succeed without thawing, a most interesting structure is produced, the nature of which is made evident by the subsequent removal by blizzards of the unconsolidated snow. It then presents a smooth, polished surface exhibiting a mosaic appearance, produced by clear ice prisms separated by lines of white ice containing air cavities.

As refrigeration goes on in the lakes, the saline contents are gradually concentrated in the residual liquid and a continuously increasing cold is required to freeze each succeeding separation. Ultimately a meshwork of ice and cryo-hydrate crystals is formed at the bottom of the lakes. As some of the lakes are very saline, this cryohydrate often bulks large. Some of it freezes at as low a temperature as 50° Fahr. below freezing-point.

In the case of sea-ice the first stages differ somewhat from those of lake-ice. On account of the wave movements, the regular growth of vertical prisms is seldom possible in the initial stages. Instead, delicate glancing scale-like crystals of ice form on the surface and are seen floating about, even at the depth of several feet. The single scales are but an inch or less in diameter, but they soon unite to form freely floating rosettes. Eventually, as these become more abundant, a complete felt-work is produced. In this stage the ice, only one to three inches in thickness, is partially transparent and appears dark, as the colour of the sea shows through it. This ice is peculiarly flexible. Unless a heavy swell intervenes the ice is now sufficiently strong to hold together and to aid in maintaining the sea in a tranquil state. If, however, the swell increases, the ice is broken into pieces of a foot or more in diameter, depending on its thickness at the time. These fragments, jostling each other continuously, become rounded and develop a turned-up rim. This is called pancake ice. Eventually, with the advent of calmer conditions, the pancake ice may be firmly frozen together. Crystallisation now proceeds by additions from below. Thus, vertical prismatic ice similar to that found in the lakes is produced. These prisms may be half an inch in diameter, and many inches in length. A few hours of still, cold weather goes far to freeze over the sea. If, however, the air temperature rises, the active formation of ice is suspended, to be resumed when the temperature again falls. By a succession of such variations in temperature a horizontal banded appearance is produced in the sea-ice; each band representing a separate period of

PRESSURE-ICE

CRYSTALS ON SEA ICE

refrigeration. The lake-ice often showed a banded appearance resembling this, and the greater or lesser abundance of air-bubbles set free from the frozen water and included in the ice is due to a similar cause. The banding in the sea-ice is due to an alteration of more and less opaque layers of ice, of half an inch to several inches in thickness. The sea salt mechanically separates from the ice as the latter forms and is partially forced out into the sea water below, and partially included in white vertical tracts between the ice prisms. When the unfrozen sea surface is agitated by winds at very low temperatures, the spray is apt to freeze as it scuds through the air and it falls back as ice. In this way the whole surface of the sea may assume a pea-soup-like consistency. Strong winds, producing a swell, may break up the solid sea-ice and drive it along as a field of separate floating masses. This is termed pack ice, and may eventually become frozen together as an irregular surface field of ice. In cases of this kind huge ice-bergs are not infrequently found frozen in amongst the smaller ice. Where the wind drives floe ice before it, pressure ridges may be formed by the mounting of some of the ice upon that in front; fragments may be piled up to many feet in height, an operation usually accompanied by a great crunching and grinding noise.

During the autumn, sea spray, dashing on the coast, remains behind as ice. Thus a huge ice-foot develops along the coast. Grottoes are not uncommon in this ice-foot, resembling limestone caves of remarkable beauty, filled with stalactites (up to several feet in length), and stalagmites of ice. These owe their origin largely to the fact that the more saline residual water dripping from the roof is further chilled by exposure, and thus continual additions are made to the formations from which the drip has taken place. The water is highly saline and stalagmites are produced only at very low temperatures, when they may consist entirely of cryo-hydrate.

ADDITIONAL NOTES BY JAMES MURRAY

SEA-ICE.—When a crack opened out to a yard or more in width it was very quickly filled with new ice. The prisms composing this grew out horizontally from the older ice at each side. They were straight and at right angles to the edges of the crack, except where that was abruptly bent or curved, when the adjustment to the influence of the two sides of the angle produced oblique or even curved prisms. Good photographs of this horizontal prismatic ice were obtained.

Polygonal plates of thin ice accumulated round the dredging line when left undisturbed in the sea for some time, as previously noticed by Hodgson. At Cape Royds this only occurred late in the season, and seemed to be much less in amount than at Hut Point, probably on account of the higher temperatures which we experienced.

PRISMATIC ICE.—It has been explained (by Mawson) how the vertical prismatic structure of the irregular surfaces of hardened snow found on the lakes has originated in the building up of additions to the prisms of lake ice. Further light on the formation of prismatic ice was afforded by observations in summer when he was absent.

The fresh-water ice found under tranquil conditions was clear and transparent, and the prismatic structure was not evident. In colder weather it became revealed as a very fine hexagonal reticulation. On the surface of hardened snow-drifts prismatic structure was developed while the lower layers of the drift remained as snow, and therefore independently of the lake ice prisms as a foundation.

The most interesting formation of prismatic structure

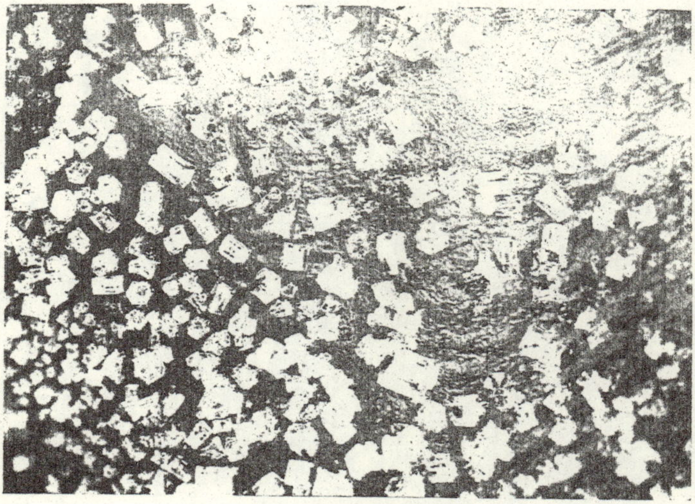

Rectangular Crystals on ice

Ice Crystals formed on a line

was observed at Clear Lake. On November 3, 1908, a large heap of chips of ice were thrown out from a shaft. The chips were of hard clear ice, and of all sizes up to several inches in diameter. On February 24, 1909, it was found that prismatic structure had developed in the heap, making it quite like the general surface of the lake, which we supposed to have been derived from snow. The prisms were most distinct at the surface, while within the mass traces of the original chips were still visible.

The mosaic design showing on the surface of prismatic ice was not uniform, nor yet was it irregular. It built up patterns, the prisms arranging themselves round centres, as well shown in Professor David's photographs. As the lake surfaces were more and more smoothed by ablation the mosaic changed in character. The prismatic structure became less distinct, and at some depth gave place to rows of bubbles arranged in curvilinear designs, the relation of which to the original hexagonal mosaic was not apparent.

About midwinter a curious optical effect was noticed at Blue Lake. The surface of the lake was almost entirely prismatic, but had been much smoothed by ablation, and was marked by ripples like those of the sand of the seashore, but much larger. On walking across the lake the moonlight was constantly glancing apparently from very large plane crystalline surfaces, several inches in diameter, inclined at a low angle to the horizontal. On touching these spots it was evident that there were no large crystals on the surface, but only the smooth-rounded ripples. The internal structure of vertical prisms made it unlikely that these broad crystalline surfaces would exist within the ice.

In the heat of summer the prismatic layer on the surface of Blue Lake became greatly eroded, large holes as much as two feet in depth being formed. Parts of the lake were unsafe to walk on, the loosened prisms crumbling away under foot.

Large Crystals in Crevasses

Similar crystals to those referred to elsewhere by Mawson, several inches in diameter, were found in the shafts sunk in the lakes. These were thin triangular plates, growing from the wall, to which they were attached, sometimes by a side, sometimes by an angle. They were regularly striated in lines parallel to one side. Some were twinned, two series of striæ meeting in the centre of a symmetrical plate. Designs were built up by the growth of new crystals from the free edges of others, or even from angles.

Ice-flowers occurred on fresh-water ice at Clear Lake and Blue Lake. They were on the ice rapidly and tranquilly formed in the trenches sunk for the observation of temperature. They were much smaller than those on the sea-ice, being only half an inch or less in diameter.

In the height of summer the combined action of the sun and air on compacted snow-drifts caused deep erosion of the snow. A kind of stratification resulted which appeared to have no relation to any original stratification of the snow. Thin flat layers of ice were formed, separated by cavities. These dipped at a gentle angle to the south, that is to say, their edges were directed towards the sun at the time of day when it is highest. These ice plates were so fragile that they collapsed in multitudes as we walked over the drifts, and a slight breeze whirled quantities of them along, often rolling on their edges.

Temperatures of Lake-ice

During the winter a series of shafts were cut through the ice of most of the lakes, in order to observe the temperature at different depths. The first trenches at Clear Lake were the work of the whole scientific staff. After

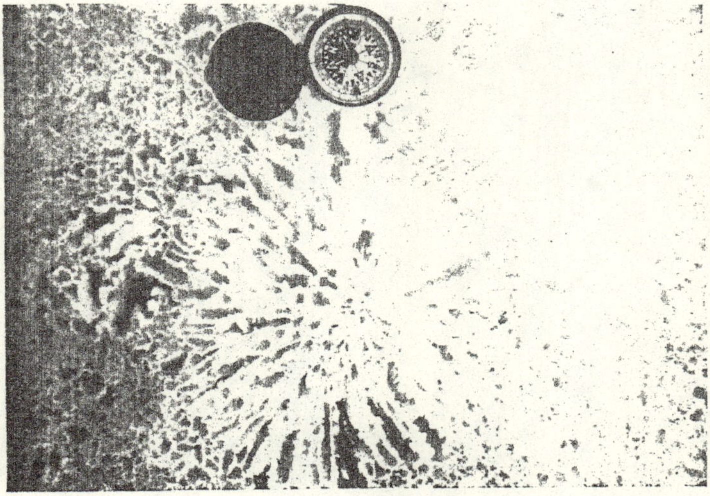

*

ICE CRYSTALS
(The Reproduction of the Compass on the upper picture is of natural size)

that the entire work devolved upon Priestley till the end of winter. Then Brocklehurst cut one in Blue Lake which proved the deepest of all reaching water at 21 ft.

A single series of observations at each shaft was of but little value, especially as the admission of air to the exposed surfaces caused very rapid changes. One shaft at

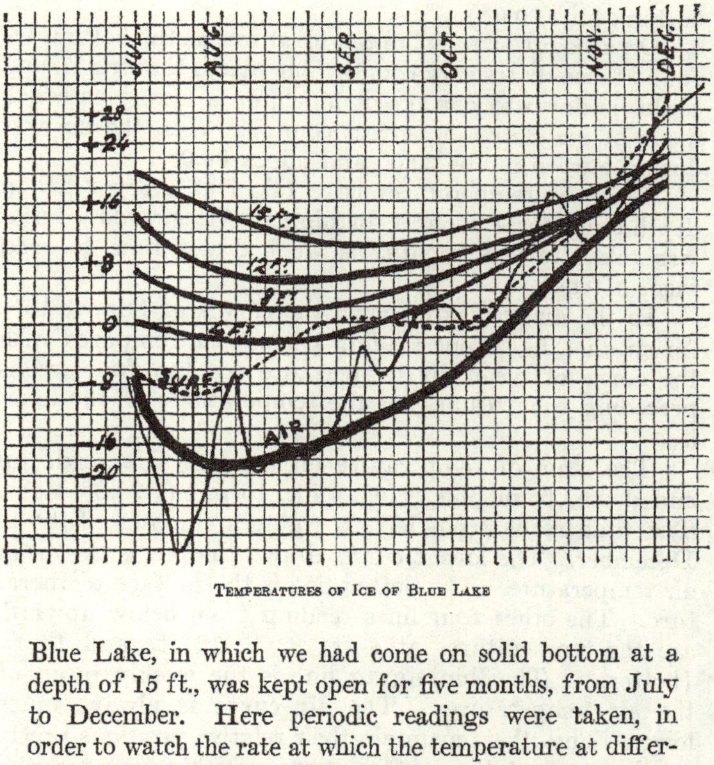

TEMPERATURES OF ICE OF BLUE LAKE

Blue Lake, in which we had come on solid bottom at a depth of 15 ft., was kept open for five months, from July to December. Here periodic readings were taken, in order to watch the rate at which the temperature at different depths changed in correspondence with the air temperatures. After some experience the shaft was covered with sacking to prevent it filling up with snow. This would also reduce the circulation of air.

On the first day, when the shaft was sunk to a depth of 5 ft., there was a difference of 23° between the surface (minus 21.0° F.) and the bottom (plus 2.0° F.). When the bottom at 15 ft. was reached a week later the whole range from top to bottom was 26° (top, minus 6.0°; bottom, plus 20.0°).

On account of the high temperatures at the bottom we found these shafts very comfortable places to work in, and could lie down to partake of lunch, on a luxurious couch made of ice chips, in perfect comfort, when the air was down between minus 30.0° and minus 40.0°.

Series of temperatures were taken at every 2 ft. in depth daily for a about a fortnight. Afterwards a series was taken once a month, as we had not time to read them oftener.

In the diagram are shown the curves of temperature for six months, at one reading per month, compared with the curve of mean air temperature for the four weeks preceding each reading of the temperatures of the lake-ice.

The scale of temperature on the left side reads up and down, from plus 28.0° Fahr. to minus 20.0° Fahr. The time reads from left to right, in months, July to December. The lowest curve, drawn thicker, is the mean air temperature. The dotted line is the surface temperature. The other four lines reading from below upward, are the temperatures at 4 ft., 8 ft., 12 ft., and 15 ft. (bottom). The thin zigzag line is the weekly mean of the air temperatures. The air curve is always much lower. The others maintain their relative positions pretty steadily, except the surface curve, which fluctuates, and becomes highest of all in December.

These curves show some points of interest. The

similarity and uniformity of all except the surface curve are remarkable. We cannot build much theory on such curves, as the curves of ice-temperature represent single observations in each month, while the air curve is a real mean deduced from twenty-eight daily means. Still the similar course traced by each curve cannot be entirely chance. If there were no direct relation between the air curve and the others we would expect greater differences at different depths.

The monthly mean is selected as giving the nearest estimate as to the rate at which the temperatures within the ice follow the air temperatures. The daily mean of air temperature would not be expected to correspond at all with the slower changes within the ice, but the curve was drawn, and fluctuated extremely. Then the mean for a week before each observation was taken, also the weekly means for the entire period, and they showed no obvious relation to the ice-temperatures. Even with the means for a fortnight before each monthly series the correspondence is far from close.

When the curve of the monthly means is drawn beside the others it is at once evident that it takes a place in the series, but that its range is much greater. All the curves converge steadily after August, and approximate very closely in November and December.

MINERALS AND CHEMISTRY

NOTES BY DOUGLAS MAWSON, B.Sc., B.E.

A large variety of minerals, chiefly rock-forming types, were met with by the expedition. The minerals included felspar, pyroxenes, amphiboles, micas, garnet, &c. Among the most notable features was the occurrence of idiomorphic felspar crystals a couple of inches in length, found abundantly scattered about the old crater of Erebus. These had apparently been expelled by the explosive force of steam from the molten lava in which they occurred.

Epidote, actmolite, tourmaline and calcite in the form of marble, were abundantly developed in the vicinity of Cape Bernacchi. In the marble cubes of graphite and iron pyrites, together with some tetrahedra of copper pyrites, were observed. In a moraine in this vicinity also a boulder of reef quartz containing iron pyrites was observed. Natrolite was found in seams in boulders of basic lava in a moraine near Mount Larsen. Titanium minerals appeared to be abundant in the eruptive rocks and schists met with between Granite Harbour and Mount Larsen.

An important occurrence of mirabilite near Cape Barne was noted by Priestley. This he found in rough masses several pounds in weight piled up at the northern end of one of the lakes. It is at a situation some height above the present lake level, but no doubt owes its origin to salts originally contained in the lake water. Gypsum was found by Joyce in fissures amongst the kenyte at the Penguin Rookery, Cape Royds. Mixtures of magnesium and and sodium sulphates, apparently originally derived from the sea, are to be found under most of the loose stones in the neighbourhood of Cape Royds. Sea spray and blown saline snow has no doubt carried these salts to their present resting-place.

* Ice Crystals formed on the line of a fish trap

OPTICS

METEOROLOGICAL OPTICS

NOTES BY DOUGLAS MAWSON, B.Sc. B.E.

MIRAGE.

Wonderful exhibitions of mirage were of daily occurrence, especially in the early morning hours. In summer time, travelling over the sea-ice on the Magnetic Pole journey, it was usually impossible to make theodolite observations between the hours of 1 A.M. and 6 A.M. on account of the extreme distortion of distant objects due to mirage effects. This was attributable to the fact that, at about this hour, a large body of cold dense air descends from the great plateau of South Victoria Land, flowing down the glacier valleys and mingling with the warmer air over the sea-ice. For the same reason the western mountains observed from Cape Royds, always loomed larger in the early morning. Distant capes viewed over the open water often appeared to be hung up in the sky.

The type of illusion known as *Fata Morgana* was of very frequent occurrence in the case of distant floating ice rafts; the warmer stratum of the air in proximity to the sea causing the slight irregularities on the ice surface to appear as lofty pinnacles.

RINGS AND CROSSES ROUND THE SUN AND MOON

These phenomena, proceeding from the refraction of the light of the sun and moon, were numerous and varied. Both large and small rings were observed. Usually only those portions of the ring appeared which neighboured on the horizontal line through the sun or moon, normal to the line of sight. Parhelia and parselene of this kind were of common occurrence.

In summer-time on the plateau we observed the most complicated and gorgeous phenomena of this kind. These were always best seen through coloured glasses.

During the winter a fine example of a parselene appeared between us and Mount Erebus. The image must have been formed within a mile or two of us.

In winter when the atmosphere was cold, clear and still, similar rings could be artificially formed near the face by breathing towards the moon. The moisture in the breath freezes instantly on leaving the body, and the optical effect is produced in the cloud of tiny floating ice particles.

Rings, coloured like the rainbow, closely investing the moon, were of frequent occurrence during the winter night. On one occasion a magnificent exhibition of this kind appeared as a series of three coloured rings; that nearest the moon showed the colours of the first order in Newton's scale; the second and third rings showed the second and third orders respectively. The effect was similar to that seen when viewing a uniaxial crystal in convergent polarised light along the principal axis.

Other Colour Effects

At certain periods of the year certain clouds are seen very brightly coloured. This colouring is strong, and all the colours of Newton's scale are seen as in the rainbow. This succession of colours increases in a direction away from the sun. The orders of colour increase successively with a corresponding reduction in distinctiveness, until too faint for observation. An isolated patch of cloud illuminated in this way resembles a fragment of a mineral like olivine viewed through an analyser under polarised light. This phenomenon was strongly shown only for the few days preceding the departure and the arrival of the sun respectively before and after the winter night.

Especially at the intermediate seasons of the year the advent and departure of the sun each day was accompanied by prismatic sunset and sunrise effects. Mount Erebus was often bathed in a delicate pink light.

Purple lights are apt to be produced on snow surfaces when obliquely lighted.

Cavities in snow formations appear of a wonderful azure blue colour. Those in ice, on the other hand, appear bluish-green, or greenish.

Earth Shadows

The earth shadows, or dark shadow bands crossing the sky, seen when the sun is very low on the horizon, were observed in a variety of forms. Some of these certainly bore a relation to the relative positions of Mount Erebus and the sun. When on top of Mount Erebus we remarked the great conical shadow it threw at sunrise over McMurdo Sound and even as far as the western mountains. It was noted later on that a relationship existed between some of the earth shadows seen in the sky from Cape Royds and this conical shadow of Erebus.

Other forms of the shadows are not so easily explained. On one occasion when the sun was low on the northern horizon near noon, just after its return, we observed the sky overhead crossed by six parallel earth shadow beams, directed from the sun supplying a Noah's Ark appearance.

THE HEART OF THE ANTARCTIC

Additional Notes by JAMES MURRAY

The sun, prime source of all the optical phenomena referred to, was seen for the last time before the long winter night on April 27, 1908. One-third of its disc was above the horizon at noon. It was again seen for the first time on August 17, 1908. The entire disc was above the horizon, and the bottom edge one quarter of the diameter from the horizon, so that it could probably have been seen a day or so earlier if the weather had been clear. The long night was therefore of 111 days (or less). It is supposed that the night at latitude 77° 30′ considered astronomically, should be several days longer, and that the sun was seen later and earlier on account of refraction.

The limits of the long day could not be so readily determined (by observation) as the sun went behind the mountains. The first year we had only a small part of the long day, and we supposed that the sun began to set on February 22, the day the *Nimrod* left us. The second year we experienced the entire long day, from about October 17 to about February 22.

There was perpetual daylight on each side of the long day for several weeks, and similarly the long night was tempered by very good twilight during the day for some weeks at the beginning and end.

Some of the most striking optical phenomena, as the earth shadows, iridescent clouds, &c., were only exhibited during the short periods when the sun rose and set each day. Others, like the prismatic sunrises and sunsets, continued long into the night.

In addition to the optical effects familiar in polar regions, such as mock suns and moons, halos and crosses, there were observed a number of optical phenomena which are not so well known.

Several of these (the prismatic sunsets and sunrises, the curved earth-shadows, and the arched ice-blink of the Barrier) are characterised by their arched form, their upper boundary being an arc of a great circle.

Prismatic Sunrises and Sunsets

The phenomenon to which we gave this name consisted in a high arch, coloured in bands like the solar spectrum, which appeared in the sky opposite to the sun, before sunrise and after sunset. It was of daily occurrence in clear

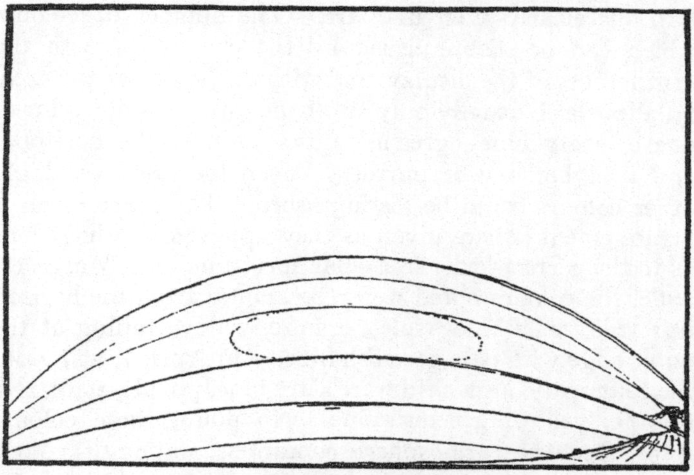

Diagram of Prismatic Arch, March 13

weather, during the whole of the two periods when the sun was rising and setting. Even during the long night it was common at noon for nearly a month after the sun had set, and it recommenced a month before the sun reappeared.

It began instantly when the sun set and got higher as the sun sank further below the horizon, at least it appeared so, perhaps through the arch becoming more distinct as the darkness deepened. Some low arches which were measured were 25° and 30° above the horizon, and it was estimated that at its highest the arch reached to 45°. Round the horizon it extended for about 90° or more.

The accompanying figure gives an idea of its size and of the breadth of the bands of colour.

Though the rainbow colours are unmistakable, the bands are very delicate, and their limits, as well as the boundary of the whole arch, are very obscure, though the latter is clearly a circular arc. The number of colours which can be distinguished by the eye varies with the distinctness of the display, and with the height of the arch. In a low arch usually only two bands were visible, a lower one of slaty blue (greenish quite close to the horizon), and a higher one of purple. When the arch was high, other colours could be distinguished. The order of their arrangement is here given as they appeared, without bias as to the correct order in a solar spectrum. On March 13, 1908, the colours noted were (reckoning from the horizon upwards) pale blue, violet, orange, yellow, fading at the outer edge with pale greenish blue. On April 7, at 7 A.M., the colours distinguished were slaty blue, purple, pink, red, orange, yellow, greenish-blue. No doubt these colours were affected by atmospheric conditions, and by the colour of clouds in the background, for the arch could often be seen against clouds.

Apparently the colours were not concentric bands, but those near the horizon formed arcs of larger circles than the upper ones. If not so the blue and purple bands would be very small, but they have really the greatest horizontal extension. Sometimes the reds and yellows did not appear to form arched bands like the lower colours, but were limited to an elliptical area, as indicated by dotted lines in the diagram.

On October 2 the blue zone rose to 15° above the horizon, and the purple zone to 25°.

In all cases the whole arch, from the summit to the horizon, was filled with bands of colour, thereby differing from rainbows and halos.

Simultaneously with the appearance of a prismatic arch, opposite to the sun there were frequently to be seen, over the sun's position, brilliant sunset colours, which also formed an arch in which the colours were in inverse order to those of the prismatic arch, ranging from orange at the horizon, through yellow and green to blue. The bands were less distinct than in the prismatic arch, but the colours were much brighter. The prismatic arch often occurred without this complementary display being visible. Though not confined to the Antarctic region this phenomenon appears to be exceptionally distinct and frequent there.

Ice-blink

The ice-blink of the Great Barrier, viewed from some little distance, was always in the form of a low arch. On January 27 and 28, 1908, we coasted the Barrier all day long, but out of sight of it. All the time the arched ice-blink accompanied us, unvarying except in height as we receded or approached, and apparently always bounded by a true circular arc. When close in to the Barrier the white glare seemed to penetrate to a very limited distance through the air, which was unaffected at a higher level.

*

Mock Suns and Moons, &c.

No illustrations are available of the more complicated displays of this kind observed on the western plateau. There are some sketches of the simpler occurrences witnessed at Cape Royds. Most commonly there were simply two bright patches equi-distant on each side of the sun or moon. These were not round mock suns and moons, as

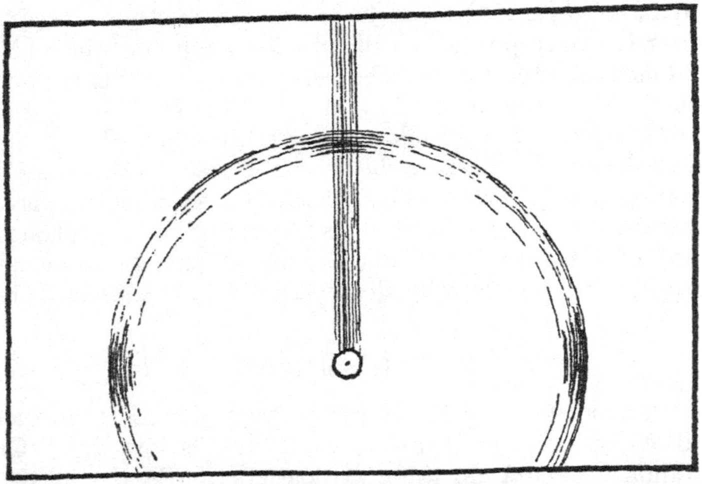

Lunar Halo, Bright Patches and Beam, June 12

frequently reported in the arctic regions, but were bits of a ring concentric with the sun or moon. Less frequently a similar patch was visible right over the sun or moon. Generally the ring connecting these patches could not be seen, but occasionally it was visible, and on June 12 there was a complete ring round the moon, accompanied by the three bright patches, and the lower part of the ring was between the observers and Mount Erebus. All these parts, which are associated alike with sun and moon, are shown in the figure. There is an additional vertical beam shown.

After the lateral sun-dogs the commonest phenomenon was this vertical beam of light, which rose from the sun or moon, and passed beyond the ring, if one were present. In the case of the sun the beam of light was yellow, that from the moon was often decidedly red. It could not readily be seen when the whole disc was in sight, but often

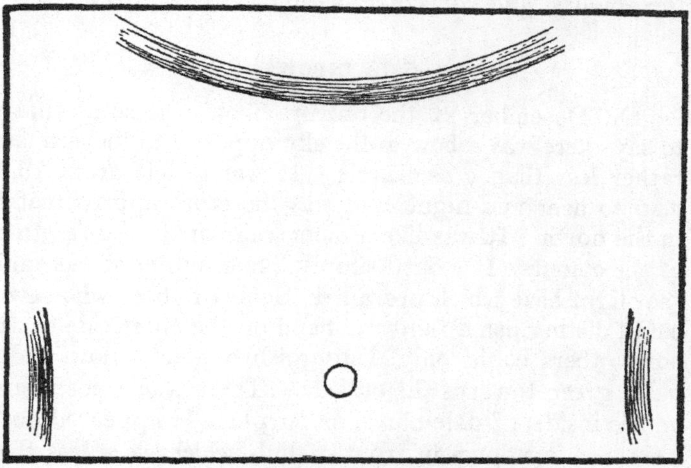

INVERTED RINGS ROUND THE SUN, FEBRUARY 7, 1909

the yellow beam was seen rising from behind a hill, the sun itself being entirely hidden. It could never be traced below the sun or moon, and we have no notes of a horizontal beam of the same kind.

A complete halo round the sun or moon was usually of uniform brightness, without brighter spots or straight beams.

On February 7, 1909, there were seen portions of three rings, convex towards the sun. The one over the sun was nearly in the zenith, and more than an eigth part of the circle was visible. The lateral ones were quite short, and like the ordinary sun-dogs, but the centres of the circles away from the sun. This display is here figured.

The iridescent colours of these sun-dogs were not disposed in broad bands as in the rainbow, but in a succession of minute coloured streaks, each repeating all the colours, and concentric with the whole ring.

Rainbow

On December 22 the only rainbow was seen, that is to say there was a bow in the sky opposite to the sun, and rather less than a semicircle. It was visible from 10.30 P.M. to near mid-night, and was therefore approximately in the north. It was like a moon rainbow in the faintness of the colours. It looked simply like a lighter streak amid the slight haze which prevailed. Some of those who saw it could distinguish a faint red band on the outer side of the bow, others could only distinguish a pale yellow and a paler green towards the outside. There was a deep band on the inside, of pale bluish or purple. It appeared quite near, coming down in front of hills which were less than a quarter of a mile away. The essential difference between such a bow as this, and the prismatic arch seen at sunrise and sunset, is that it is a band of narrow limits, while the spectrum of the other extends from the top of the arch down to the horizon.

Iridescent Clouds

The sketch is intended to show the bands of colour (vivid purple, orange, green, &c.) on the margins of the cloud, and the central patch. Each cloud has a broad central tract of deep clear blue, which looks exactly like a bit of ordinary blue sky, seen through a gap in the cloud. A great bank of grey stratus passing behind the wisps of illuminated cloud, proved that there was no blue sky present.

The colours of the iridescent clouds were brighter than in any of the other phenomena of the same class, such as the prismatic sunrises and sunsets. They were brighter than any rainbow and only to be compared with the spectrum of sunlight shown by a glass prism.

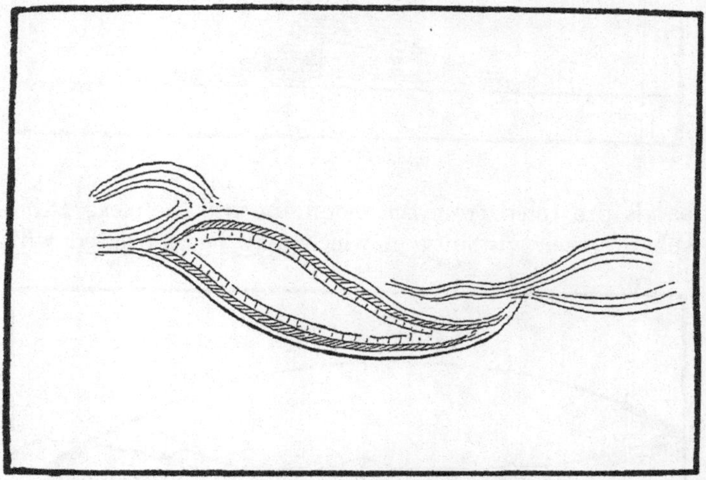

IRIDESCENT CLOUDS, AUGUST 22

EARTH SHADOWS

The name is that used by Captain Scott, though perhaps their appearance may be better suggested if they were called aerial shadows. These shadows appear to be generally from mountain peaks, but their source is often hidden. They most commonly take the form of straight

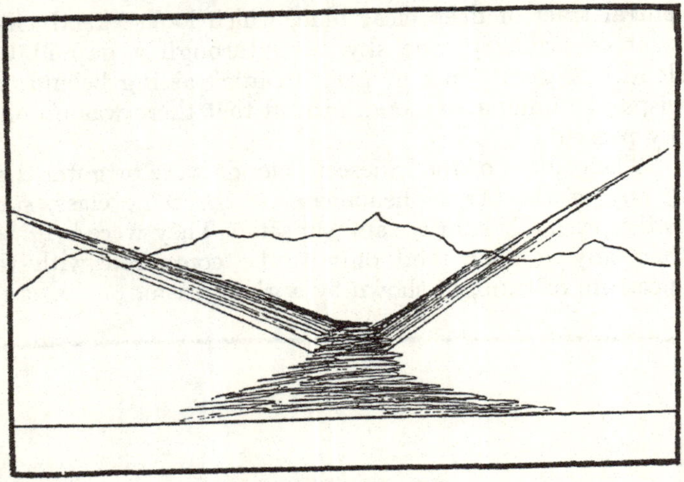

Earth Shadows, April 12

bands projected from the mountain into the clear atmosphere (their visibility in which may be associated with

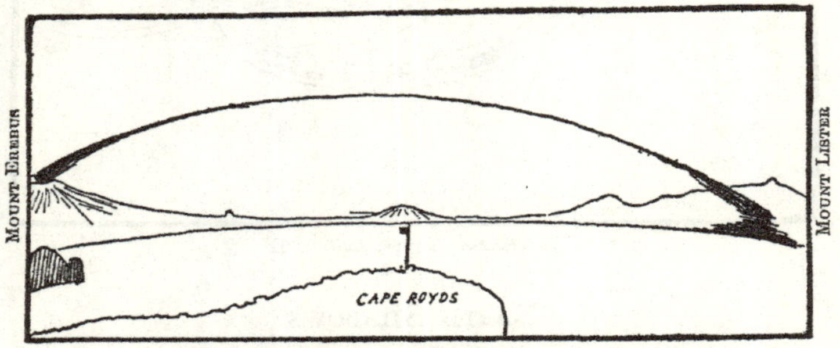

Curved Line Joining Erebus Summit and its shadow on the Western Mountains, April 12

the presence of invisible particles of ice). These bands are in the beginning sufficiently darker than the sky to be called shadows, but they fade out upwards till there remains merely a fine line which cannot be called either lighter or darker than the sky.

The figure shows one of the first of these shadows noticed on the western mountains, April 12, 1908. Two bands rose up at different angles. As the sun went round the band on the left became more nearly horizontal, the other went steeper, keeping the same relation to one

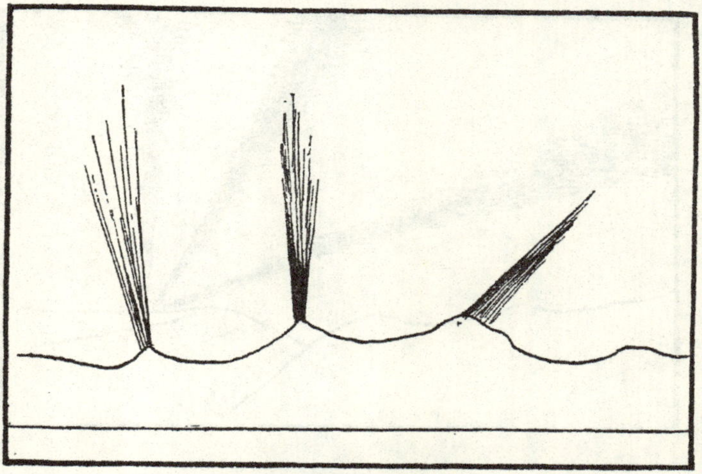

EARTH SHADOWS, OCTOBER 15

another. The line on the left could be traced without interruption to the summit of Erebus as shown in the next figure, that on the right died out altogether. The line on the left was curved, but the fact was not remarked upon at the time.

SHADOWS FROM MOUNT EREBUS

The most remarkable "shadows" were those from Erebus. As early as April 12, 1908, we noticed that the two shadows rising from Mount Lister joined the ordinary shadow of Erebus cast upon the foot-hills of Lister. There was at that time no suggestion made as to the nature of these shadows, and it was not till later detected that some of them were not straight.

On September 2 the curious curved "shadow" was first noticed, and was afterwards observed frequently.

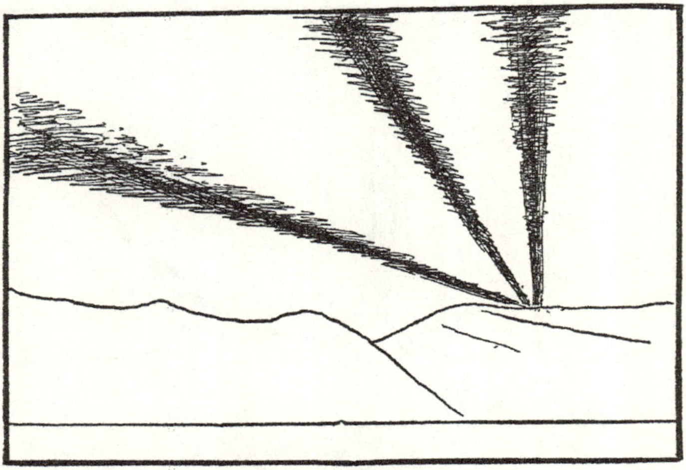

Earth Shadows, September 17

To illustrate this remarkable "shadow" it is necessary to make a diagram in which perspective is everywhere violated. When looking at the shadow on Mount Lister, Erebus was almost behind us, and the whole curve could only be seen by turning the head.

In the picture both mountains are shown, Erebus on the left, Lister on the right. The shadow of Erebus falls up on Lister, and the curved line joins them, broadening out near each end. In the foreground is outlined

* A Photograph of the Aurora Australis. The Lights in the Sky indicate the Position of the Streamers

Flagstaff Point, Cape Royds, near which the observer stood.

While the broad ends of these curved " shadows " are appreciably darker than the sky, the central portion is different. It is an exceedingly fine line, neither lighter nor darker than the sky above and below, yet perfectly well defined. In some instances the darker ends have a bright line outside, and a blue zone inside, which broadens at the ends. Usually the observer was in the shadow of Erebus when these effects were seen, but similar shadows could be detected from the outside, as on September 29, when a curved shadow crossed the Ferrar Glacier, low down, the sun being in the south-east. On the same day the curved shadow from Erebus was measured, 30° above the horizon (to the north). On this occasion it did not reach to the mountains, but ended on the sea ice.

The two previous figures show shadows from Erebus. The following one shows three peaks of the western mountains, with shadows projected from them, the sun being behind them.

Almost as puzzling as the curved " shadows " are those instances when several of the shadows diverge from one point as shown in the figure below. On this occasion the sun was behind the observer.

THE HEART OF THE ANTARCTIC

MAGNETIC OBSERVATIONS

THE MAGNETIC POLE AND THE AURORA

NOTES BY DOUGLAS MAWSON, B.Sc., B.E.

THERE are two * localities where the lines of magnetic force stand perpendicular to the earth's surface. One of these is situated in the Northern Hemisphere to the north-west of Hudson Bay and is called the North Magnetic Pole; the other, in the Southern Hemisphere, in the northern part of South Victoria Land, is known as the South Magnetic Pole.

The lines of magnetic force are imaginary lines passing through any place in the direction along which a freely suspended magnet will align itself. In the vicinity of the magnetic poles such a magnet stands vertically and at intermediate positions assumes an angle intermediate between a vertical and a horizontal position. The south-seeking end of the magnet dips downwards and is attracted towards the South Magnetic Pole in the Southern Hemisphere, whilst in the Northern Hemisphere it is the north-seeking end of the magnet which dips.

Magnets mounted as compasses are balanced on a vertical pivot and consequently they are free to swing in a horizontal circle only. They are controlled by the horizontal components of the earth's magnetic force at the spot where any observations is made and consequently, if used at the magnetic poles, where the whole of the magnetic force is vertical, they are unaffected and useless.

---

* Observations of dip in the Northern Hemisphere indicate the existence of two magnetic poles of unequal strength. The stronger of these is regarded as the North Magnetic Pole, the other is situated in Siberia and is generally referred to as the Asiatic focus.

The dip circle is the instrument used for measuring the vertical component of magnetic force, and consequently it is a very important instrument in the polar regions. It consists of a magnetised needle swinging on a horizontal axis, and the readings are taken in degrees from the vertical.

The magnetic poles, or ends of the magnetic axis of the earth, do not bear any necessary relation to the geographical poles, which are the extremities of the rotation axis of the earth. They are not diametral, but are unsymmetrically placed. In this connection one authority says: " In natural magnets the points at which attraction takes place, otherwise called poles, are generally unsymmetrically placed and depend entirely on the internal structure of the magnet as well as on the irregularities of its surface."

The magnetic poles are not fixed spots but are constantly travelling onward, executing an unknown path and apparently completing a cycle in a period of many hundreds of years. Besides this onward movement of a few miles per year, there is a lesser daily oscillation. The North Magnetic Pole was reached in 1831 by Sir James Clark Ross, who afterwards visited the Antarctic in the hope of securing the double event, but he was successful only in locating the South Magnetic Pole by observations made on his ship at a distance. In the interval between 1841, when these observations were made, and 1902, when the *Discovery* expedition again located the South Magnetic Pole, it had moved about two hundred geographical miles to the eastward.

Observations of magnetic declination and dip were taken at intervals along the route to the South Magnetic Pole. Those taken on the coast, when compared with values determined by the *Discovery* expedition, indicate that the magnetic pole has, in the interval, moved in a northerly and westerly direction. This fact was further ascertained by actually sledging inland from the Drygalski Barrier, following as nearly as possible the magnetic meridian, until the dip readings showed approximate verticality. Here the flag was hoisted. The determination of the exact centre of the magnetic polar area could not be made on the spot, as it would involve a large number of readings taken at positions surrounding the Pole. The execution of such observations under conditions of such low temperature and prevalent high winds is a matter of very great difficulty, and when it is borne in mind that many days would be necessary for the operations, the impossibility of such a course for sledging-parties such as ours is obvious.

The Aurora Australis

The aurora was first observed during the evening of March 26, 1908. Earlier in the season the daylight overpowered the light of the aurora, and therefore observations were not possible. After October 4, 1908, likewise, observations were not possible on account of the continuous daylight. Nevertheless, the aurora was probably in the heavens during the summer-time, as observations made by the ship's party in lower latitudes showed. As the *Nimrod* travelled north into regions where dark nights prevailed, auroral displays were observed both in the latter part of February 1908 and early in March 1909. Between the dates of March 26 and October 4 scarcely twenty-four hours passed without some display. At times the auroral lights were present in the heavens for many days together, though of course at full moon the brilliancy of the light obscured the more delicate auroral effects. Certain hours of the day were attended by greater displays than others. About half-past seven in the evening a brilliant display was to be expected and this continued with little reduction in intensity throughout the evening hours. The effect increased in brilliancy at about four o'clock in the morning, and died away towards 7 A.M. At one period of the year we experienced bright auroræ frequently about three o'clock in the afternoon. Very little colour was observed in connection with these auroræ other than the usual yellowish-green tints, but at some times the luminosity showed yellower than at others. In some of the most brilliant displays the curtains were bordered below by a narrow zone of deep crimson colour. The displays were usually in the form of arches, which showed minor convolutions and appeared as beautifully draped curtains. These were sharply defined below but merged insensibly into the heavens above; their depth appears to have been many thousands of feet. Besides the curtain auroræ, diffused nebulous lights were frequently observed, often in connection with clouds. On still cold evenings a faintly luminous mist enveloped Ross Island, and this seemed to have some connection with the nebulous auroræ. In fact, very early in the winter a nebulous type of aurora was seen on one occasion to descend between us and the slopes of Erebus, apparently only about five or six thousand feet in height above us.

When at their greatest brilliancy the displays were powerful enough to throw shadows but were yet insufficiently strong to allow of their being photographed. We obtained impressions on photographic plates after about ten minutes' exposure, but as the curtains had altered their shape during the interval, the result was of little value. With regard to the curtain auroræ, when once outlined in the sky they experienced spasmodic kindling, the waves of light travelling usually in one definite direction. This has the effect of producing to the eye the appearance of ripples of luminescence traversing the curtain at a very rapid rate. The curtains, as a whole, slowly drift in a determined direction, generally towards the magnetic pole. The displays, however, were very seldom observed in that part of the heavens situated towards the magnetic pole; they usually appeared in the north, through east to south. The arches sometimes travelled towards us from the south-east. Observations of electric potential showed nothing remarkable during displays of the aurora.

I am informed by the Chief of the Telegraph Department of South Australia that during September 12, 1908, the telegraph lines, both north and south and east and west across Australia, were much affected by earth currents. At the same time we experienced considerable auroral displays. Since then, on May 14 and 15, 1909, the same authority reports further disturbed earth currents, and it is interesting to note that brilliant auroral displays were observed in the Southern Indian Ocean by a passing steamer.

NOTES ON THE AURORA AUSTRALIS

By JAMES MURRAY

The different forms exhibited by the aurora could be much better understood from a few pictures than from much description. To depict the delicate and evanescent beams of auroral light by harsh black lines is very unsatisfactory, but as some illustration is desirable, these sketches attempt to give the general forms. Marston's coloured picture gives a good idea of a curtain aurora.

Curtain Aurora.—At Cape Royds this was by far the commonest type of aurora. It consists of broad ribbon-like bands made up of transverse parallel fibres. It hangs in folds like those assumed by heavy drapery, and looks very like the fringe of separate threads bordering a curtain. The folds and convolutions of these curtains are sometimes very complicated, as shown in the figure.

The picture shows part of a magnificent display of curtain aurora, which was seen on April 28, 1908. The whole sky was covered by a series of curtains, from horizon to zenith, with very little blank space. The curtains were much folded and convoluted. Similar displays occurred on May 24, and on several other occasions.

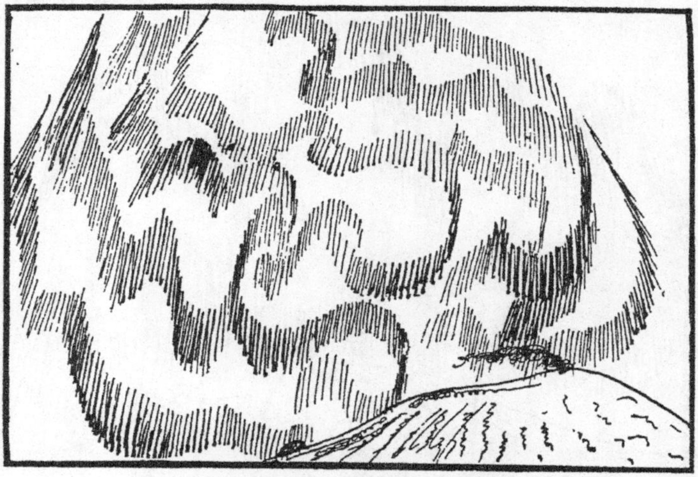

PART OF CURTAIN DISPLAY, APRIL 28

Sometimes a single curtain formed a complete band, more or less folded, which usually encircled the summit of Mount Erebus at some distance. Such a band is shown in the figure of the aurora of May 26.

Similar complete rings occurred which did not go behind the mountain.

Curtains sometimes had the beams separated, or little groups of beams separated from one another. These were spoken of as disjointed curtains. Some curtains were not folded at all, but stretched across the sky as simple bands, which were sometimes called arches. An entirely different kind of arch could, however, be distinguished.

ARCH AURORA.—This consists of a series of bright patches, arranged in a large arc or circle, and of long tapering beams going upwards from each patch. Examined closely there is a sharp distinction between the narrow

CURTAIN BAND CIRCLING EREBUS, MAY 26

tapering beam, and the brighter expanded portion. The arch in its typical form is much less common than the curtain. The rays are often seen coming from beyond the horizon, or from behind hills, while the basal arc is hidden. Similar beams often rise far apart, but still apparently radiating from a centre. Others are unconnected, and incline at various angles. These form a transition to the searchlight type.

The figure shows a typical arch, with a bright curtain beneath it, witnessed on June 19, 1908.

SEARCHLIGHT AURORA.—Long bright expanding beams suggest a resemblance to searchlights, not only by their form, but by the rapidity of their movements. They swing about in the same manner, and shoot out suddenly from the horizon or from a patch of nebulous aurora, and as suddenly disappear. They often accompany other

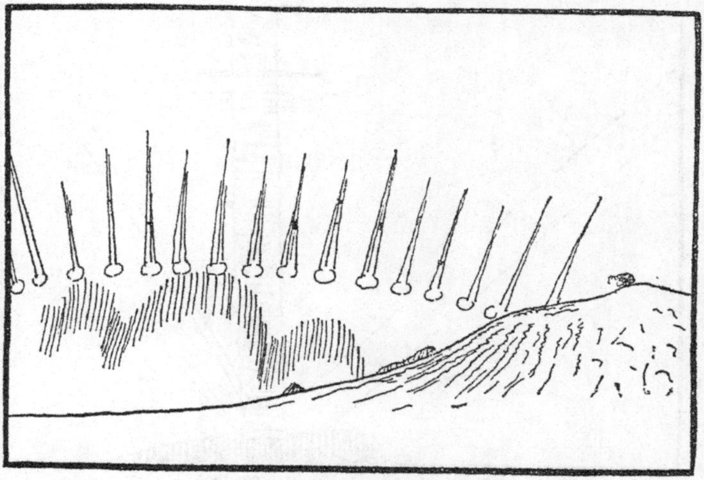

ARCH AND CURTAIN, JUNE 19

forms of aurora. Some searchlight beams are shown in the figure of the very peculiar aurora of May 23, 1908.

NEBULOUS AURORA.—A nebulous auroral haze is constantly associated with clouds, occurring on the upper surface and showing at the edges. Sometimes the nebulous light accompanied haze so thin that it would not be seen without the illumination, and in such cases it might be seen drifting along with the wind.

OTHER AURORÆ.—All the definite types of auroræ

readily change into other types. Curtains lose their form and become nebulous; the hazy border of a cloud gives out beams like an arch or searchlights; any type will suddenly concentrate into a single beam or into a shapeless patch. Some auroræ occurred which could not be classed with any of the types recognised, but they were not common enough to get special names. Two of the most striking auroræ are figured.

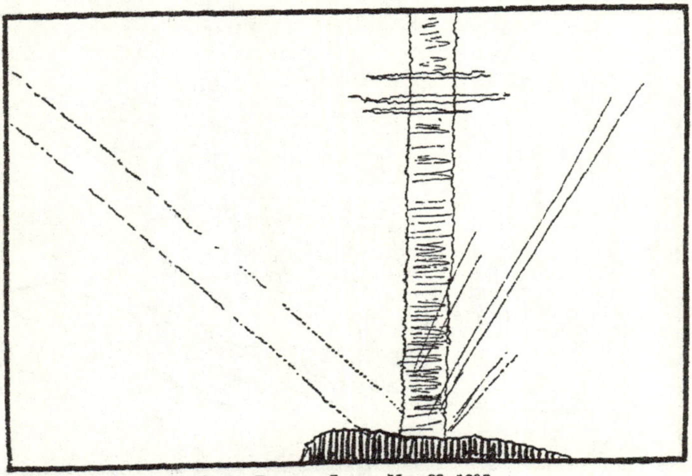

GREAT BEAM OF LIGHT, MAY 23, 1908

On May 23 at 11 A.M. a great broad beam of light appeared from behind a stratus cloud in the west from Cape Royds, and grew upwards till it reached the zenith. It was at first accompanied by some searchlight beams (shown inclined to the right) which soon disappeared. The broad beam illuminated some strips of cirrus in its path, which were scarcely visible to right and left of it.

It swung slowly down to the left, the lower end keeping in the same apparent position, till it made an angle of forty-five degrees with the horizon, when it faded. The whole display lasted twenty minutes. The dotted line in the figure shows the last position of the beam. It differed from the ordinary searchlights in the great breadth and parallel sides of the beam as well as in its steady motion.

This was one of the few auroræ somewhat in the direction of the Magnetic Pole, but it did not appear to be very distant, to judge from its illumination of cirrus clouds at a great height.

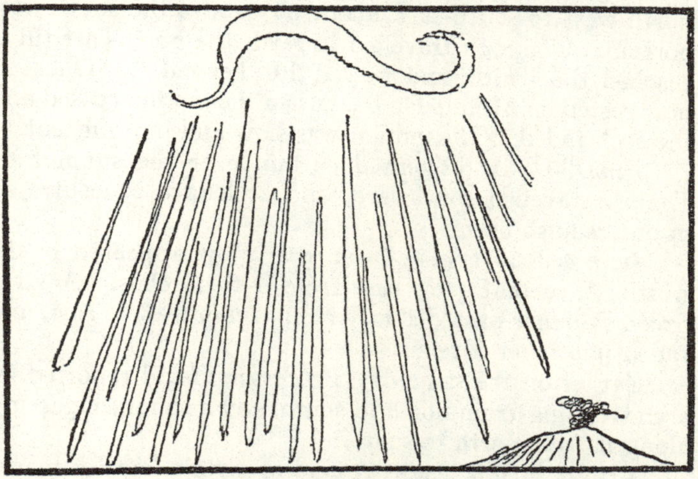

Aurora on August 31

On August 31 was seen the aurora figured below. Almost in the zenith there was a large tract of sigmoid shape. From near this there diverged very long faint streamers over the whole quarter of the sky from east to north. Though pretty close together the streamers were not joined as in a curtain, and were of unequal lengths. On several other occasions similar displays, converging on bright tracts in the zenith were seen, which approached the corona type of aurora, as figured by Peters.

MOVEMENTS OF AURORA.—Is there any general direction in which auroræ move, or in which the shimmer or kindling passes along the curtains, &c.? Some notes bearing on the question are available.

Those auroræ associated with clouds must travel simply with the wind. Auroræ not connected with clouds may have their motions determined by other causes.

May 24.—An even arch appeared, trending from north-west to south-east, about 30° above the south-west horizon. The arch travelled towards the north-east till it reached the zenith, and passed 20° beyond it. One end now rested on Mount Erebus, when the motion ceased and the arch faded. The motion was slow and intermittent.

June 30.—A disjointed curtain over the summit of Erebus travelled southward, and at length concentrated in one reddish beam.

June 4, 8.30 P.M.—An auroral glow appeared in the west. A curtain grew east from it to Erebus. A vivid green shimmer passed slowly along from west to east, and faded just over Erebus.

Earlier on the same day Professor David reported an arch trending from north to south, and a shimmer passing along it from north to south.

August 26, 6.15 P.M.—A typical curtain low down over Erebus, its ends bearing north-east and south-east. Waves of brighter light passed along it from north to south.

TIDES AND CURRENTS

BY JAMES MURRAY

The coast-line, always encumbered by the ice-foot, would render littoral tide observations in Antarctica difficult or impossible. Fortunately the fast ice, filling all

The Tide Gauge

Murray and Mawson at a hole made in one of the frozen lakes

confined waters during the winter and spring months, gives a unique opportunity for such observations, impossible in open waters. The ice-surface affords a platform on which to erect any tide-observing instrument, which, while rising and falling freely with the tides or other dishevelling agencies, is undisturbed by any lateral

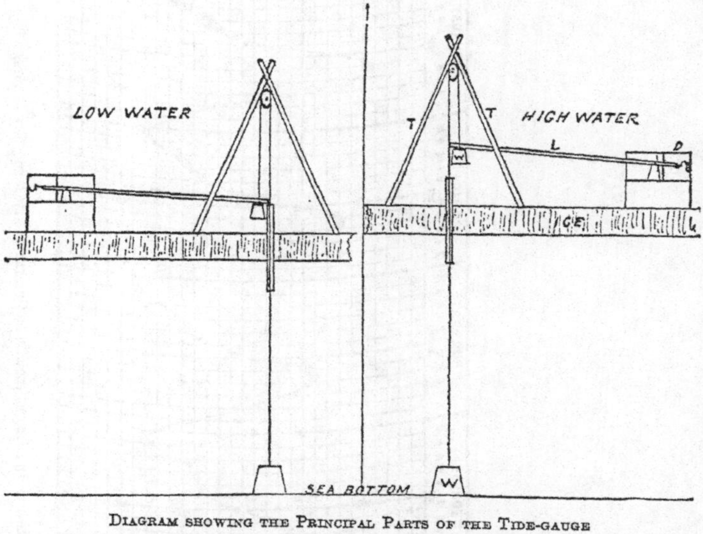

DIAGRAM SHOWING THE PRINCIPAL PARTS OF THE TIDE-GAUGE

W = the weights, the larger one as anchor, lying on the sea-bottom, the smaller one (the object of which is to keep the line taut at all times) on the free end of the line, below the lever; L = the long bamboo lever for reducing the scale; T = two legs of the bamboo tripod, supporting the pulley over which the wire passes; D = the drum on which the record is made. The recording part of the apparatus is more complicated than the diagram indicates. It will be described in detail in the scientific publications.

movements. By the combined efforts of most of the scientific staff, and the skilful hands of our engineer, Day, a form of tide-recording instrument was devised and made. It was set up on the ice of Backdoor Bay, about one hundred yards from shore, by Dr. Mackay and Professor David.

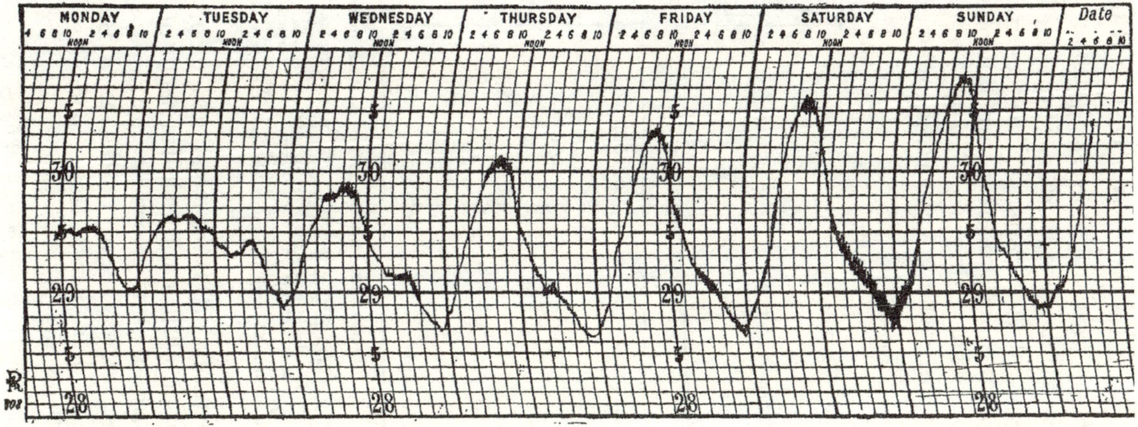

Tracing of Tide Record from Third Quarter to New Moon (New on 28th)

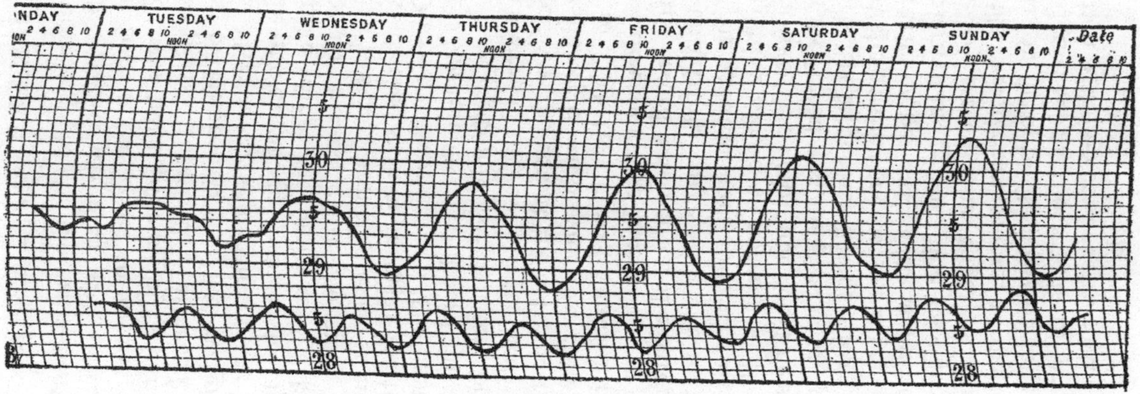

July 20 to July 27, 1908

The recording part of the instrument was a modified barograph. A box weighted with stones was put down through a hole in the ice as an anchor. The wire from this anchor was taken over a pulley hung from a tripod of bamboo poles. It was then attached to one end of a long lever of bamboo, which was weighed to keep the wire taut. By the arrangement of the lever the record of the tide movement was reduced to one twentieth and thus brought within the limits of the barograph drum. The parts above the ice are shown in the photograph. The wire in passing through the ice is enclosed in a tube which is kept filled with oil, as used for the same purpose on the *Discovery* expedition, The instrument worked well, and a continuous record was got for about three months interrupted only for half an hour weekly, when the papers were changed. At the end of that time the wire broke from the box at the bottom of the sea, and the tripod was blown down during a severe blizzard. The ice was then so thick that it was found impossible to cut a hole to put down another anchor.

The tide record obtained on the barograph drum was a simple undulating curve with one maximum per day, attaining the greatest amplitude at full and new moon, and diminishing almost to nothing at the quarters, when shorter waves of less amplitude could be seen.

When the record was analysed it was resolved into two undulations, the larger one having the period equal to the lunar day, the smaller one having a period of half a day. As one maximum per day of the lesser tide coincides with the maximum of the greater tide, its effect on the original record is to increase the apparent amplitude in that phase, while the other maximum of the lesser tide causes a flattening in the opposite phase of the greater one.

This figure shows the results of an analysis of the same portion of tide-record shown in the previous figure.

The upper curve is the larger side, and shows how it diminishes at the first and third quarters (end of line to the left) and increases at new and full (to right). The lower curve is the smaller tide, and it is almost uniform through all phases of the moon, though it is not always so uniform as in the portion figured.

SEICHES.—The tide record shows what is known as a "festooning," due to lesser undulations. These are seen chiefly during blizzards, but they are known to occur also in calms. The small scale on which the record is made prevents any analysis of these undulations, but they were more clearly seen on the first tide-measuring instrument which we used. Before the recording tide-guage was devised Dr. Mackay built an instrument for measuring tides. In this there was simply a weight sliding on an inclined plane which was marked as a scale. A pencil working in a straight line gave the amount of the tides, but to get the time of the different phases it had to be observed at regular intervals. It was noticed that over and above the steady rise and fall of the tide, the weight was constantly rising and falling at intervals of one or a few minutes, and sometimes to the extent of four inches or more.

It would have been possible to plot a detailed curve of these oscillations by observing at intervals of half or quarter minutes, but the cold was too great to permit uncovering the hands to record the observations. These oscillations are considered to be of the nature of seiche waves. Seiche was the name given to certain free oscillations of enclosed bodies of water, first observed in Swiss lakes by Professor Forel. Whatever the origin of the inequalities of level which must precede the oscillation, that continues as a seiche, diminishing till a state of rest is again reached.

The period of the seiche (*i.e.*, the time between the recurrences of the same phase of the oscillation) has a definite relation to the dimensions of the body of water. Seiches are best observed in enclosed basins of water, where they are not interfered with by tides, but oscillations of the same nature are of frequent occurrence in deep bays on the sea coast. These may be set up by the ordinary lunar tides, but the recurring tidal waves may often obscure them. The seiches at Cape Royds might originate in this way. The Ross Sea is a deep bay of large size in which secondary oscillations might be expected, and McMurdo Sound is a similar bay of small dimensions. The seiches observed, with periods of only a few minutes, and small amplitude, might be supposed to be readily set up by winds.

Currents in McMurdo Sound

There were many indications of a permanent current setting south past Cape Bird towards Cape Royds. About half a mile north of Cape Royds, at Black Sand Beach, it was by some influence deflected, and left the land, passing away southward for some miles. The readings of the current indicator set up near Cape Royds seem to show that the deflecting influence is another current coming from the southward.

The first indication of the southward setting current was a pretty one. On March 16, when the Sound was entirely open and no large ice was in sight, a strong southerly wind was blowing. This brought a quantity of fine broken ice with it which drifted along the shore from Cape Royds northwards, forming a zone about one hundred yards wide. In the figure the dark shading shows the band of ice fragments, the small arrows the direction of the wind, and the long arrow the currents from the north which stopped the drift of the ice.

At Black Sand Beach the band of ice left the shore and swung round and went away across the Sound in a direction south of west. The north edge of the ice band was as clean as if it had stopped against a solid, while the spindrift flying northward showed that there was no change in the wind. The line followed by this band of ice fragments became approximately the limit of the fast ice later on.

After the permanent ice formed in the Sound, its edge extending in a line south-west from Black Sand

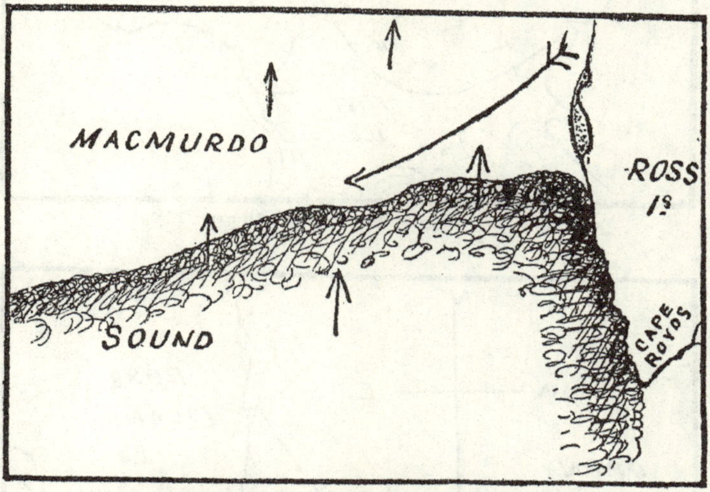

WIND-DRIVEN ICE STOPPED BY CURRENT FROM NORTH, MARCH 16

Beach, every gale cleared the ice out from the sea north of this line. When calm weather returned, pack ice and bergs floated down the east side of the Sound, till they struck the fast ice, when they began to roll along its edge away south-westward.

The drift of the *Nimrod* from January 7 to 16, 1909, when held in the pack, extended our knowledge of this current, and proved a vast eddy from Cape Bird round

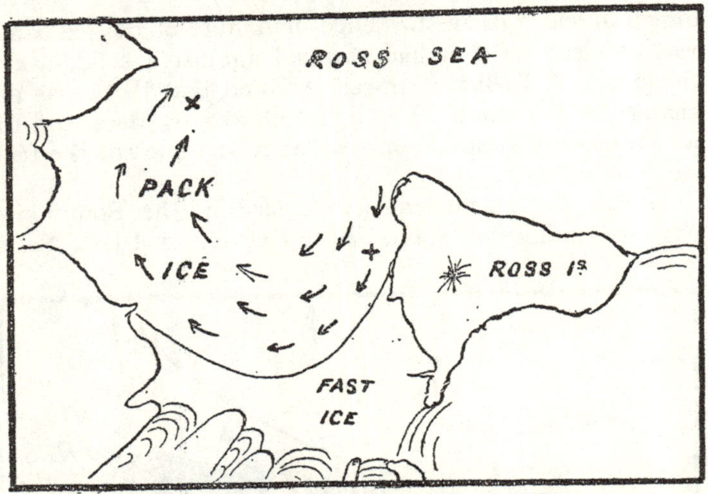

Eddy of Pack in McMurdo Sound, January, 1909

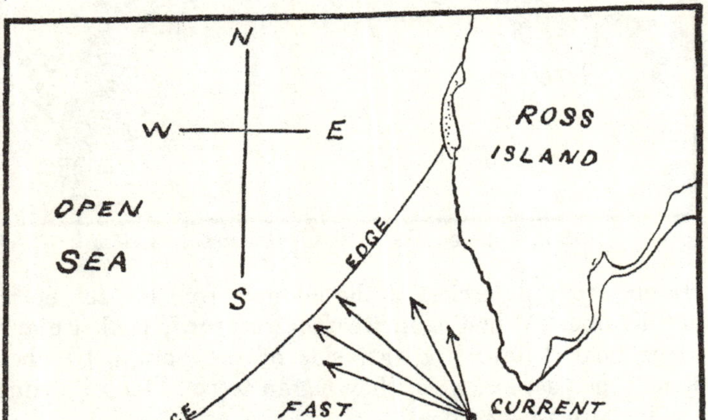

Variation in Direction of Current, June 30 to July 23, Indicated by the Arrow by Cape Royds and Granite Harbour, to the Nordenskjold Barrier.

The arrows indicate roughly the drift of the ice. The cross (+) marks the spot where the *Nimrod* was caught in the pack, and the other cross (×) about where she was released. The line below the arrows is the edge of the fast ice.

CURRENT INDICATOR.—This was devised by Shackleton, and set up on the ice a short distance off Cape Royds. It was put in charge of Brocklehurst, who visited it several times a day when practicable, from June 30 to August 18, when it was finally out of action. Only the earlier readings are taken into account, as it is suspected that latterly the vane beneath the ice became encumbered by the growth of the ice, and the readings are therefore not so trustworthy.

These readings bring out two things: *first,* a current nearly at right angles to that coming from Cape Bird; *secondly,* an absence of any indication of tidal influence. From June 30 to July 25, the current was steadily between north and west, and most frequently north-west. The arrows show the amount of variation in the direction.

Appendix Five

METEOROLOGY

A SUMMARY OF RESULTS

By Professor T. W. EDGEWORTH DAVID, B.A., F.R.S.,
and
Lieutenant ADAMS, R.N.R.,
Meteorologist to the Expedition, 1907–1909

SYSTEMATIC meteorological observations were kept on the voyage of the *Nimrod,* commencing on January 1, 1908, from Port Lyttelton down to winter quarters at Cape Royds. These observations were taken hourly. On the return voyage from Ross Island to Port Lyttelton observations were taken during every watch. On the return voyage of the *Nimrod* from Lyttelton to Ross Island, on her cruises in the Antarctic, and the return voyage to Lyttelton, meteorological observations were taken at intervals of four hours.

At winter quarters, Cape Royds, systematic observations were taken by one of us (Lieutenant Adams) during the day, from March to October 1908, observations at night being taken by whoever happened at the time to be night watchman. These observations were two hourly. From October 1908 till February 1909, the biologist, J. Murray, who had previous experience of meteorological work, was in charge of the meteorological observations and records.

Attempts were made to ascertain the amount of annual snowfall in the Antarctic in the neighbourhood of Ross Island. This task was, of course, beset with the same difficulty which the meteorologists of all other Antarctic expeditions experienced, viz., that of distinguishing between fresh falling snow and old snow drifted through the air by blizzards. We did our best to make this discrimination, but the results can only be set down as empirical and provisional. Our general conclusion is, that at Cape Royds the snowfall from February 1908 to February 1909 was equal to about 9½ in. of rain.

Mackintosh, when on the Southern Supporting-party's expedition, in charge of Joyce, in January 1909, found as the result of excavations made at the old Depôt A of the *Discovery* expedition near Minna Bluff, that the snowfall there for the past six years had been equal to an annual rainfall of 7½ in.

During the whole time of our residence in the Antarctic, from February 1908, till the beginning of March 1909, no rain whatever fell. The snow usually came with a blizzard. These blizzards blew from a general southerly direction; at Cape Royds they were mostly from the south-east. It is clear that the snow brought by the blizzards is in part drift snow, in part new falling snow. On several occasions we noticed that, whereas in the earlier part of the blizzard the snow was largely redistributed old snow in the form of drift, towards the end of a blizzard fresh new falling snow would be deposited. As, at the time of the blizzard, the wind was travelling very rapidly from the south at the rate of perhaps sixty to seventy miles an hour, we argued that this new falling snow was probably produced by moisture carried by the upper currents. The temperature of the atmosphere invariably increased considerably from the beginning of a blizzard towards its end. This rise was very marked, for whereas the initial temperature, at the beginning of a blizzard, would be perhaps minus 30° Fahr., at the end of a blizzard, after a lapse of possibly twenty-four to thirty hours, the temperature would have risen to plus 12° or plus 15° Fahr.

This rise in temperature may have been due to causes of which the more important are: First, the usual föhn effect, the temperature of the air being raised through compression as the air descends from higher levels to lower. This compression effect ought theoretically to make itself strongly felt at the atmospheric South Polar vortex. Secondly, the latent heat set free when aqueous vapour in the atmosphere is passing into the form of snow which, of course, tends to raise the atmospheric temperature. It might also be suggested that as the atmospheric circulation during a blizzard is immensely accelerated, probably the upper winds under these conditions may transfer relatively warm air from tropical regions polewards. If this be so, it is quite possible that some of the snow which falls towards the close of a blizzard was formed out of the moisture generated in warmer climates.

The Southern Party specially studied the question of whether much snow fell far south, and the Northern Party, who went to the South Magnetic Pole, also paid special attention to this point of whether the snowfall increased in proportion as one receded from the South Pole. As the result of the observations of the Southern Party it was clear that within ninety-seven geographical miles of the South Pole there were still very strong south-south-east winds, bringing with them a quantity of snow. During the time the Southern Party were on the plateau no falls of fresh snow were observed; but there was nothing to suggest that the annual snowfall was less than at winter quarters.

Temperature.—The lowest temperature we experienced was minus 57° Fahr., near White Island on the Great Ice Barrier, on August 14, 1908. We may refer to

Cloud Spirals above Mount Erebus

* Cloud Spirals in Upper Currents of Air near Erebus

Mr. Murray's notes for a comparison of the temperatures observed by us and by other Antarctic expeditions.

Winds.—These may be divided into surface winds and high-level winds.

As regards surface winds, we found in Ross Sea that these were controlled to a great extent by the presence or absence of ice over Ross Sea. Once McMurdo Sound and the Ross Sea to the north of it became firmly frozen over, we found that we enjoyed calmer weather conditions than we did when the sea was open. Evidently the presence of a large surface of comparatively warm water at plus 28° Fahr. acts as a disturbing factor in the local atmospheric circulation. The surface winds at our winter quarters were either gentle northerly winds, whose speed seldom exceeded twelve miles an hour, or gentle winds from the south-south-east or south-east. If these latter winds become strongly developed they pass over into a definite blizzard. One of the rarest winds at Cape Royds was a north-westerly.

On the southern journey it was noted that south-south-east winds predominated on the surface of the Great Barrier. These sometimes swung round to between south-south-west and west-south-west. The Southern Party experienced a violent south-south-east blizzard at a point just beyond latitude 88° South. At the furthest point south attained by them, latitude 88° 23′, the sastrugi were large and high, and trended from south-south-east to north-north-west. There was much soft snow on this part of the plateau at an altitude of over 10,000 ft.

On the journey of the Northern Party to the South Magnetic Pole it was found that the chief winds on the coast, as well as on the high plateau, are from south-south-west to west, with occasional blizzard winds from the south-south-east and south-east. Both the Northern and Southern Party kept specially careful records of the direction of the dominant sastrugi. Maps of these will be published in the Meteorological Memoir. It is certain that a good deal of the westerly wind experienced by the Northern Party in their journey along the sea-ice near the coast was practically a land breeze. It used to spring up soon after midnight, and keep on blowing pretty freshly until about 10 A.M. on the following day. It was observed that occasionally in the neighbourhood of Mount Nansen a breeze would spring up from the north-east off Ross Sea, carrying dense cumulus clouds inland.

HIGH-LEVEL ATMOSPHERIC CURRENTS.—On January 14, 1908, Messrs. Leo. Cotton, Douglas Mawson and T. W. Edgeworth David were able to get some observations of the direction and rate of movement and height of the upper wind current, in latitude 69° 53′ South, longitude 179° 47′ West. We estimated that the mean height of the mackerel sky which seemed to be formed at the base of the anti-trade wind was on this occasion between 13,000 and 14,000 ft., and we determined that the rate of movement of the mackerel clouds was about twenty-miles per hour in a south-easterly direction. This does not mean necessarily, of course, that the upper wind was not moving at a still more rapid rate, but the figure may be looked upon as the minimum speed for that current.

At Mount Erebus our winter quarters were situated in an exceptionally favoured position for observing the upper currents of the atmosphere. Not only had we the great cone of Erebus to serve as a graduated scale against which we could read off the heights of the various air currents as portrayed by the movements of the clouds belonging to them, but we also had the magnificent steam column in the mountain itself, which by its swaying from side to side indicated exactly the direction of movement of the higher atmosphere. Moreover, during violent eruptions like that of June 14, 1908, the steam column

Dumb-bell Cloud above Erebus

rose to an altitude of over 20,000 ft. above sea-level. Under these circumstances it penetrated far above the level of a current of air from the pole northwards, so that its summit came well within the sweep of the higher wind blowing in a southerly direction, the result being that the steam-cloud in this region was dragged over powerfully towards the south-east. On such occasions one usually saw evidence of two high-level currents, the one coming from a northerly direction, its under limit being about 15,000 ft. above sea-level; and the other, or middle current from a southerly quarter, usually blowing towards the east-north-east, meeting its upper limit at 15,000 ft. normally while its lower limit was between 6000 and 7000 ft. above sea-level. While these two currents were blowing strongly, there would frequently be a surface current blowing gently from the north. This would bring up very dense masses of cumulus cloud from off Ross Sea. The cumulus would drift up to the 6000 or 7000 ft. level on the north-west slopes of Erebus, and then the tops of the cumulus would be cut off by the lower edge of the northward flowing middle current. Wisps of fleecy cloud would be swept along to the east-north-east torn from the tops of these cumulus clouds by the middle current. The whole appearance is illustrated in the accompanying diagram.

It is of especial interest to note the effects of blizzards on the direction of movement of the high-level currents, as well as on their altitude. As the result of our ascent of Erebus we ascertained that the whole of the snowfall lying within the rim of the second great crater, at an altitude of from 11,500 to over 12,000 ft., is strongly ridged with sastrugi, which trend from about west-south-west to east-north-east in the direction of the prevalent middle-air current. The sharp points of these sastrugi are directed towards the west-south-west, the quarter from which of course the prevalent wind blows.

Our actual experience of a heavy blizzard, at a level of over 5000 ft. on Erebus, as well as our subsequent observations of the height to which the blizzard wind extended, showed that during blizzards the whole atmosphere from sea-level up to at least 11,000 ft. moves, near Cape Royds, from south-east to north-west, and the speed of movement is from forty up to over sixty miles an hour.

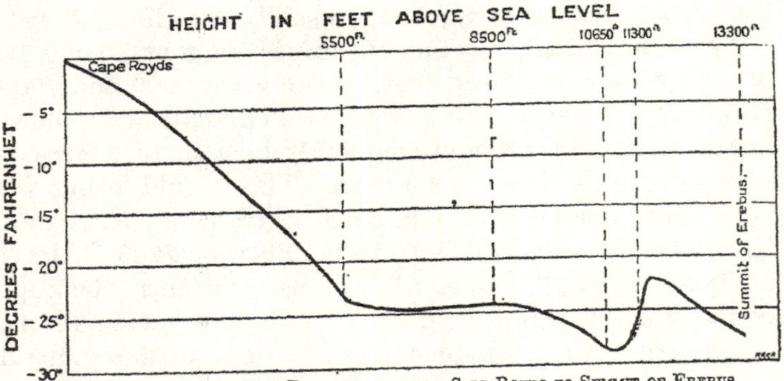

THE CURVE OF ATMOSPHERIC TEMPERATURE FROM CAPE ROYDS TO SUMMIT OF EREBUS

The day that we reached the summit of Erebus, March 10, 1908, we found ourselves at the level of over 13,000 ft. within the lower limit of the upper wind. Subsequent observations by us of the point in the steam-cloud over Erebus, where the bend took place at the junction of the lower limit of this current, with the top limit of the middle current, showed that after and during the blizzard the middle-air current, normally blowing from the west-south-west, is temporarily abolished, being absorbed by the immense outrushing air stream of the south-east blizzard. We noticed that usually in winter time, especially when a blizzard was impending, strong cloud radiants were developed towards the north-west. These radiants were produced by the apparent convergence, due to perspective, of long belts of cirro-stratus clouds; they could be observed swinging round for several hours from north-west to true north, and even east of north. When this was the case a blizzard was certain to be impending. Unfortunately, during a blizzard, the air was generally so thick with snow that we were unable to see the top of Erebus. At the end of a blizzard the air current over Erebus became suddenly reversed, the steam-cloud swinging round from the south to the north. After a time, following on the conclusion of a blizzard, a high-level current was seen to be floating the cirrus clouds from the south-east towards the north-west, and the steam of Erebus would stream out towards the north-west. We could not account for this high-level south-easterly current. It looked like a reversal of the usual upper wind, and it appears to be a fact new to meteorological science.

As regards the sequence of events during a blizzard, they would seem to be as follows: First, there would be gentle northerly winds at Cape Royds for perhaps one or two days; temperatures would be low, and it would appear as though the air flowing south was coming to supply the void which otherwise would be caused by the contraction of the atmosphere near the pole. Then would follow two or three days of absolute calm, the temperature meanwhile continually falling. We may suggest hypothetically this meant that a great mass of air near the pole was constantly becoming heavier and denser as the result of the increasing cold. Sooner or later, perhaps a week after the northerly wind ceased to blow, this heavy mass of cold air would seem to force a passage for itself equatorwards. It would commence rushing out as a south-easterly blizzard, and as soon as this rapid current was started, and even before it had got to the latitude of Cape Royds, 77° 32′ South, an acceleration of the upper current above it had already set in. This led to the curving of the upper current into a direction more or less parallel to meridians as it rushed in to take the place of the cold wind escaping from the neighbourhood of the pole. The increased speed of circulation of the upper current increased the normal föhn effect at the pole, and this, combined with the more rapid transfer of warm air from the north to the south by the medium of the upper current, together with the latent heat from the snow formed combined to raise the polar atmospheric temperature, and so temporarily to make that region a region of relatively low pressure instead of relatively high, that is, relatively, as compared with the normal atmospheric pressure there. As soon, however, as this warming of the polar air had become general conditions for a blizzard wind ceased, and a period of calm supervened. At the moment of the cessation of the wind, conceivably, a species of hydraulic ram effect made itself felt in the suddenly checked anti-trade wind current above, which led to the stoppage of that great air stream temporarily, and its resurging back equatorwards, thus producing a curious high-level current frequently seen by us after a blizzard in the neighbourhood of Erebus.

To return from this theoretical digression to observed facts, an occasional precursor of the cessation of a blizzard was the veering of the wind from south-east through east-south-east to east.

That these blizzard winds occasionally blow right across the Antarctic Circle, and reach the shores of Australasia, is indicated by the experience of the *Nimrod*. When she left Cape Royds on February 23, 1908, she left us in a light south-easterly wind, the survival of a previous blizzard, and the following four days it was still blowing a blizzard. Then, after a partial cessation of the

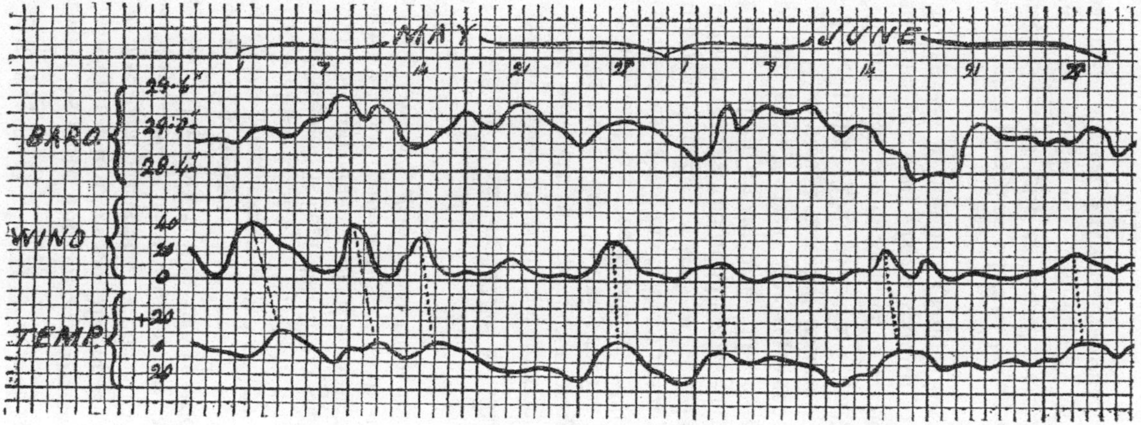

Diagram showing Daily Means of Barometric Pressure, Wind and Air Temperature wind for two days more, the blizzard freshened again on March 1, continuing to blow on March 2. The *Nimrod* experienced these south-easterly winds all the way back, from Cape Royds to Port Lyttelton in New Zealand, arriving at the latter port within twelve and a half days after she left Cape Royds.

It is, of course, premature as yet to generalise on the result of these observations. It is hoped that when the Meteorological Memoir is completed that the observations, especially on the movements of the higher atmosphere, will contribute to our knowledge of meteorological conditions and the laws which control them in the Southern Hemisphere.

Additional Notes By JAMES MURRAY

In regard to the relation of barometric pressure and temperature to wind-storms, the accompanying diagram shows the curves traced by the daily means of barometric pressure, wind, and air temperature. The curves are drawn to such scales as will make them readily comparable.

The uppermost curve is that of the daily means of the barometric readings, the middle curve is the wind, and the lowest one is the temperature. The dates are marked at the top, and each vertical line represents an interval of one day. The period selected includes the whole of May and June 1908. The scales are marked vertically at the left-hand side, the barometer in inches and tenths, the wind in miles per hour (from zero up to 40), the temperature in degrees Fahrenheit (from plus 20° to minus 20°).

The midwinter period is selected because at that time two features characteristic of the Antarctic climate are best developed. The first is the absence of constant and definite correspondence between barometric movement and wind; the second is the constant relation of temperature and wind.

It is not meant that wind-storms never accompany or follow great changes of pressure, but that the relation often fails, and is not constant and reliable. Severe wind-storms may occur when the barometer is steady or shows only gentle movements, as in the first few days of May 1908. On the other hand, and more commonly, very great rises and falls of the barometer may occur in comparatively calm weather, as between June 1 and 12, in which period there was only a few hours of moderately fresh breeze on the 4th. June 19 to 22 is a more striking example, when a very rapid rise of barometer occurred, without any decided wind-storm. But the wind-storms between May 5 and 15 accompanied considerable fluctuations of the barometer.

The relation of wind and temperature is shown in the remarkable parallelism of the two curves. A wind-storm is almost invariably followed by a rise of temperature, as indicated by the dotted lines. The only exception in the period illustrated is the moderate storm of May 20 to 21, when there was no appreciable rise of temperature.

Cold blizzards have been noted. The worst of all was that of July 25 to 27, when the daily means ranged from minus 22.0° to minus 29.0°, and as low as minus 35.0° was recorded.

The rise of temperature during a blizzard amounted to as much as 32° on July 9, and even greater rises have been noted.

The intimate relation between wind and temperature continues throughout the cold weather, but almost disappears in the summer months. The relation of wind to barometric pressure (possibly on account of the proximity of open sea) is also much more regular in the summer.

ANTARCTIC TEMPERATURES

The accompanying diagram was drawn while in the south in order to make a comparison of the temperatures experienced by all the Antarctic expeditions (of which records, giving monthly means for a period of one year or more, were available). The results of plotting the temperature curves were so curious and interesting that the diagram is here reproduced.

The months read from left to right, and the temperatures, in degrees Fahrenheit, from above downwards.

The two uppermost curves are Dr. Bruce's records (*Scotia* expedition) dot and dash —.— = his first year (1903), dots . . . . = his second year (1904).

The thin plain line = Borschgrevinck's record (1899) (This curve is taken from Mr. Armytage's book.)

The short dashes - - - - - = the *Discovery* records for the first year (1902), the long dashes — — — = the second year (1903).

The thick plain line gives the curve for our expedition which only covers ten months of the year 1908.

Dr. Bruce's temperatures were recorded much further north than any of the others and outside the Antarctic Circle. They are much higher throughout except in the summer months at which season all the records approximate pretty closely. The remarkable feature in these two series is the alternating of higher and lower temperatures in the two years so that the two curves when plotted together make a regular chain.

Borschgrevinck's curve is much higher than those of the *Discovery* except in July and September which were equalled in the second year of the *Discovery*.

The *Discovery* curves for the two years show a peculiarity similar to those of the *Scotia*. The second year was colder throughout, till October, after which it

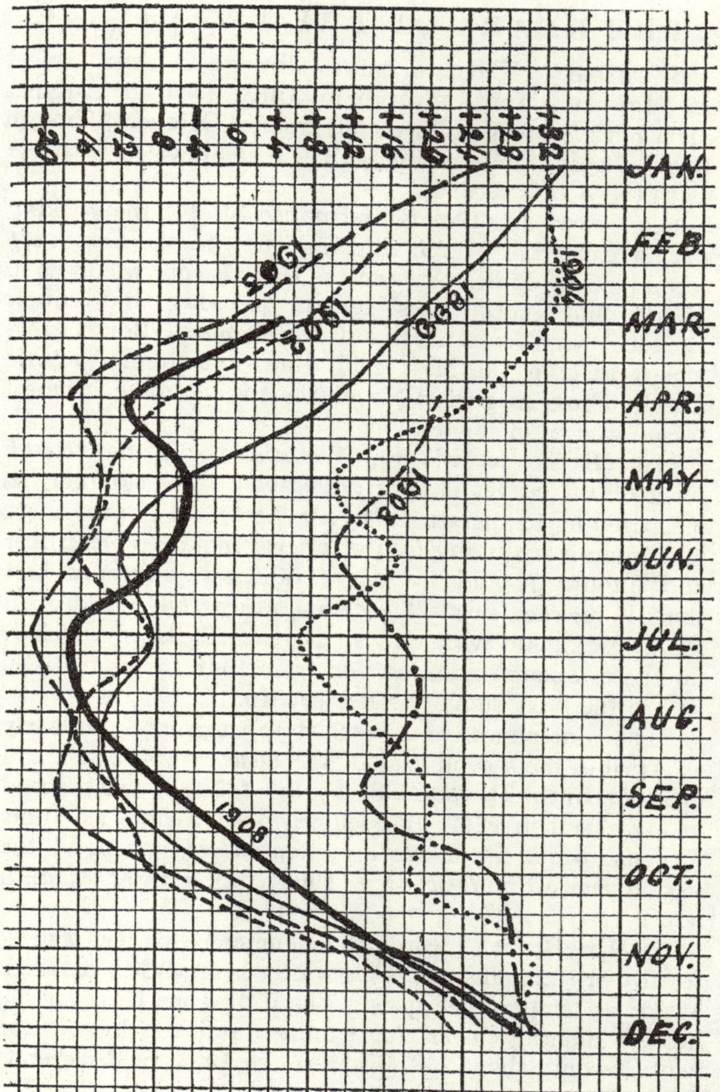
JAN.
FEB.
MAR.
APR.
MAY
JUN.
JUL.
AUG.
SEP.
OCT.
NOV.
DEC.
1908

was warmer. The two curves diverge in April, meet in May and June, diverge in July, meet in August, diverge in September, meet and cross in October. Though there is only one crossing the chain is almost as distinct as in the *Scotia* records.

Our curve for 1908 is first noticeable for the striking parallelism from March till August, with the *Discovery* second year. It lies between the two *Discovery* curves in March and April, rises greatly in May and equals Borschgrevinck's. In June it is well above Borschgrevinck's, making the least cold June known in Victoria Land. From August to November it is again the warmest season known in this region.

In all the records for Victoria Land the trough form of the curves is very noticeable. The summer, as measured by temperature, is very short: the approximate maximum is maintained only for two months (December and January), and the curve falls away steeply and steadily on each side. The temperature falls rapidly till the approximate winter mean is reached, after which it fluctuates about this mean for from four to six months, before the next summer's rise sets in.

The winter mean was reached approximately as early as April in the second year of the *Discovery* (1903), and in 1908. In the *Discovery* first year (1902) it was reached in May, and in 1899 (Borschgrevinck) as late as June. The rise began in all cases in September, except in the second year of the *Discovery* (1903), when it was a month later.

In Dr. Bruce's records the summer is longer, lasting for five or six months, and the winter is much less marked, as far as temperature goes. The difference between the highest monthly mean for summer and the lowest for winter is only about 26° (Fahr.). In 1908 the difference was 47°.

METEOROLOGY

An important characteristic of Antarctica is the cold summers, with the monthly mean rarely above freezing-point. The diagram brings out the close approximation (within 10°) of all the means recorded for December and January, from regions so far apart in latitude as the wintering stations of the *Discovery* and the *Scotia.*

NOTE ON THERMOMETERS FOR POLAR WORK

BY JAMES MURRAY

For work in regions where temperatures below the freezing point of mercury are to be expected it is customary to trust entirely to spirit thermometers. We had reason to regret doing so. Whatever may be their behaviour in temperate climates, in polar regions the error of the spirit thermometers is apt to fluctuate in a puzzling and irritating manner, especially in thermometers which have to be carried about from place to place.

For example, several thermometers were tested in the sea in winter, when the temperature was just a small fraction above the freezing-point of sea water. They showed errors varying from one or two to many degrees. When tested immediately afterwards in water from melting fresh-water ice, they read quite correctly, that is to say, they showed errors no greater than those indicated on the Kew certificates.

To make corrections in the readings only for the certified error would be most misleading. To attempt to restore the thermometers by getting the displaced spirit back into the column before beginning a series of observations was of no avail, as the thermometers might go wrong at any moment.

The only means of giving the readings of spirit thermometers any credibility at all was to test them constantly in comparison with a mercury thermometer, which was itself tested occasionally in melting fresh-water ice. Fortunately we had half a dozen mercury thermometers, taken for other than meteorological purposes. Unfortunately they were only graduated down to zero (Fahr.). Had the scale gone down to near the freezing-point of mercury, they might have served for all our observations. Whenever a series of temperatures (as in a shaft sunk through the ice) came partly below and partly above zero (Fahr.), the whole series was taken with spirit thermometer, but every one above zero was repeated with a mercury thermometer, and the whole series corrected in correspondence with its readings. This practice at least minimised the chance of change in the error of the spirit thermometer, by allowing the least possible time for it.

In the meteorological screen a mercury thermometer was hung beside the spirit one used for the dry-bulb readings. No variation was ever detected in this one, which was undisturbed. The maximum and minimum frequently went wrong, and had to be corrected by observing the position at which they rested when shaken down.

Greely remarks on the unreliability of spirit thermometers. It is desirable for polar expeditions to be provided with mercury thermometers graduated down as low as possible. We have no reason to believe that our thermometers were not of excellent quality. The defect is inherent and largely due to climate.

CLOUD FORMS

By JAMES MURRAY

Without treating them from the meteorological standpoint, some notes and sketches of the more striking forms of cloud associated with Mount Erebus may be of interest.

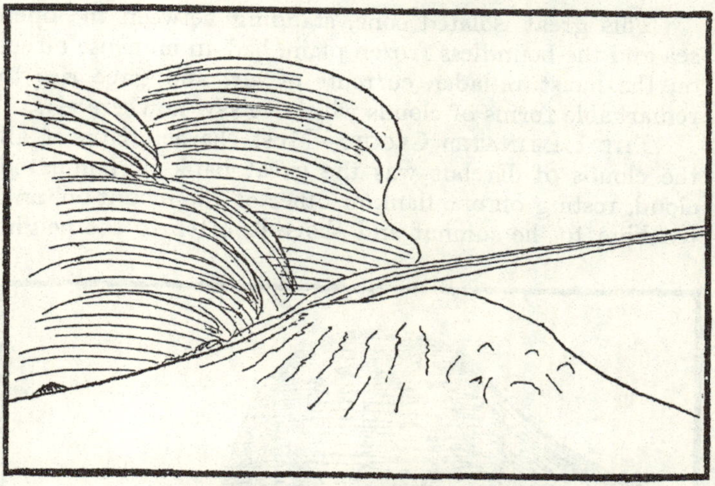

Stratified Cloud on Erebus, October 13

Spiral Cloud, July 23

This great isolated cone, standing between the open sea and the boundless frozen plain, had an immense effect on the moisture-laden currents of air, and gave rise to remarkable forms of clouds which we saw nowhere else.

THE LAMINATED CLOUD.—Most characteristic of all the clouds of Erebus was the great bank of laminated cloud, resting on one flank or other of the mountain, and reaching to the summit and above it, often to the height

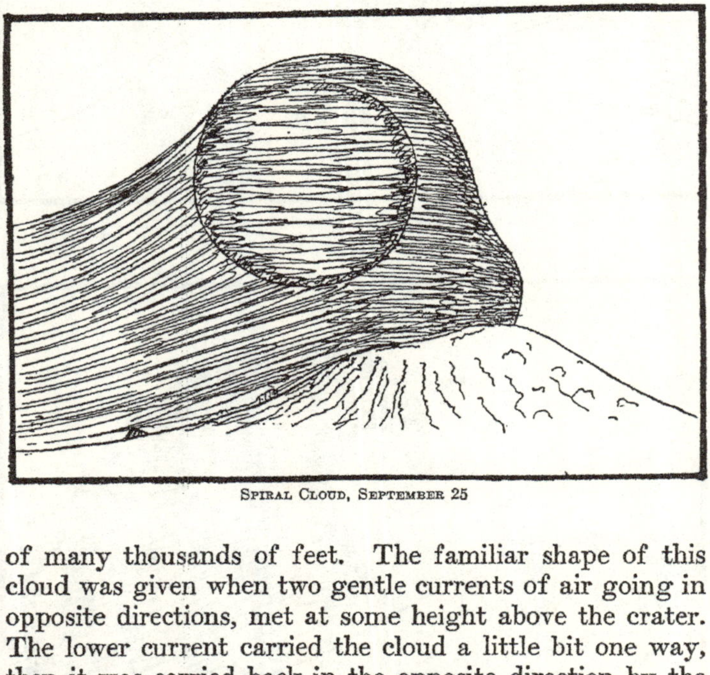

SPIRAL CLOUD, SEPTEMBER 25

of many thousands of feet. The familiar shape of this cloud was given when two gentle currents of air going in opposite directions, met at some height above the crater. The lower current carried the cloud a little bit one way, then it was carried back in the opposite direction by the upper current, as shown in the illustration.

The disposition of the clouds shown in this sketch is unusual, as there is a thin stratified band (not the smoke-cloud) trailing off to the south, while the great bank hangs on the north. Generally this great cloud rests on the north slope, but occasionally the position is reversed.

Spiral Cloud.—A modification of the great cloud-bank, produced by the meeting of very gentle currents, resulted in the rolling up of the cloud into a spiral form.

On July 23 a spiral was formed from a bank on the south side of the mountain, as figured below. The upper

Whale-backs, September 16

current on this occasion was from the south. A similar spiral cloud is figured, right over the mountain, in the picture of saw-edged stratus, October 1.

The most remarkable spiral cloud observed was that of September 25, 1908. The bank of cloud lay on the north flank. The upper current was from the south, as on July 23. Its action was so regular that the spiral could be traced for more than a complete turn.

Whale-back Clouds, September 16

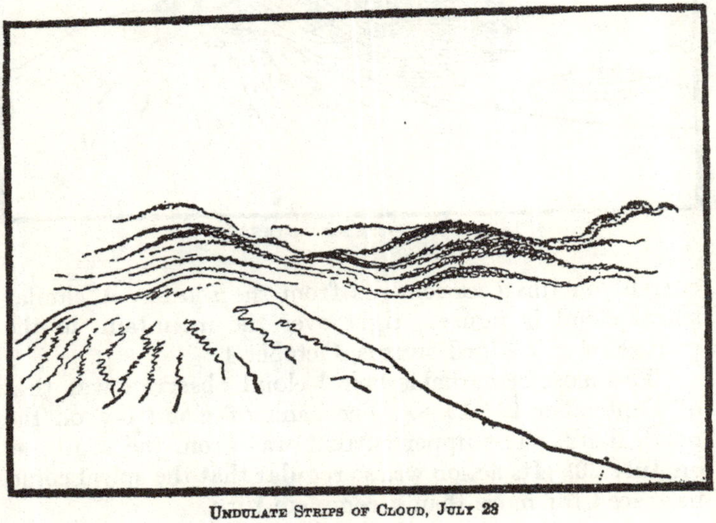

Undulate Strips of Cloud, July 28

Interlacing Clouds, September 16

Saw-edged Stratus and other Clouds, October 1, 1908.

WHALE-BACK CLOUDS.—The most striking of all the cloud forms on Erebus were those to which we gave the name of whale-backs. They were small patches of cloud, isolated or two or three together, one over another, with very evenly rounded upper surfaces, the lower surfaces more indefinite, often rounded too. A very good development of them occured on September 16, 1908. The largest stratified group of them formed an object like a comet, the others were isolated. This display went through many changes, slowly and gradually. Two stages are figured. The first figure shows the clouds at an early stage, in which they appear imbricated in a characteristic manner, observed on many other occasions They were then stratified, showing generally three layers in each group.

At a later stage, shown in the next figure, the whale-backs were isolated, except the one large comet-like mass.

While usually evenly rounded, the curves were frequently reversed at the ends, as shown in the lower goup in the first figure.

Whale-backs were usually seen after a blizzard, occasionally before one. During the blizzard the mountain was generally hidden, so that we could not tell if they were formed during the storm.

UNDULATE CLOUDS.—Narrow lines of cloud were often arranged in regular or irregular undulations. The first figure shows fairly regular undulations, over the summit of Erebus, on July 28, 1908. The second figure shows a modification of this form, in which the wisps of cloud are intricately interlaced. This occurred on the south side of Erebus, on September 16, at the same time as well-developed whale-backs on the north side.

Sometimes, especially after storms, a great variety of clouds were piled up over and around Erebus, including whale-backs, cumulus undulate clouds, with perhaps a spiral cloud over all.

One such accumulation is figured, in which there were, cumulus, just below the summit, two banks of stratus at different levels, the upper one with deeply serrated edge, and a spiral cloud. Over all was a Noah's Ark of cirro-stratus, not figured, and beyond the influence of Erebus. Saw-edged stratus was seen on one or two other occasions, but was not common.

Appendix Six

REPORT ON THE HEALTH OF THE EXPEDITION

By Dr. ERIC MARSHALL, M.R.C.S., L.R.C.P.

THE fact that there was no case of scurvy during the period of the expedition's residence in the Antarctic may be attributed to the fact that the utmost care had been taken in provisioning the expedition with foods of the best quality obtainable, in that variety which is essential under polar conditions. Bottled and preserved fruits were used liberally during the long winter, and when the spring approached and the preparations for the spring and summer sledging involved an increasing amount of physical work, the allowance of fresh meat (penguin, seal and mutton) was increased. When the spring depôt party started south on September 22, 1908, all the members of the expedition had been on a liberal allowance of fresh meat for a month. During the whole winter all the men took daily exercise in the open air, this routine being interrupted only by the most severe blizzards. There was no case of sickness.

We found, in the matter of clothing, that heavy pilot-cloth garments and furs were not essential provided that wind-proof suits were worn, and that the body temperature was maintained by a full diet. On the southern journey, when the rations had been reduced to the minimum, and our clothing was worn and torn, so that it no longer kept out the biting wind, our temperatures were subnormal. At the end of a long day's march and in the face of a blizzard wind, when our altitude was about ten thousand feet, our temperatures were on several occasions reduced to 94° Fahr., rising to 97° or 98° after we had eaten a hot though scanty meal. Frost-bites were more frequent at these times, and it was more difficult to restore the part attacked.

It is an interesting fact that the members of the expedition did not suffer from colds during their stay in the Antarctic save in August 1908, when a bale of new clothing was opened in the hut, and all the men were at once seized with acute nasal catarrh. The symptoms were quickly dispelled when we took exercise in the open, and those who remained in the hut recovered after two or three days.

On the return of the expedition to New Zealand the *Nimrod* laid up for one day at the mouth of Lord's river, Stewart's Island, and a number of the staff went ashore to bathe and fish, &c. All who went ashore suffered considerably from the inflammation caused by the bites of sand flies, yet it was only those members, who, on arrival at Lyttelton and Christchurch, New Zealand, who were not immediately seized with colds.

The expedition was not entirely free from accidents, for on arrival at the ice in January 1909, A. L. A. Mackintosh was struck in the right eye by a hook while unloading cargo. The accident necessitated the immediate removal of the eye. His recovery was extremely satisfactory, so that on the fourth day he was able to get about. This, however, prevented him from remaining with the shore staff, as it was deemed necessary that he should return to Australia.

During the ascent of Mount Erebus, Brocklehurst, while wearing ski-boots, was frost-bitten in both feet, eight toes being affected. Under treatment seven recovered, but the great toe of the left foot showed no signs of improvement, and ultimately, dry gangrene having set in, I amputated the last joint a month after the accident. Recovery was slow owing to the limited amount of healthy tissue for the posterior anterior flap. The ultimate result was satisfactory. B. Day, while tobogganing, fractured the base of the third metatarsal of his right foot. These accidents, together with a few septic fingers, were all that arose in the surgical line.

Weights and measurements were taken regularly, but although slightly on the increase during the winter, did not vary much from month to month, although one member has now well-marked linear albicantes on both upper arms as a memento of the adiposea of the south.

During one time or another all the members of the Southern Party suffered more or less severely from dysentery. Some of the pony meat was not wholesome, and as the supply of oil was small, it was either eaten raw or warmed to about 100° Fahr., with the result that we were unable to digest it. An acute enteritis resulted and prostrated us from time to time. At this time we were almost entirely dependent on the pony meat, the starch food available being of the scantiest.

A considerable quantity of Easton Syrup tabloids were used on the plateau and were found of assistance. Only on two or three occasions did any one suffer from snow-blindness, and on each occasion the snow goggles provided had not been worn.

The deep amber glasses were a sufficient protection, as they cut out all the actinic rays, and had a very pleasing tone. A combined flash red and worked green glass, giving an orange-brown tone, was also provided. The glasses themselves entirely eliminate the violet and ultra-violet rays and were an absolute protection against snow-blindness, but a more complete system of ventilation in the vulcanised fibre cylinders was required.

Appendix Seven

SOUTHERN JOURNEY DISTANCES

THE following table gives detailed information regarding the distances travelled day by day on the southern journey. The geographical miles given in the first column cover the period from November 15, when the party left Depôt A, until January 9, when the furthest south point was reached. The distances have been taken from the chart after all corrections have been made, and represent a direct line from camp to camp.

In the second column will be found the noon latitudes, calculated from observations taken as opportunity offered. The observations have been checked by the officers of the Royal Geographical Society in London.

The last column shows the distances travelled day by day according to sledge-meter, and these figures of course take into account all deviations and *détours,* so often rendered necessary by the condition of the surface. That the sledge-meter was reliable is proved by the fact that on the homeward journey we were able to calculate our positions without taking latitude observations. We took only one observation during the journey back to the coast (January 31, noon position 82° 58′ South), and on that occasion the theodolite confirmed the record of the sledge-meter.

Observations for variation were taken whenever we took a latitude observation, and the results will be found recorded on the chart.

The latitude observations noted in this table were taken with a three-inch theodolite, which was carefully

Date.	Geographical miles.	Noon latitudes.	Statute miles.	Yards.	Relay.
1908					
October 29	....	....	14	880	....
October 30 (Hut Point)	....	....	9	880	....
October 31 (back to Royds)	....	....	23	....	....
November 1 (to Hut Point)	....	....	23	....	....
November 2 (blizzard)	....	....	no march	....	
November 3	....	....	12	300	....
November 4	....	....	16	500	....
November 5	....	....	9	1200	....
November 6 (blizzard)	....	....	no march	....	....
November 7	....	....	1	....	....
November 8 (blizzard)	....	....	no march	....	....
November 9	....	....	14	600	....
November 10	....	....	15	1550	....
November 11	....	....	15	....	....
November 12	....	....	15	1650	....
November 13	....	....	15	1550	....
November 14	....	....	15	100	....
November 15	7.39 (from noon)	79° 36′ S.	12	1500	....
November 16	14.91	....	17	200	....
November 17	13.3	....	16	200	....
November 18	13	....	15	500	....
November 19	13.7	....	15	200	....
November 20	13.6	....	15	800	....
November 21	13.3	....	15	500	....
November 22	16	....	15	250	....
November 23	14	....	17	1650	....
November 24	15.4	....	17	680	....
November 25	14.6	....	17	1600	....
November 26	13.2	82° 12′ S.	16	1700	....
November 27	15.5	....	16	1200	....
November 28	13.6	82° 39′ S.	15	1500	....
November 29	11.7	....	14	900	....
November 30	11	....	12	150	....
December 1	10.5	....	12	200	....
December 2	10.3	....	11	1450	....
December 3 (Mount Hope)	....	....	20	....	....
December 4	10.5	83° 33′ S.	10	....	....
December 5	3.1	....	5	....	4
December 6	4.1	....	....	....	3
December 7	9.1	....	10	570	....

SOUTHERN JOURNEY

Date.	Geographical miles.	Noon latitudes.	Statute miles.	Yards.	Relay.
1908					
December 8	7.7	....	12	150	....
December 9	9.8	84° 2′ S.	11	1450	2
December 10	9.8	....	11	860	....
December 11	7.2	....	8	900	....
December 12	3.1	....	3	500	6 1000
December 13	4.5	....	5	....	6
December 14	8	....	7	880	2
December 15	11.5	....	13	200	....
December 16	12	84° 53′ S.	13	1650	....
December 17	9.1	....	12	250	1
December 18	3	....	6	600	12
December 19	7.4	....	10	....	1 880
December 20	10	85° 19′ S.	11	950	1
December 21	7	....	6	....	3
December 22	7	....	4	....	6
December 23	6.2	....	13	....	....
December 24	9.2	....	11	250	....
December 25	9.2	....	10	650	....
December 26	11.4	....	14	480	....
December 27	12	....	14	930	....
December 28	11.7	....	14	450	....
December 29	10.1	....	12	600	....
December 30	3.7	....	4	100	....
December 31	8.5	....	11	....	....
1909					
January 1	9.7	86° 59′ S.	11	900	....
January 2	9.1	....	10	450	....
January 3	12.6	87° 22′ S.	11	1680	....
January 4	12.2	....	14	660	....
January 5	13.4	....	15	480	....
January 6	13.2	(88° 7′ camp)	15	313	....
January 7 (blizzard)	....	....	no march	....	....
January 8 (blizzard)	....	....	no march	....	....
January 9	16.5	88° 23′ S.	18	704 from camp	....
		(furthest south)	18	704 back	....
			4	40 to camp	....
January 10	....	....	21	308	....

Date.	Geographical miles.	Noon latitudes.	Statute miles.	Yards.	Relay.
1909					
January 11	....	....	19	1580	....
January 12	....	....	14	100	....
January 13	....	....	15	1560	....
January 14	....	....	20	1600	....
January 15	....	....	20	....	....
January 16	....	....	18	800	....
January 17	....	....	22	850	....
January 18	....	....	26	900	....
January 19	....	....	29	....	....
January 20	....	....	15	800	....
January 21	....	....	17	....	....
January 22	....	....	15	900	....
January 23	....	....	14	100	....
January 24	....	....	16	....	....
January 25	....	....	26	....	....
January 26	....	.... }	16	{	....
January 27	....	.... }		{	....
January 28	....	....	14	890	....
January 29 (blizzard)	....	....	2	....	....
January 30	....	....	13	....	....
January 31	....	82° 58	13	850	....
February 1	....	....	13	1400	....
February 2	....	....	13	900	....
February 3	....	....	5	900	....
February 4 (dysentery)	....	....	no march	....	....
February 5	....	....	8	....	....
February 6	....	....	10	....	....
February 7	....	....	12	880	....
February 8	....	....	12	....	....
February 9	....	....	14	900	....
February 10	....	....	20	300	....
February 11	....	....	16	1320	....
February 12	....	....	14	450	....
February 13	....	....	12	....	....
February 14	....	....	11	1400	....
February 15	....	....	12	440	....
February 16	....	....	13	....	....
February 17	....	....	19	200	....
February 18	....	....	15	400	....
February 19	....	....	14	440	....
February 20	....	....	14	....	....

Date.	Geographical miles.	Noon latitudes.	Statute miles.	Yards.	Relay.
1909.					
February 21	....	....	20	....	....
February 22	....	....	20	800	....
February 23	....	....	14	500	....
February 24	....	....	15	....	....
February 25 (blizzard)	....	....	no march	....	....
February 26 (left A. and M.)	....	....	24	....	....
February 27	....	.... }	39	{	....
February 28	....	.... }		{	....
March 1	....	.... }		{	....
March 2	....	.... }	63	{	....
March 3	....	.... }	30 out	{	....
March 4	....	.... }	33 back	{	....

adjusted before the start for the southern journey. An observation taken on the return journey, in February, when the position was known from bearings, showed that the instrument was correct. The observations were only taken with the theodolite "face left," but as the instrument was in good adjustment this was sufficient.

On the outward journey the last latitude observation was taken in latitude 87° 22′ South. The remainder of the distance marched towards the south was calculated by sledge-meter and dead reckoning. The accuracy of the sledge-meters used was proved by the fact that on the return journey we were able to pick up the depôts without taking observations. The " slip " was ascertained by careful tests before the start of the journey.

The chronometer watches taken were rated before leaving and on the return, and the error was only eight seconds. All bearing, angles and azimuths were taken with the theodolite. Variation was ascertained by means of a compass attached to the theodolite, and the steering compasses were checked accordingly. At noon each day the prismatic compasses were placed in the true meridian, and checked against the theodolite compass and the steering compasses.

The total distance marched, from October 29 to March 4, as recorded on the sledge-meters, was 1755 miles, 209 yards statute, this including relay work and back marches.

Appendix Eight

CONSUMPTION OF STORES AT WINTER QUARTERS

Fifteen Men

Week ending July 27, 1908

	LBS.
Mutton	30
Seal meat	15
Penguins (six Adelies)	12
Bacon (one side)	30
Tinned fish	5
Tinned meat	8
Tinned soups	36
Tinned and dried vegetables	19
Sugar	20
Syrup	6
Honey	5
Jam	15
Chocolate	4
Crystallised fruits	5
Dried fruits	2
Tinned and bottled fruit	40
Dried milk	56
Coffee	1
Tea	6
Oatmeal and Quaker Oats	6
Eggs (fresh and dried)	6
Butter	20
Suet (tinned)	6
Cocoa	2
Rice	4
Biscuits	10
Flour	56
Currants and raisins	8
	433

Such articles as salt and pepper were issued as required, and on special occasions there were extra dishes such as tinned roast reindeer, roast black cock, marrow pudding, lobscouse and fish-cakes.

The amount of food consumed per man per day in this week was about 4.12 lb.

Appendix Nine

THE *NIMROD'S* HOMEWARD VOYAGE. IN SEARCH OF DOUBTFUL ISLANDS

THE homeward voyage of the *Nimrod,* after the members of the shore-party had been landed at Port Lyttelton, was made interesting by a search for some charted islands, the existence of which was doubtful. J. K. Davis, who had been first officer, was in command at this time, and he had under him the members of the ship's staff, all the members of the shore-party proceeding from New Zealand by the ordinary passenger routes. The *Nimrod* went first to Sydney, where the naval authorities very generously assisted in effecting certain necessary repairs. The appended report by Captain Davis deals with the voyage from that point.

"MONTE VIDEO, *July* 8, 1909.

" Leaving Sydney on May 8, I steered south on the 151st meridian against moderate southerly winds. On May 12, when the position was latitude 43° South, longitude 151° East, the wind came away fresh from the south-east, and as I did not consider that I was far enough to windward of Macquarie Island to allow me to stand east in that latitude, I stood west and decided to carry out your instructions regarding the Royal Company Island if the wind continued south-easterly. This was a fortunate decision, for the wind continued from the south and east for four days, so that on May 17 I was only ninety-seven miles from the position of the islands as given on the chart. At noon, after ascertaining our position, I took a sounding in 2430 fathoms. The bottom specimen was lost through the wire parting while heaving in. On May 18, at 2 P.M., in fine clear weather, we sailed over the position assigned to the Royal Company Island, with nothing in sight. I stood east till 4 P.M., and then south, but saw nothing to indicate the existence of land in the vicinity.

"On May 24, when 190 miles off Macquarie Island, we encountered a heavy north-west gale, which the vessel weathered with very little damage, though deeply loaded. This was followed on the 25th by a heavy gale from the south-west, but at 11.45 P.M. of this day we sighted the island and managed to get to leeward of it till daylight, when the wind and sea were less violent. My instructions were to visit Macquarie Island for the purpose of making zoölogical and geological collections to connect those already obtained in the Antarctic with the life and rocks of Australia and New Zealand, and also of observing whether any Antarctic birds or penguins migrated there in the winter months, as they (the penguins particularly) leave the far south when the sea freezes over in April, and do not return until the following summer; where they migrate to has so far not been discovered. We sighted the island on the night of May 26, and stood off till daylight, when we were soon able to make out its distinctive features. Approaching from the eastward one is at once struck by the rugged boldness of the coast-line, which rises sheer out of the water to a height of nearly 1500 ft. in places. At the south-east extremity there is a reef of dangerous ragged rocks on which the sea breaks heavily, and as we got nearer we were able to see that the mountain slopes were green, and to trace the course of several waterfalls. A wide bay or rather curve in the coast-line forms the southern anchorage called Lusitania

Road; close inshore is a line of rocks, and it is through a break in these rocks that the landing-place is reached.

"At 10 A.M. we anchored in eight fathoms about a quarter of a mile from the shore. From the ship we could see two huts situated on the lower ground at the foot of the hill. A large rookery of penguins and some sea elephants appeared the only life visible. A boat was soon lowered and sent ashore in charge of the chief officer; the party succeeded in landing through the heavy surf and the boat was hauled up into a bed of kelp. The men scattered to collect specimens, &c. They found the slopes of the hills covered with a long coarse grass. There are no trees or even shrubs on the island. A small river running down a valley formed by the hills made the low ground swampy. Of the two huts the larger was evidently a boiler-house for rendering down the blubber of the sea elephants, and the smaller was the one in which the sealers lived while engaged in this work during the season. Both wore a very neglected and forlorn appearance. The sea elephant, which is like a big seal and in some cases over thirty feet long, is an awkward, clumsy animal and apparently spends most of its time asleep in the long grass near the water. It has large teeth which somewhat resemble tusks, but although it appears very fierce is not dangerous. The penguins, of which there were a great number, were of the King genus, and they keep up a continuous squeaking. The young birds were just fledged and were nearly ready to take to the water.

"We remained at anchor here for the night and at day-break, which was not till eight o'clock, we steamed along the coast northwards. There was a fresh north-west wind, and it came down the hill-sides in violent gusts, called by the whalers 'willywaws,' raising a sheet of foam on the water. About six miles up the coast there is a break in the hills at a place called Green Gorge, which is a wide valley running across the island. Further on we could make out Nugget Point, from which extends a reef of rocks for some distance seaward. As we drew nearer to this point we could make out two huts on the shore and also the wreck of a vessel high and dry on the beach. Suddenly, to our surprise, a column of smoke rose from the smaller of the two huts. As we had heard nothing of any one living on the island this was extraordinary. Presently with the glass we could make out the figure of a man standing at the door of the smaller of the two huts watching our approach. We came to an anchor, and the boat was lowered and headed for the shore. The man who had been watching us from the hut now walked down to the beach accompanied by two little dogs. There was a heavy surf, but our Crusoe-like friend, after pointing out the best landing-place, walked into the water and assisted in beaching the boat. Every one at once wanted to know what he was doing there, was he alone, how long had he been there, &c., and the following information was gathered in less time than it takes to write. Mr. W. McKibbon had arrived at the island last March in a small vessel called the *Jessie Nicoll,* belonging to a company that has the lease of the island for the purposes of obtaining sea elephant and penguin oil. This vessel brings down a number of men to work on the island during the short season trying down blubber on shore, &c. When the previous season was over and the ship filled with barrels of oil, instead of returning with the other men, our friend had elected to remain behind on the island for the winter by himself, and to collect oil for the next season.

" The next day I went ashore myself; we were met by our friend of the day before who escorted me to his little hut, which was very warm and comfortable and scrupulously clean. It consisted of two rooms, both of which had fires; one of them was used as a work-room and store, the other as a sleeping-place. I learnt, as I sat in front of a comfortable little fire, that McKibbon was a native of Carrick-on-Shannon, and had been in the navy for twenty years; he was a typical sailor, clean-shaven, and despite of his fifty years of active life was as keen and alert as a boy. 'I never had an illness in my life,' he said when I asked him what would happen to him if he fell ill. He was nearly as surprised to see us as we were to see him. 'I thought at first it was the New Zealand Government light-house tender; then when you came closer I saw you were a wooden vessel and put you down as a surveying ship.' When he heard that we had just returned from the Antarctic he told us that he had been in the Arctic in the paddle frigate *Valorus,* which vessel took up stores for the expedition under Nares and Markham.

"We spent four days at Macquarie Island, and obtained a good collection of specimens. We left the island on May 30, and as soon as we got from under its lee encountered a fresh westerly gale with high seas. Running before this we passed over the charted position of Emerald Island at 8 P.M. on the 31st. It was a clear night, three days from full moon, and if there had been even low land in the vicinity we should have seen it. I decided to stand on, as it was blowing hard, and a high westerly sea made soundings impracticable. On June 9 we arrived in the vicinity of the position assigned to the Nimrod Islands, and at 1.15 A.M., in fine clear weather, passed over this spot with nothing in sight. The weather here again was much against us, a very high sea, with fresh westerly wind and squally, and the barometer at 28.20. After steaming east sixteen miles from the position, it being now dark, I hove to and attempted to sound. This operation, being performed in a kind of blizzard by lamplight, with the ship rolling heavily all the time, was not a great success. We slacked out 1140 fathoms but obtained no bottom, so hove in again as the wind was coming away in fierce squalls from the south-west. I found that we were setting fast to leeward, so decided to continue our voyage.

"At noon on June 17 we were up by dead reckoning to the position of Dougherty Island, as given by Captain Dougherty, but as the weather was overcast could not be sure of our position. Captain Keates places the island in the same latitude thirty-four miles further east. I therefore continued eastward on the parallel over this position (by dead reckoning). As it was now dark and the weather moderate, I stood back again to the westward, hoping to get sights at daylight and did so. Good star observations were confirmed at noon, when the island, according to Captain Dougherty's position, should have borne west distance four miles. No land was in sight from the masthead in clear weather. I stood east again, and at 4 P.M., when darkness was just setting in, the island according to Captain Keates' position should have borne east four miles; nothing in sight. At 4.30 we passed over this position and continued eastward along the parallel of 59° 21′ South, but saw no indication of land. It is just here that we met with ice during our passage, and I am inclined to think Dougherty Island has melted. The darkness was sixteen hours out of the twenty-four; it was, in fact, to quote from the 'Directory,' one long starless desolate night, a perpetual gloom which the sun seems never to penetrate, and the conditions made a search of this kind more or less doubtful. Although I will not say these islands do not exist, I can with confidence say they do not exist in the locality laid down or anywhere near it. When in latitude 59° 31′ South, longitude 107° West, we encountered north-easterly winds, which drove us into 61½° South, where we met moderately warm weather and continual rain. As the sun had only an altitude of 5° at noon there was almost constant darkness. At 10.45 P.M. on June 27 we sighted the Diego Ramirez Island right ahead at a distance of fourteen miles. We made an extremely good landfall, and this satisfied me that our chronometers, despite the changes of temperature to which they had been subjected, were reliable. It was a cloudy moonlight night, and the fact that we saw these islands, or rather rocks, for the highest point is only 587 ft. above sea-level, so far off convinces me that if the other islands had existed anywhere near the localities laid down for them we could not have failed to see them."

The *Nimrod* touched at Monte Video and arrived at Falmouth on August 26. Four days later she was berthed in the Thames, thus completing the most adventurous cruise of her eventful career. She had been away from the Thames for nearly twenty-five months.

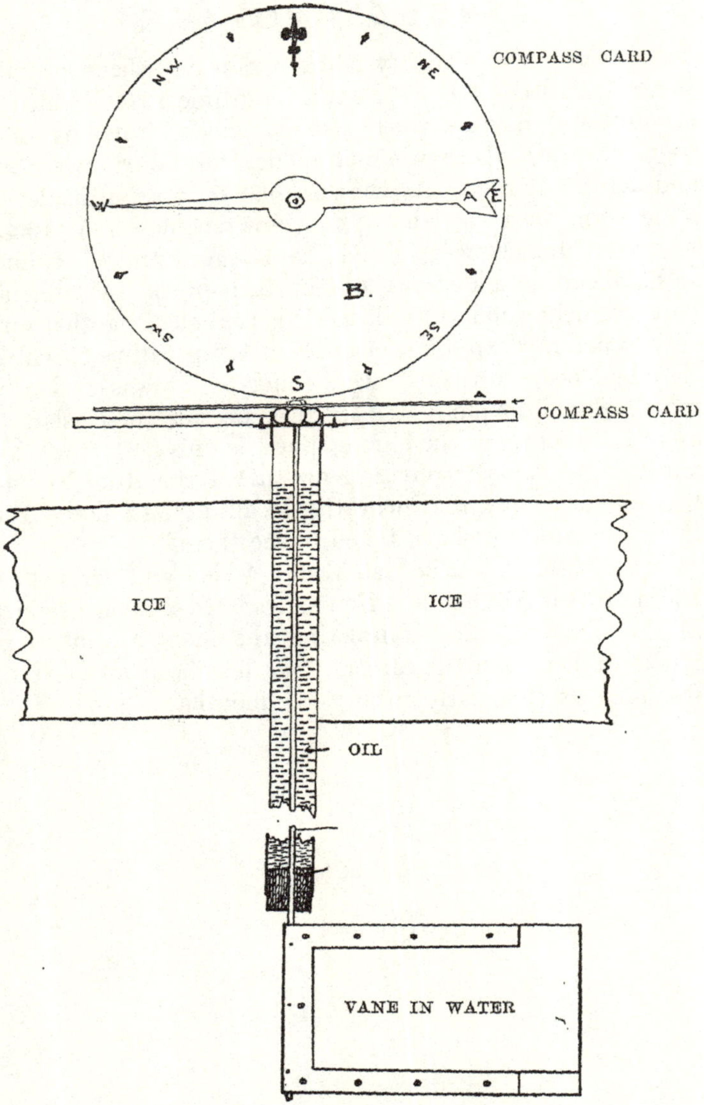
COMPASS CARD
NW
NE
W
A
E
B.
SW
SE
S
COMPASS CARD
ICE
ICE
OIL
VANE IN WATER

TIDE GAUGE

Appendix Ten

ESTIMATED COST OF THE BRITISH ANTARCTIC EXPEDITION 1907 UP TO AUGUST, 1909

	£	s.	d.	£	s.	d.
Purchase price of *Nimrod*	5000	0	0			
Cost of improvements, alterations, and repairs in dry dock, London	2550	0	0			
Cost of improvements, alterations, additions and repairs at Lyttelton, New Zealand	740	19	9			
Cost of repairs at Lyttelton, New Zealand, after her return from the Antarctic on her first voyage when she was damaged during heavy weather	957	3	3			
				9248	3	0
Equipment for ship, engine, carpenters, deck and galley stores, &c.				2540	4	8
Relief equipment ship				490	6	7
Food-supplies, ship				1821	6	11
Wages: captain's, officers', and crew of ship				5161	3	3
Coal account for ship at London, Torquay, St. Vincent, Cape Town, Lyttelton, New Zealand (three supplies) and Monte Video				1520	8	8
General expenses of ship, including labour, pilotage, shipbrokers' charges, &c.				1365	15	1
Insurance on ship, &c.				863	12	0
Equipment shore-party for sixteen men for two years				4296	18	4
Relief equipment shore-party				328	16	11
Food-supplies shore-party, sixteen men for two years				2005	14	7
Relief food-supplies for whole expedition, forty men for one year				1807	10	7
Salaries shore-party				6055	10	4
Cost of fifteen Manchurian ponies, including purchase price, freight from China to New Zealand, harness, two years' forage, &c.				1517	17	8
Cost of nine Esquimau dogs and food				126	0	0
Interest account at bank on guarantees, &c., and bank charges				2021	11	5
Passage of shore-party and manager of expedition to New Zealand and back				1016	15	10
Travelling expenses				443	4	9
General expenses: manager's salary, office rent, telegrams, postages, typewriting correspondence, exhibition of equipment, wages district messenger-boy, telephone, rent, &c., &c.				1484	19	7
Cost of cablegrams between London and Australia, and New Zealand before the expedition sailed and on its return				264	14	7
				£44,380	14	9

INDEX

INDEX

INDEX

INDEX

INDEX

This page intentionally left blank

Made in United States
Orlando, FL
05 November 2026

90229684R00360